CAMBRIDGE TRACTS IN MATHEMATICS

General Editors

132 Mixed Hodge Structures and Singularities

Valentine S. Kulikov
Moscow State University of Printing

Mixed Hodge Structures and Singularities

CAMBRIDGE
UNIVERSITY PRESS

PUBLISHED BY THE PRESS SYNDICATE OF THE UNIVERSITY OF CAMBRIDGE
The Pitt Building, Trumpington Street, Cambridge CB2 1RP, United Kingdom

CAMBRIDGE UNIVERSITY PRESS
The Edinburgh Building, Cambridge CB2 2RU, United Kingdom
40 West 20th Street, New York, NY 10011-4211, USA
10 Stamford Road, Oakleigh, Melbourne 3166, Australia

First published 1998

Printed in the United Kingdom at the University Press, Cambridge

Typeset in 10pt Times New Roman PS

A catalogue record for this book is available from the British Library

Library of Congress Cataloguing in Publication data

Kulikov, Valentine S., 1948–
Mixed Hodge structures and singularities/Valentine S. Kulikov.
p. cm. – (Cambridge tracts in mathematics ; 132)
Includes bibliographical references and index.
ISBN 0-521-62060-0 (hb)
1. Hodge theory. 2. Singularities (Mathematics) I. Title.
II. Series.
QA564.K85 1998
516.3'5–dc21 97-11978 CIP

ISBN 0 521 62060 0 hardback

Contents

Introduction

The aim of this book is to introduce and at the same time to survey some of the topics of singularity theory which study singularities by means of differential forms. Here differential forms associated with a singularity are the main subject as well as the main tool of investigation. Differential forms provide the main discrete invariants of a singularity as well as continuous invariants, i.e. they make it possible to study moduli of singularities of a given type.

A singularity is a local object. It is a germ of an algebraic variety, or an analytic space, or a holomorphic function. However, the majority of the ideas and methods, used in the theory under consideration, originated in the 'global' algebraic geometry. Therefore we first give a very brief and schematic description of these ideas.

The idea of using differential forms and their integrals to define numerical invariants of algebraic varieties goes back to the classic writers of algebraic geometry. It will be important for us that holomorphic and algebraic forms can be used to calculate the singular cohomology of a smooth algebraic variety over $\mathbb{C}$. Developing the ideas of Atiyah and Hodge (1955), Grothendieck (1966) showed that $H^i(X^{an}, \mathbb{C}) \simeq H^i_{\mathrm{DR}}(X/\mathbb{C})$, where $H^i_{\mathrm{DR}}(X/\mathbb{C})$ is the De Rham cohomology. Grothendieck defined $H^i_{\mathrm{DR}}(X/\mathbb{C})$ as the hypercohomology $\mathbb{H}^i(X, \Omega^{\cdot}_X)$ of the complex of sheaves of holomorphic differential forms on X. The comparison theorem enables us to calculate the cohomology of the complement to a hypersurface in projective space by means of the cohomology classes generated by rational differential forms.

Algebraic differential forms have also proved to be useful in the study of the monodromy of a family of complex varieties, using the Gauss–Manin connection. The monodromy transformation is the transformation of fibres (or their homotopic invariants) of a locally trivial fibration corresponding to a loop in the base. This notion appears when studying the multivalued analytic function, where it corresponds to the notion of the covering or sliding transformation. Very often the monodromy transformation appears in the following situation. Let $f\colon X \to S$ be a proper holomorphic map of an analytic space to the disk in the complex plane. Let X_t be the fibre $f^{-1}(t)$, $t \in S$, $S' = S\backslash\{0\}$ and $X' = f^{-1}(S')$. By reducing the radius of S, if necessary, we can make the fibration $f'\colon X' \to S'$ a locally trivial C^∞-fibration. The monodromy transformation T associated with the loop in S' surrounding 0 is called the monodromy transformation of the family f. The action of the monodromy T on the vector space $H^*(X_t)$ is obtained by the parallel displacement of the cohomology classes in the fibres of the locally constant fibration $\underline{H} = \bigcup_{t\in S'} H^i(X_t, \mathbb{C}) = R^i f_* \mathbb{C}_{X'}$. Grothendieck also defined the relative De Rham cohomology sheaves $\mathscr{H}^i_{\mathrm{DR}}(X/S) \simeq \mathbb{R}^i f_*(\Omega^{\cdot}_{X/S})$. If $f\colon X \to S$ is a smooth proper morphism of algebraic varieties/$\mathbb{C}$, then from Grothendieck's theorem it follows the existence of the canonical isomorphism of coherent analytic sheaves $\mathscr{H}^i_{\mathrm{DR}}(X/S)^{\mathrm{an}} \simeq R^i f_*{}^{\mathrm{an}}(\mathbb{C}) \otimes_{\mathbb{C}} \mathscr{O}_{S^{\mathrm{an}}}$. The presence of the locally constant sheaf $\underline{H} = R^i f_*(\mathbb{C})$ in $\mathscr{H}^i_{\mathrm{DR}}(X/S)$ defines a topological connection on the sheaf $\mathscr{H}^i_{\mathrm{DR}}(X/S)$. Katz and Oda (1968) gave an algebraic definition of the canonical connection (the Gauss–Manin connection) on the sheaves $\mathscr{H}^i_{\mathrm{DR}}(X/S)$ such that the sheaves of its horizontal sections are $R^i f_*(\mathbb{C})$. They calculated this connection explicitly and showed that it reduces to the definition orginally given by Manin (1958) for the case in which X/S is an algebraic curve over the field of functions.

The family of varieties X_t defined by the morphism f degenerates at the point $0 \in S$, and the Gauss–Manin connection has a singularity at the point 0. This singularity is regular. The notion of a regular connection generalizes the classical notion of a differential equation with a regular singular point, and is the subject of Deligne's book [D1]. Katz gave an algebraic proof of the regularity of the Gauss–Manin connection (1970). Analytic proofs were given by Griffiths (1971) and by Deligne [D1]. The regularity theorem is related to the monodromy theorem. When the space X as well as all fibres X_t, $t \neq 0$, is smooth, the monodromy theorem states that T is quasi-unipotent on $H^*(X_t, \mathbb{Q})$, i.e. there exist positive integers k and N such that $(T^k - 1)^N = 0$. There are several proofs of this basic theorem (Clemens 1969; Katz 1971; Griffiths & Schmid 1975). Many of the characteristic features of the degenerate family $f\colon X \to S$ become apparent in the properties of the monodromy. The monodromy of the family f is closely connected with the mixed Hodge structure (MHS) on the cohomology

$H^*(X_0)$ and $H^*(X_t)$. The Hodge structure and the period map of a family of algebraic varieties defined by the Hodge structure on the cohomology of fibres $H^*(X_t)$ are the second most important notions used in this book.

The concept of a pure Hodge structure is a formalization of the structure of the cohomology groups of a compact Kähler manifold. From the theory developed by Hodge in the 1930s it follows that $H^n(X, \mathbb{C}) = \oplus_{p+q=n} H^{p,q}$, $\overline{H^{p,q}} = H^{q,p}$, where the cohomology vector space is identified with the vector space of harmonic forms, and harmonic forms, and consequently the cohomology, can be decomposed into the direct sum of components of the type (p, q). Under a variation of a variety X in a family X_t the variation of subspaces $H^{p,q}(X_t)$ is not complex analytic. For this reason it is more convenient to study the Hodge filtration $F^p H^n(X, \mathbb{C}) = \oplus_{r \geqslant p} H^{r,n-r}$, which does depend analytically on the parameter t. One can give an equivalent definition of the Hodge structure in terms of the Hodge filtration. Using the isomorphism $H^n(X, \mathbb{C}) \simeq H^n_{\mathrm{DR}}(X)$, we can express the Hodge filtration in terms of the stupid (obvious) filtration $\sigma_{\geqslant p}\Omega^{\cdot}_X = \{0 \to \Omega^p_X \to \Omega^{p+1}_X \to \ldots\}$ on the De Rham complex. One of Griffiths's discoveries [Gr] was that the Hodge filtration is related to the pole order filtration: if ω has a pole of order not greater than $k+1$ along a non-singular projective variety X, then the residue $\operatorname{Res}\omega$ is a sum of forms of the type (p, q) with $q \leqslant k$. This gives a purely algebraic definition of the Hodge filtration.

The theory of MHS developed by Deligne [D2, D3] is used more and more. The definition of the MHS includes, besides a decreasing Hodge filtration $F^{\cdot}$, an increasing weight filtration W. Deligne showed that the cohomology of any algebraic variety (possibly non-proper and singular) has a natural MHS. The MHS also appears in investigations of degenerations of algebraic varieties. Let a morphism $f\colon X \to S$ define a family of non-singular projective varieties over the punctured disk S', and let $X_0 = f^{-1}(0)$ be the degenerate fibre. Schmid [Sm] and Steenbrink [S1] investigated the question of what happens with the Hodge structure on $H^n(X_t, \mathbb{Z})$, when t is limited to the point 0. The limit object appears to be a MHS. The Hodge filtration of the limit MHS is in a sense the limit of the Hodge filtration on $H^n(X_t)$, and the weight filtration is related to the monodromy.

The study of period maps goes back to the investigations of Abel and Jacobi on integrals of algebraic functions. A tempestuous development of the theory of periods of integrals begins after Griffiths's papers (1968). Griffiths studied the properties of integrals in terms of the notions of the period matrix space and the period map which he introduced. Let us consider a family X_t of non-singular projective varieties depending on a parameter $t \in S$, and defined by a proper morphism $f: X \to S$. Using the connection on the fibration $\underline{H} = \bigcup_{t \in S} H^n(X_t)$, we can displace the Hodge structures on $H^n(X_t)$ to the cohomology space $H = H^n\ (X_{t_0})$ of a fixed fibre. Considering the Hodge filtration $F^p H^n(X_t)$ only, we can associate a flag $F^\cdot(t)$ in H to every point $t \in S$,

$$H = F^0(t) \supset F^1(t) \supset \ldots \supset F^{n+1}(t) = \{0\},$$

and hence obtain a point $F^\cdot(t)$ of the manifold of flags of given type. We obtain the period map $\Phi: S \to \mathscr{F}$. In coordinates the period map is given by the periods of integrals (we have to choose a basis in the homology space consisting of continuous families of cycles, and to define the subspaces $F^p(t)$ by bases of differential forms). In fact the definition of Φ is not correct, since as t moves round a loop in S, the identification of $H^n(X_{t_0})$ with itself need not be the identity. So we have to consider either a map $\tilde{\Phi}: \tilde{S} \to \mathscr{F}$ of the universal cover of S, or a map Φ of S to a quotient of $\mathscr{F}$ by some group.

Griffiths found that to first order, F^p is deformed only into the subspace F^{p-1}. In terms of the Gauss–Manin connection this can be interpreted as $\nabla F^p \subset \Omega^1_S \otimes F^{p-1}$ (the horizontality theorem). In fact, Griffiths considered polarized Hodge structures on the primitive cohomology groups $P^n(X_t)$. He constructed the period matrix space D of all possible polarized Hodge structures of given type, which is a submanifold of the flag manifold distinguished by the Hodge–Riemann bilinear relations. It turns out that D is an open homogeneous complex manifold, there is a naturally defined properly discontinuous group Γ of analytic automorphisms of D such that $M = D/\Gamma$ is an analytic space, and then we obtain the period map $\Phi: S \to M$.

The period map can be used for the description of the moduli of algebraic varieties. Here there are problems about Torelli-type theorems. The global Torelli problem is the question of whether the period map $\Phi: \mathscr{M} \to M$ of the moduli space of algebraic varieties of a given type is an embedding, i.e. whether the period matrix uniquely characterizes the polarized algebraic variety. The affirmative answer to this question in the case of algebraic curves is the usual Torelli theorem. The local Torelli problem is one of deciding when the Hodge structure on $H^*(X_t, \mathbb{C})$ separates points in the local moduli space (Kuranishi space) of X_t. The infinitesimal Torelli problem is one of deciding when the tangent map $d\Phi$ to the period map of the universal family is a monomorphism. The criterion for the infinitesimal Torelli theorem to hold, which was obtained by

Griffiths (1968), stimulated the appearance of many papers on this theme. We are not able in this introduction to go in detail into the problems touched upon above, and we refer the reader to one of the surveys on the theory of Hodge structures and periods of integrals, e.g. to [K-Ku] and [B-Z], where one can find other references.

The aim of this book is to transfer the ideas and notions, described above, to the local situation – to the case of isolated singularities of holomorphic functions $f\colon (\mathbb{C}^{n+1}, 0) \to (\mathbb{C}, 0)$. Again we have a morphism $f\colon X \to S$, but now the fibres are local analytic hypersurfaces in an open set $X \subset \mathbb{C}^{n+1}$. In Chapter I we introduce the main personage of this book – the meromorphic Gauss–Manin connection $\mathscr{M}$ of a singularity f and the Brieskorn lattice $\mathscr{H}^{(0)}$ in it. We prove the regularity of the singularity of the Gauss–Manin connection and the monodromy theorem. The discussion in Chapter I is based on the classical papers of Brieskorn [Br] and Malgrange [M]. In Chapter II we consider the limit MHS appearing on the vanishing cohomology $H^n(X_t, \mathbb{C})$ of an isolated singularity f. In the main we follow the development as it occurred historically. Here the main contribution comes from the papers of Steenbrink, Varchenko and Scherk. Initially the MHS on the vanishing cohomology $H^n(X_t)$ was constructed by Steenbrink [S3] following a suggestion of Delinge. He used an embedding of the morphism $f\colon X \to S$ in a projective family $Y \to S$ and the limit MHS in the case of a degeneration of projective varieties [S1]. Then Varchenko [V2, V3] proposed and accomplished the direct introduction of the limit Hodge filtration $F^{\cdot}$ on $H^n(X_\infty, \mathbb{C})$ (by means of asymptotics of integrals), without using an embedding to a projective family. Following this idea Scherk and Steenbrink [Sc-S] introduced the filtration $F^{\cdot}$ in a different way. They showed how the filtration $F^{\cdot}$ is obtained from the embedding of the Brieskorn lattice $\mathscr{H}^{(0)}$ in the meromorphic connection. Finally, in Chapter III we consider the period map of a μ-const deformation $f_y(x)$ of isolated singularities parametrized by points $y \in Y$ of a non-singular manifold. The basis of this chapter is the papers of Saito [Sa8], Karpishpan [Ka2] and Hertling [He1, He2]. First we consider the period map defined via embedding of Brieskorn lattices but not the one defined via the limit MHS on $H^n(X_t, \mathbb{C})$ as it should be according to the ideology presented at the beginning of this introduction. Instead it is connected with the following. Firstly, the limit MHS is determined by the embedding of Brieskorn lattice, and, secondly, the limit MHS on $H^n(X_t, \mathbb{C})$ is a rougher invariant than the embedding $\mathscr{H}^{(0)} \subset \mathscr{M}$ is (because the filtration $F^{\cdot}$ is defined not by the embedding $\mathscr{H}^{(0)} \subset \mathscr{M}$ but by the embedding of adjoint objects $Gr^{\cdot}_V\mathscr{H}^{(0)} \subset Gr^{\cdot}_V\mathscr{M}$). We give examples of the explicit calculation of the period maps for the deformations of unimodular singularities and prove the horizontality theorem and the infinitesimal Torelli theorem.

Now we give a detailed description of the contents of this book. In §1 of Chapter I we recall the definition of the Milnor fibration [Mi] $f\colon X \to S$ of an isolated singularity $f\colon (\mathbb{C}^{n+1}, 0) \to (\mathbb{C}, 0)$. The restriction of f over the punctured disk S' is a smooth locally trivial fibration. We introduce the cohomological fibration $\underline{H} = \bigcup_{t\in S'} H^n(X_t, \mathbb{C})$ defining the topological Gauss–Manin connection ∇_{top} on the sheaf of sections $\mathscr{H} = \underline{H} \otimes \mathscr{O}_{S'} = R^n f_* \mathbb{C}_{X'} \otimes \mathscr{O}_{S'}$. In §2 we develop elements of the theory of connections on locally free sheaves, the presentation of which one can find in Deligne's book [D1]. We pay special attention to the relation between ∇ and the dual to it connection ∇^*. For Gauss–Manin connections, determined by the homological and cohomological fibrations, this leads to the period matrix of a singularity f being a solution of system of linear differential equations – the Picard–Fuchs equation of the singularity.

In §3 we introduce the De Rham cohomology sheaf $\mathscr{H}^n_{\text{DR}}(X/S) = \mathscr{H}^n(f_*\Omega^{\cdot}_{X/S})$, which is a natural extension of the sheaf $\mathscr{H}$ to the whole disk. This sheaf establishes the connection between topology and algebra and analysis, and reflects the analytic nature of the singularity f. §4 contains Brieskorn's calculation of the Gauss–Manin connection of the sheaf $\mathscr{H}^{(-2)} = \mathscr{H}^n_{\text{DR}}(X/S)$ in terms of differential forms. This calculation naturally leads to the sheaf $\mathscr{H}^{(-1)} = f_*\Omega^n_{X/S}/d(f_*\Omega^{n-1}_{X/S})$ which contains $\mathscr{H}^{(-2)}$ and is also an extension of the sheaf $\mathscr{H}$. We introduce the notion of the meromorphic Gauss–Manin connection on the sheaf $\mathscr{M} = \mathscr{H}^{(-2)} \otimes \mathscr{O}_S[t^{-1}]$. We also explain how to give a more conceptual description of the calculation of the Gauss–Manin connection on $\mathscr{H}^{(-2)}$ as a connecting homomorphism in an exact cohomology sequence. In §5 the main personage of this survey appears. This is the Brieskorn lattice $\mathscr{H}^{(0)}$, the third natural extension of the sheaf $\mathscr{H}$ to the whole disk S. The lattice $\mathscr{H}^{(0)}$ is defined in terms of $(n+1)$-forms. The identification of $\mathscr{H}^{(0)}$ and $\mathscr{H}^{(-1)}$ on S' is realized by means of the Leray derivative (the Poincaré residue), $\omega \mapsto \omega/\mathrm{d}f = \operatorname{Res}[\omega/(f-t)]$. All three lattices $\mathscr{H}^{(-i)}$, $i = 0, 1, 2$, are the terms of an increasing filtration on $\mathscr{M}$. The correlation between them is contained in the diagram shown in (I.5.3.4). In §6 from Malgrange's theorem, which claims that the periods of an n-form ω have the limit $\lim_{t\to 0}\int_{\gamma(t)}\omega = 0$, we obtain a result of Sebastiani about the absence of torsion in the sheaves $\mathscr{H}^{(-i)}$, and hence we obtain that these sheaves are locally free $\mathscr{O}_S$-modules of rank μ.

In §7 we recall the classical definition of a regular singular point of a system of linear differential equations and Sauvage's theorem on the regularity of a simple singular point. We give the definition of a connection with regular singularity and of the residue of such a connection wrt a saturated lattice. In §8 we prove a fundamental theorem on the regularity of the Gauss–Manin connection. We give two proofs of this important fact. Firstly, we prove that periods of holomorphic forms give solutions of the Picard–Fuchs equation. Then from Malgrange's theorem it follows that horizontal sections of the Gauss–Manin connection have a moderate growth. Secondly, using a resolution of the singularity and the sheaves of differential forms with logarithmic poles, we construct a saturated lattice in the meromorphic Gauss–Manin connection. In §9 from the regularity of the Gauss–Manin connection we deduce the monodromy theorem claiming that all eigenvalues of the monodromy are roots of unity. We give a beautiful proof of this due to Brieskorn, based on the positive solution of the seventh Hilbert problem. The part of the monodromy theorem concerning the size of Jordan blocks is proved in (II.3.5.9). We obtain corollaries of the monodromy theorem about the decomposition into series of the periods of integrals of differential forms.

In §10 we consider non-isolated hypersurface singularities. Starting from the construction of the Gauss–Manin connection as a connecting homomorphism in an exact sequence of complexes and in the spirit of [Sr], we obtain a natural generalization of Brieskorn lattices $\mathscr{H}^{(i)}$ to the case of non-isolated singularities. We give Van Straten's criterion on the absence of torsion in the sheaves $\mathscr{H}^{(i)}$ and its application to the non-isolated singularities with a one-dimensional critical set.

In Chapter II we consider the limit MHS appearing on the vanishing cohomology $H^n(X_t, \mathbb{C})$ of an isolated hypersurface singularity. In §1 we recall very briefly the necessary basic definitions from the theory of MHS developed by Deligne. In §2 we introduce the limit MHS constructed by Schmid [Sm] for a variation $(\underline{H}, \mathscr{F}^{\cdot})$ of pure Hodge structures $F_t^{\cdot}$ on a vector space H, parametrized by points of the punctured disk S'. The limit Hodge filtration $F^{\cdot}$ appears on the zero fibre $\mathscr{L}/t\mathscr{L}$ of the canonical extension $\mathscr{L}$ of the sheaf $\mathscr{H} = \underline{H} \otimes \mathscr{O}_{S'}$ to the point 0. The weight filtration $W_{\cdot}$ is the weight filtration $W(N)$ of the nilpotent operator $N = -(1/2\pi\mathrm{i}) \log T_u$. In §3 we introduce the limit MHS on the vanishing cohomology $H = H^n(X_\infty, \mathbb{C})$ (X_∞ is the canonical fibre of the Milnor fibration) according to Steenbrink. Steenbrink used an embedding of the Milnor fibration $X \to S$ to a projective family $Y \to S$. He introduced the limit MHS on $H^n(Y_\infty, \mathbb{C})$ using a resolution of singularities of the zero fibre and the complex of relative differential forms with logarithmic poles. We give Steenbrink's construction only very schematically, without going into technical details. The limit MHS on $H^n(X_\infty, \mathbb{C})$ constructed by Steenbrink can be considered as a quotient of the MHS on $H^n(Y_\infty, \mathbb{C})$. From here it follows the symmetry of Hodge numbers of the MHS on $H^n(X_\infty, \mathbb{C})$, and also the monodromy theorem on the size of Jordan blocks.

In §4 we consider the Hodge theory of a non-singular hypersurface Y in a non-singular manifold X developed by Griffiths and extended by Deligne to the case of divisors with normal crossings. This theory enables us to calculate the cohomology of Y and of the complement $X \setminus Y$ by means of differential forms on X with poles on Y, and also relates the Hodge filtration $F^\cdot$ to the pole order filtration $P^\cdot$. When Y has singularities, the filtrations $F^\cdot$ and $P^\cdot$ do not, generally, coincide [Ka1, D-Di]. We apply this theory for the description of $H^n(X_t)$, where the hypersurface $X_t \subset X$ is the fibre of the Milnor fibration. The relative variant of this theory enables us to obtain a natural extension $\mathscr{H}_X$ of the sheaf $\mathscr{H}$ to S. Then $\mathscr{H}_X$ is described as the cohomology sheaf $\mathscr{H}_X = \mathscr{H}^{n+1}(K^\cdot, \underline{d})$ of the bicomplex $(K^{\cdot\cdot}, \mathrm{d}_1, \mathrm{d}_2)$, the terms of which are differential forms with poles, and the degrees are defined by the order of a differential form and the order of its pole. In the bicomplex $K^{\cdot\cdot}$ the Poincaré complex and the Koszul complex are intertwined, and its differentials are exterior differentiation $\mathrm{d}_1 = \mathrm{d}$ of differential forms and exterior multiplication by the form df. The Gauss–Manin connection and the Hodge filtration $F^\cdot$ on $\mathscr{H}_X$ are defined in terms of this bicomplex. In fact $\mathscr{H}_X$ is a D-module, or a differential system, which appears in the papers of Pham [Ph1, Ph2], Scherk and Steenbrink [Sc-S]. The study of this is extended in the papers of M. Saito [Sa1–Sa10] in the frames of the theory of D-modules and of the 'monstrous' theory of mixed Hodge modules developed by him. This latter theory raises the level of abstractness by at least one more order, and moves further from the soil, in which all this began to grow. The 'new' homological algebra, derived cathegories, perverse sheaves, the Riemann–Hilbert correspondence, etc. are essentially used in it. Under Saito's influence the theory becomes more and more abstract and technical. One of the aims of this book was not to follow this line but to evolve the theory using the 'traditional' (i.e. habitual to algebraic geometricians) language of sheaves, connections, spectral sequences etc. and to avoid using the language of the theory of D-modules, the theory of mixed Hodge modules etc. This is possible because of the following circumstance. As is shown in §5 the operator ∂_t is invertibl $\mathscr{H}_X$, $\mathscr{H}_X$ contains the Brieskorn lattice $\mathscr{H}^{(0)} = F^n\mathscr{H}_X$ as well as canonical lattice $\mathscr{L}$, $\mathscr{H}^{(0)} \subset \mathscr{L} \subset \mathscr{H}_X$. That is the pair $\mathscr{H}^{(0)} \subset \mathscr{L}$ tains all the information which is of interest to us. The localization of differential system $\mathscr{H}_X$ leads us to the classical frames of the meromorphic connection $\mathscr{M} = (\mathscr{H}_X)_{(t)}$, and as is explained in the supplement to §7, the pair $\mathscr{H}^{(0)} \subset \mathscr{L}$ happily stands localization and removes to $\mathscr{M}$, $\mathscr{H}^{(0)} \subset \mathscr{L} \subset \mathscr{M}$. All this enables us to work in the classical frames of the meromophic connection $\mathscr{M}$.

In §6 we study the structure of the meromorphic connection $\mathscr{M}$, its decomposition into the sum $\mathscr{M} = \hat{\oplus}_\alpha C_\alpha$ of root subspaces of the operator $t\partial_t$, the $V^\cdot$-filtration on $\mathscr{M}$. We determine the isomorphism ψ: $H \rightrightarrows \mathscr{L}/t\mathscr{L} = \oplus_{-1<\alpha\leq 0} C_\alpha$ between the canonical fibre of the fibration $\underline{H}$ and the zero fibre of the canonical lattice $\mathscr{L} = V^{>-1}\mathscr{M}$, and it enables us to introduce the MHS on H in terms of the lattice $\mathscr{L}$.

In §7 we introduce the MHS on $H = H^n(X_\infty, \mathbb{C})$, according to Varchenko [V3] and Scherk and Steenbrink [Sc-S]. At first we define the limit Hodge filtration $F^\cdot$ according to Scherk and Steenbrink, who transformed the approach of Varchenko and gave the definition of $F^\cdot$ in terms of the differential system $\mathscr{H}_X$ and the embedding $\mathscr{H}^{(0)} \subset \mathscr{L}$, and then we define $F^\cdot$ according to Varchenko. Such an order of exposition is connected with the fact that, in our opinion, the construction of Scherk and Steenbrink has genetically a more natural motivation. In the beginning we observe the sequence of steps leading to the construction of $F^\cdot$ by Scherk and Steenbrink.

In §8 we study the main discrete invariant of a hypersurface singularity – its spectrum Sp(f), and a more detailed invariant Spp(f) the set of spectral pairs. To give Spp(f) is equivalent to giving Hodge numbers $H_\lambda^{p,q}$. The spectrum Sp(f) is a set of μ rational numbers $\alpha_1, \ldots, \alpha_\mu$, where $\alpha_j = -(1/2\pi \mathrm{i})\lambda_j$ are logarithms of the eigenvalues λ_j of the monodromy T. It codes the relation between the semisimple part of the monodromy and the limit Hodge filtration $F^\cdot$. We study the properties of spectrum. We show that $Sp(f) \subset (-1, n)$ and that the spectrum is symmetric wrt the centre of this interval $(n-1)/2$. We develop the techniques for the calculation of the spectrum. We explain: how to find the spectrum of a (semi)quasihomogeneous singularity; how to find the spectrum in terms of the Newton filtration defined by the Newton boundary; and in particular, how to find the negative part of spectrum, the degree of which, $\sum_{-1<\alpha\leq 0} n_\alpha = p_g$, is equal to the geometric genus of the singularity [S3]; how to find the spectrum of the join of isolated singularities, and in particular, how the spectrum changes under adding squares of new variables. This technique enables us to find spectra of all the simple, uni- and bi-modal singularities, which we gather together in a table. Finally, we study variations of MHS of families of hypersurface singularities. We deal with this more widely in Chapter III, but in §8 we consider the behavior of discrete invariants under deformations. We give the results of Varchenko and Steenbrink on the semicontinuity of spectrum and on its stability under μ-const deformations. At the end of §8, following Steenbrink, we define the spectrum of non-isolated singularities and give a theorem on the relation of the spectrum of a singularity with a one-dimensional critical set and the spectra of isolated singularities of its Iomdin series.

In §1 of Chapter III we begin to study μ-const deformations $f_y(x)$ of isolated singularities parametrized by the points $y \in Y$ of a non-singular variety. We explain how to glue the objects, associated earlier with an 'individual' singularity, to a family parametrized by Y. In particular, we obtain the family of Milnor fibrations $X(y)$, the family of cohomological fibrations $\underline{H} = \bigcup \underline{H}(y)$, the family of meromorphic connections $\mathscr{M} = \hat{\oplus}_\beta C_\beta$, etc. In §2 we obtain the formula for differentiating wrt parameter y geometric sections $s[\omega](t, y)$, defined by a holomorphic $(n+1)$-form $\omega = g(x, y)\,dx$. The same formula is used for differentiating the root components $s(\omega, \beta)$ of the geometric sections $s[\omega](y) = \sum_{\beta > -1} s(\omega, \beta)(y)$. In the case of upper diagonal deformations of a quasihomogeneous singularity $f(x)$ from this formula we obtain a formula for decomposition of $s(\omega, \beta)(y)$ into Taylor series in degrees of y. In §3 first we define the period map Φ: $Y \to \Pi$ defined by the embedding of Brieskorn lattices. For all Brieskorn lattices $\mathscr{H}^{(0)}(y)$ we have inclusions $V^{>-1} \supset \mathscr{H}^{(0)}(y) \supset V^{n-1}$. The period map takes a point y to the subspace $\mathscr{H}^{(0)}(y) \bmod V^{n-1}$ in the finite-dimensional vector space $V^{>-1}/V^{n-1}$, i.e. to the point in the Grassman manifold Π. Then we define the period map $\overline{\Phi}$ defined by the MHS on the vanishing cohomology. This period map takes a point y to the Hodge filtration $F^{\cdot}(y)$ of the limit MHS, i.e. to the point in the flag manifold. We give, following Hertling, explicit calculations of the period maps of universal families of unimodular singularities. In §4 we calculate the tangent map of the period map and prove the horizontality theorem and the infinitesimal Torelli theorem for the period map Φ. We compare the period maps Φ and $\overline{\Phi}$ of the miniversal μ-const deformation of a quasihomogeneous singularity. From this comparison it follows that the Torelli theorem for the period map $\overline{\Phi}$ in general is false. Finally, in §5 we consider the 'global' Torelli problem. Varchenko introduced the notion of the Picard–Fuchs singularity PFS(f) of a singularity f in terms of framed Picard–Fuchs equations. Hertling reformulated this in terms of embeddings of Brieskorn lattices. We interpret it as a point in the quotient Π/G_1 of the period space by the group $G_1 \subset GL(H_{\mathbb{Z}})$ of automorphisms commuting with the monodromy. Then the Torelli problem for the family f_y, $y \in Y$, concerns the injectivity of the map Φ_1: $Y/\sim \to \Pi/G_1$, where '$\sim$' is the equivalence relation induced by the R-equivalence of singularities. We give Hertling's results [He1, He2].

The reader is expected to have the knowledge and training usual in algebraic and analytic geometry. This includes knowledge of sheaf theory and the technique of spectral sequences.

The list of references reflects the development of singularity theory. It includes: firstly, the original papers of E. Brieskorn, B. Malgrange, J.H.M. Steenbrink, J. Scherk, A.N. Varchenko, F. Fham, M. Saito, Ya. Karpishpan and C. Hertling, directly connected with the considered topic; secondly, some general papers on MHS and periods of integrals (P. Griffiths, P. Deligne, W. Schmid); thirdly, some books and surveys [Mi, AGV, Ph1, D1, Di3, B-Z]. Some other references are also included.

Now about numbering and cross-references in this book. Each of three chapters is divided into sections, and each section is divided into subsections. In a subsection all claims, remarks and displayed formulae are numerated successively in a uniform way by three numbers, the first of which is the number of the section and the second the number of the subsection. For example, the tag (1.3.2) means claim (or formula) 2 in subsection 3 of section 1. To refer to a claim etc. we use three numbers within a given chapter, and four numbers, the first of which is a roman numeral, to refer to other chapters. Thus (II.7.5.2) refers the reader to claim 2 of subsection 5 of section 7 of chapter II.

The preparation of this book was partially supported by Grant. No. 95-01-01575 from the Russian Foundation for Fundamental Research and Grant No. 4373 from the INTAS.

I

The Gauss–Manin connection

1 Milnor fibration, Picard–Lefschetz monodromy transformation, topological Gauss–Manin connection

1.1 Milnor fibration

(1.1.1) Let $f\colon (\mathbb{C}^{n+1}, 0) \to (\mathbb{C}, 0)$ be a germ of holomorphic function $t = f(x_0, x_1, \ldots, x_n)$. We assume as a rule that f has an isolated critical point at $0 \in \mathbb{C}^{n+1}$, in other words, f has (or is) an isolated singularity. Let $B = \{|x| < \varepsilon\} \subset \mathbb{C}^{n+1}$ be a ball of radius ε, and let $S = \{|t| < \delta\} \subset \mathbb{C}$ be a disk of radius δ. Put $X = B \cap f^{-1}(S)$ and let f also denote the restriction of f onto X, $f\colon X \to S$. Let $S' = S \backslash \{0\}$ be the punctured disk, $X_t = f^{-1}(t)$ be the fibre over the point $t \in S'$, and $X_0 = f^{-1}(0)$ be a singular fibre, $X' = X \backslash X_0$. Denote by f' the restriction of f on X'

$$\begin{array}{ccccc} X & \supset & X' & \supset & X_t \\ f\downarrow & & \downarrow f' & & \\ S & \supset & S' & \ni & t. \end{array}$$

As is shown by Milnor [Mi], if ε and $\delta \ll \varepsilon$ are sufficiently small, then f' is a smooth locally trivial fibration, the diffeomorphism type of which only depends on the germ of f at 0. Usually the fibration $f'\colon X' \to S'$ is called the Milnor fibration. In the following it will be convenient for us to call the whole morphism $f\colon X \to S$ the *Milnor fibration*. Any fibre X_t, $t \in S'$, is called a *Milnor fibre*. We can think of the singular fibre X_0 as a degeneration of a family of manifolds X_t. Replacing the fibres X_t by their (co)homology we get a 'linearization' of the family X_t.

1.2 Cohomological Milnor fibration

(1.2.1) A Milnor fibration f' defines a vector bundle $\underline{H} \to S'$ on S', or a locally constant sheaf (or a system of local coefficients, in different terminology)

$$\underline{H} = \bigcup_{t \in S'} H^p(X_t, \mathbb{C}) = R^p f_* \mathbb{C}_{X'}$$

and it also defines a dual vector bundle

$$\underline{H}_* = \mathrm{Hom}_{\mathbb{C}_{X'}}(\underline{H}, \mathbb{C}_{S'}) = \bigcup_{t \in S'} H_p(X_t, \mathbb{C}).$$

We call $\underline{H}$ the *cohomological*, and $\underline{H}_*$ the *homological Milnor fibration* of a singularity f.

1.3 Topological Gauss–Manin connection

(1.3.1) Denote by

$$\mathscr{H} = \underline{H} \underset{\mathbb{C}_{S'}}{\otimes} \mathscr{O}_{S'} = R^p f_* \mathbb{C}_{X'} \underset{\mathbb{C}_{S'}}{\otimes} \mathscr{O}_{S'}$$

the locally free sheaf of sections of the fibration $\underline{H}$. A local section w of $\underline{H}$ can be viewed as a family of cohomology classes $w(t)$ depending on parameter t. The flat system $\underline{H}$ in the sheaf $\mathscr{H}$ allows us to translate cohomology classes $w(t)$ from a fibre X_t to a nearby fibre and, consequently, allows us to differentiate the cohomology classes wrt parameter, and to define a connection on the sheaf (see the §2). This connection is called the *topological* (or transcendental) *Gauss–Manin connection* and is denoted by Δ_{top}.

1.4 Picard–Lefschetz monodromy transformation

(1.4.1) A locally constant bundle $\underline{H}$ (or $\underline{H}_*$) defines (and is defined by) an action of the fundamental group $\pi_1(S', t)$ on a fibre $\underline{H}_t = H^p(X_t, \mathbb{C})$. Let γ: $[0, 1] \to S'$ be a loop representing an element of $\pi_1(S', t)$. The inverse image of the locally trivial fibration f': $X' \to S'$ defines a locally trivial, and consequently, trivial fibration $\gamma^{-1}X' \xrightarrow{\varphi} [0, 1]$ on the segment. A trivialization of this family defines a diffeomorphism h_γ: $X_t = \varphi^{-1}(0) \to X_t = \varphi^{-1}(1)$. We thus obtain the monodromy representation

(1.4.2) $$\pi_1(S', t) \to \mathrm{Aut}\, H^p(X_t, \mathbb{C}), \quad [\gamma] \mapsto (h_\gamma^*)^{-1}.$$

Let $[\gamma] \in \pi_1(S', t)$ be the generator of the fundamental group represented by a counter-clockwise oriented circle γ around the origin. The linear transformations

(1.4.3) $$M = (h_\gamma)_*: H_p(X_t, \mathbb{C}) \to H_p(X_t, \mathbb{C})$$

(1.4.4) $$T = (h_\gamma^*)^{-1}: H^p(X_t, \mathbb{C}) \to H^p(X_t, \mathbb{C})$$

are called the (local) *Picard–Lefschetz monodromy* transformation of homology and cohomology, respectively.

(1.4.5) The Milnor fibration, the Gauss–Manin connection and the Picard–Lefschetz monodromy are the main objects associated with a singularity f and they contain a great deal of information about this singularity. These objects can be studied by topological methods. In the case of isolated singularities the most important fact ([Mi]) is that the fibre X_t has the homotopy type of a bouquet $S^n \vee \cdots \vee S^n$ of μ spheres of dimension n. In particular, the Betti numbers $b_i = \dim H^i(X_t, \mathbb{C}) = 0$ for $i \neq 0$ and n and, consequently, $\underline{H} = R^n f_* \mathbb{C}_{X'}$ is the only non-trivial fibration among the fibrations $R^p f_* \mathbb{C}_{X'}$. The number $\mu = \dim H^n(X_t, \mathbb{C}) = b_n$ (the rank of this fibration) is called the *Milnor number* of the singularity f. We refer the reader to other surveys ([AGV], ch. 1; [Di3]) for topological methods for the study of singularities. In this survey we'll be interested in the study of singularities by algebraic methods in which the main subject as well as the main tool of investigation is the differential forms associated with a singularity.

2 Connections, locally constant sheaves and systems of linear differential equations

2.1 Connection as a covariant differentiation

Let S be a complex manifold of dimension m, and $\mathscr{E}$ be a locally free sheaf of rank n on S. The presence of a connection on $\mathscr{E}$ enables us to differentiate sections of $\mathscr{E}$ along vector fields on the base S. On the other hand, the notion of a connection on $\mathscr{E}$ is a way of having an invariant definition of S of a system of homogeneous linear differential equations with n unknown functions in m variables. In the following we'll be mainly interested in the Gauss–Manin connection. First we briefly recall the necessary facts about connections. An excellent exposition of this subject can found in Deligne's paper [D1].

(2.1.1) *Definition* A *connection* on a quasicoherent sheaf of $\mathscr{O}_S$-modules $\mathscr{E}$ is a $\mathbb{C}$-linear homomorphism

$$\nabla\colon \mathscr{E} \to \Omega^1_S \underset{\mathscr{O}_S}{\otimes} \mathscr{E} = \Omega^1_S(\mathscr{E})$$

satisfying the *Leibniz identity*

$$\nabla(gs) = dg \otimes s + g\nabla s,$$

where g and s are local sections of sheaves $\mathcal{O}_S$ and $\mathcal{E}$, respectively. ∇ is called the *covariant differentiation*.

This notion originates from the corresponding classical definition. The classical definition of a connection on $\mathcal{E}$ enables us to compare infinitely close fibres of the sheaf $\mathcal{E}$ and consists in giving for any pair of points x and y, infinitely close to first order in S, an isomorphism of fibres $\gamma\colon \mathcal{E}(x) \to \mathcal{E}(y)$. To translate this into modern language we have to consider points in the categorical sense, i.e. morphisms $T \to S$. Then two T-valued points $x\colon T \to S$ and $y\colon T \to S$ are said to be infinitely close to first order, if $(x, y)\colon T \to S \times S$ factors through the neighbourhood of the first order $\Delta_1 = (\Delta, \mathcal{O}_{X\times X}/I^2)$ of the diagonal Δ in $S \times S$. In this way we come naturally to the previous formal definition [D1].

2.2 Equivalent definition: a covariant derivative along a vector field

(2.2.1) For a local vector field w on S, i.e. a local section of the sheaf of vector fields $\theta_S = \mathcal{H}om_{\mathcal{O}_S}(\Omega^1_S, \mathcal{O}_S)$, and for a section s of the sheaf $\mathcal{E}$ the *covariant derivative* of s along w is defined by

(2.2.2) $$\nabla_w(s) = \langle \nabla s, w\rangle$$

where $\langle , \rangle$ is induced by the pairing $\Omega^1_S \times \theta_S \to \mathcal{O}_S$.

∇_w defines a $\mathbb{C}$-linear homomorphism

(2.2.3) $$\nabla_w\colon \mathcal{E} \to \mathcal{E}$$

and the Leibniz identity transforms into the (Leibniz) identity

(2.2.4) $$\nabla_w(gs) = w(g)s + g\nabla_w(s).$$

We obtain one more definition of the notion of connection on $\mathcal{E}$ as an $\mathcal{O}_S$-linear homomorphism

$$\mathrm{Der}_{\mathbb{C}}(\mathcal{O}_S) = \theta_S \to \mathcal{E}nd_{\mathbb{C}}(\mathcal{E}), \quad w \mapsto \nabla_w,$$

where ∇_w satisfies the Leibniz identity.

In particular, if $\dim S = 1$ and $w = \mathrm{d}/\mathrm{d}t$ is a basis vector field on S, then giving a connection on $\mathcal{E}$ is equivalent to giving an operator $D = \nabla_{\mathrm{d}/\mathrm{d}t}\colon \mathcal{E} \to \mathcal{E}$ satisfying the Leibniz identity. The operator D will be also denoted by ∂_t.

(2.2.5) *Description of all connections on* $\mathcal{E}$ If $\nabla_1, \nabla_2 \in \mathrm{Hom}_{\mathbb{C}}(\mathcal{E}, \Omega^1_S(\mathcal{E}))$ are two connections on $\mathcal{E}$, then it follows from the Leibniz identity that their difference $\nabla_2 - \nabla_1 \in \mathrm{Hom}_{\mathcal{O}_S}(\mathcal{E}, \Omega_S(\mathcal{E}))$ is not only a $\mathbb{C}$-linear, but also an $\mathcal{O}_S$-linear homomorphism from $\mathcal{E}$ to $\Omega^1_S(\mathcal{E})$. Conversely, if ∇_0 is a connection on $\mathcal{E}$ and $\Gamma \in \mathrm{Hom}_{\mathcal{O}_S}(\mathcal{E}, \Omega^1_S(\mathcal{E}))$, then $\nabla = \nabla_0 + \Gamma$ is a connection on $\mathscr{E}$. Thus connections on $\mathscr{E}$ form a principal homogeneous space over $\mathrm{Hom}_{\mathscr{O}_S}(\mathscr{E}, \Omega^1_S(\mathscr{E}))$.

2.3 Local calculation of connections. Relation to differential equations

(2.3.1) Let $\mathscr{E}$ be a locally free sheaf of rank n. Choose in $\mathscr{E}$ a basis (locally) $e_1, \ldots, e_n$, and let $e\colon \mathscr{O}^n_S \simeq \mathscr{E}$, $e\colon (0, \ldots, 1, \ldots, 0) \mapsto e_i$. For brevity we'll use matrix notation: let $e = (e_1, \ldots, e_n)$ be a row of base vectors, and let $y = (y_1(x), \ldots, y_n(x))^t$ be a column of coordinates of a section s wrt the basis e. The section s of $\mathscr{E}$ is represented by the scalar product $s = ey$.

A choice of basis e defines a constant subsheaf $e\colon \mathbb{C}^n_S \subset \mathscr{E}$ and a connection ∇_0 on $\mathscr{E}$, $\nabla_0 s = ds$, i.e.

$$\nabla_0 s = e\,\mathrm{d}y = \sum_{i=1}^{n} \mathrm{d}y_i e_i,$$

for which the sections e_i are horizontal, $\nabla_0(e_i) = 0$. Any connection ∇ on $\mathscr{E}$ is represented as

(2.3.2) $$\nabla = \nabla_0 + \Gamma, \quad \text{where} \quad \Gamma \in \mathrm{Hom}_{\mathscr{O}_S}(\mathscr{E}, \Omega^1(\mathscr{E})).$$

In the basis e, the homomorphism $\Gamma\colon \mathscr{E} \to \Omega^1_S \otimes \mathscr{E}$, $\Gamma\colon \mathscr{O}_S e_1 \oplus \cdots \oplus \mathscr{O}_S e_n \to \Omega^1_S e_1 \oplus \cdots \oplus \Omega^1_S e_n$ is given by *a matrix of connection forms* $\omega = (\omega_{ij})$

(2.3.3) $$\Gamma(e_j) = \sum_{i=1}^{n} \omega_{ij} e_i, \quad \text{i.e. } \Gamma(e) = e\omega,$$

i.e. ω is the matrix of the linear transformation Γ in the basis e: the coordinates of vector $\Gamma(e_j)$ are in the jth column of the matrix ω. Now, if $s = ey$, then $\nabla s = \nabla_0 s + \Gamma s$, i.e.

$$e\nabla y = e\,\mathrm{d}y + \Gamma(ey) = e\,\mathrm{d}y + \Gamma(e)y = e\,\mathrm{d}y + e\omega y$$

or in coordinates

$$\nabla y = \mathrm{d}y + \omega y, \qquad \nabla \begin{pmatrix} y_1 \\ \vdots \\ y_n \end{pmatrix} = \begin{pmatrix} \mathrm{d}y_1 \\ \vdots \\ \mathrm{d}y_n \end{pmatrix} + (\omega_{ij}) \begin{pmatrix} y_1 \\ \vdots \\ y_n \end{pmatrix},$$

i.e.

(2.3.4) $$(\nabla y)_i = \mathrm{d}y_i + \sum_{j=1}^{n} \omega_{ij} y_j, \quad i = 1, \ldots, n.$$

Now let $x_1, \ldots, x_m$ be local coordinates on S. Then the connection forms $\omega_{ij} \in \Omega^1_S$ are written in the form

(2.3.5) $$\omega_{ij} = \sum_{k=1}^{m} \Gamma_{ij}^{k}\,\mathrm{d}x_k,$$

where the holomorphic functions $\Gamma_{ij}^k(x)$ are called the *connection coefficients* (wrt the basis e and coordinates x).

Denote by $\partial_i = \partial/\partial x_i$ the differentiation wrt the coordinate x_i, and let $\nabla_i = \nabla_{\partial_i}: \mathscr{E} \to \mathscr{E}$ be the covariant derivative wrt the vector field ∂_i. Then (2.3.4) turns into a system of equalities

(2.3.6) $$(\nabla_k y)_i = \partial y_i/\partial x_k + \sum_{j=1}^{n} \Gamma_{ij}^{k} \partial y_i/\partial x_k,$$
$$i = 1, \ldots, n,\ k = 1, \ldots, m$$

(2.3.7) *Definition* A local section s of the sheaf $\mathscr{E}$ is called *horizontal* if $\nabla s = 0$, i.e. $s \in \operatorname{Ker}\nabla$, $\nabla: \mathscr{E} \to \Omega_S^1 \otimes \mathscr{E}$.

The condition of horizontality of section s can be written in the form $\nabla_k s = 0$, $k = 0, \ldots, m$, i.e. in the form of system of homogeneous linear differential equations of the first order

(2.3.8) $$\partial y_i/\partial x_k = -\sum_{j=1}^{n} \Gamma_{ij}^{k} \partial y_j/\partial x_k, \quad i = 1, \ldots, n,\ k = 1, \ldots, m.$$

This is a system of nm equations with n unknown functions $y_i(x)$ in m variables $x_1, \ldots, x_m$.

If $\dim S = m = 1$, t is a coordinate on S, $\omega_{ij} = \Gamma_{ij}(t)\,\mathrm{d}t$, and $\Gamma = (\Gamma_{ij}(t))$ is the matrix of connection coefficients, then the operator $D = \partial_t$ acts on a section $s = ey$ according to the formula:

(2.3.9) $$Dy = \mathrm{d}y/\mathrm{d}t + \Gamma(t)y$$

and the condition for horizontality of s is written in the form of a system of ordinary differential equations

(2.3.10) $$\mathrm{d}y/\mathrm{d}t = -\Gamma(t)y.$$

2.4 The integrable connections. The De Rham complex

(2.4.1) A connection $\nabla = \nabla_0: \mathscr{E} \to \Omega_S^1 \otimes \mathscr{E}$ can be extended to a $\mathbb{C}$-linear homomorphism of sheaves

$$\nabla_i: \Omega_S^1 \underset{\mathscr{O}_S}{\otimes} \mathscr{E} \to \Omega_S^{i+1} \underset{\mathscr{O}_S}{\otimes} \mathscr{E}$$

by means of the equalities

$$\nabla_i(\omega \otimes s) = \mathrm{d}\omega \otimes s + (-1)^i \omega \wedge \nabla_0(s).$$

(2.4.2) The $\mathbb{C}$-linear homomorphism

$$R = \nabla_1 \circ \nabla_0 \colon \mathscr{E} \to \Omega^2_S(\mathscr{E})$$

is called the curvature of the connection ∇ on $\mathscr{E}$, so that R is a section of the sheaf $\mathscr{H}om_{\mathbb{C}}(\mathscr{E}, \Omega^2_S(\mathscr{E})) \simeq \Omega^2_S(\mathscr{E}nd_{\mathbb{C}}(\mathscr{E}))$. A connection ∇ is called *integrable* if $R = \nabla_1\nabla_0 = 0$.

One can show that for vector fields $X_1, X_2 \in \theta_S$ and a section $s \in \mathscr{E}$ one has

$$R(X_1, X_2)(s) = \nabla_{X_1}\nabla_{X_2}s - \nabla_{X_2}\nabla_{X_1}s - \nabla_{[X_1,X_2]}s,$$

and the condition of integrability of connection ∇ is equivalent to the condition $\nabla_{[X,Y]} = [\nabla_X, \nabla_Y]$ for arbitrary vector fields X and Y. If $\dim S = 1$, then any connection is integrable. A local calculation of curvature R shows that the condition of integrability of ∇ is equivalent to the classical condition of integrability of the corresponding system of linear differential equations.

(2.4.3) One can show that $\nabla_{i+1}\nabla_i(\omega \otimes s) = \omega \wedge R(s)$ (the Ricci identity). Therefore the condition of integrability of a connection ∇, $R = 0$, is equivalent to the condition that

(2.4.4) $$\Omega^{\cdot}_S(\mathscr{E})\colon 0 \to \mathscr{O}_S \underset{\mathscr{O}_S}{\otimes} \mathscr{E} \xrightarrow{\nabla_0} \Omega^1_S \underset{\mathscr{O}_S}{\otimes} \mathscr{E} \xrightarrow{\nabla_1} \dots \xrightarrow{\nabla_{m-1}} \Omega^m_S \underset{\mathscr{O}_S}{\otimes} \mathscr{E} \to 0$$

is a complex, i.e. $\nabla_{i+1}\nabla_i = 0$. The complex $\Omega^{\cdot}_S(\mathscr{E})$ is called the *De Rham complex* with values in a locally free sheaf $\mathscr{E}$ with an integrable connection ∇.

2.5 Local systems and integrable connections

(2.5.1) Let S be a connected and locally connected topological space. A locally free sheaf of vector spaces E on S, i.e. a sheaf locally isomorphic to a constant sheaf $\mathbb{C}^n$, is called a *local system* on S.

(2.5.2) The fundamental group $\pi_1(S, x_0)$, $x_0 \in S$, acts on the fibre $E_{x_0} \simeq \mathbb{C}^n$. The functor $E \mapsto E_{x_0}$ establishes an equivalence of the category of local systems on S with the category of complex-valued representations of the group $\pi_1(S, x_0)$.

(2.5.3) Now let E be a local system on a complex manifold S. Consider a locally free sheaf $\mathscr{E}$ of holomorphic sections of E, $\mathscr{E} = \mathscr{O}_S \otimes_{\mathbb{C}} E$. Then there is a *canonical connection* ∇ on $\mathscr{E}$ for which the sheaf of horizontal sections coincides with E, $\operatorname{Ker} \nabla = E$. Here ∇ is defined by the formula

$$\nabla(gs) = dg\ s$$

where g and s are sections of $\mathcal{O}_S$ and E. Clearly, the canonical connection is integrable. Conversely, one has the following:

(2.5.4) *Theorem* [D1] Let ∇ be a connection on a locally free sheaf $\mathcal{E}$. Put $E = \operatorname{Ker} \nabla = \{s \in \mathcal{E} | \nabla s = 0\}$. If ∇ is an integrable connection, then E is a local system on X and $\mathcal{E} = \mathcal{O}_S \otimes_{\mathbb{C}} E$. ■

Therefore, the functors:

(a) a local system $E \mapsto$ the sheaf $\mathcal{E} = \mathcal{O}_S \otimes_{\mathbb{C}} E$ and the canonical connection ∇; and
(b) a locally free sheaf $\mathcal{E}$ with an integrable connection $\mapsto E = \operatorname{Ker} \nabla$,

establish an equivalence of categories of local systems on E and locally free sheaves with integrable connections.

2.6 Dual local systems and connections

We intend to study the local system $\underline{H} = R^p f_* \mathbb{C}_{X'} = \bigcup_{t \in S'} H^p(X_t, \mathbb{C})$ on a punctured disk, introduced in §1. $\underline{H}$ is called the *cohomological fibration*. Geometrically the dual *homological fibration* $\underline{H}' = \bigcup H_p(X_t, \mathbb{C})$ appears in a natural way. So we consider the following general situation:

(2.6.1) Let $\underline{H}$ be a local system on, say, the punctured disk S'. Denote by $\underline{H}^* = \operatorname{Hom}_{\mathbb{C}_{S'}} (\underline{H}, \mathbb{C}_{S'})$ the dual local system, let

$$\mathcal{H} = \underline{H} \underset{\mathbb{C}_{S'}}{\otimes} \mathcal{O}_{S'} \quad \text{and} \quad \mathcal{H}^* = \underline{H}^* \underset{\mathbb{C}_{S'}}{\otimes} \mathcal{O}_{S'} = \mathcal{H}om_{\mathcal{O}_{S'}} (\mathcal{H}, \mathcal{O}_{S'})$$

be the locally free sheaves corresponding to $\underline{H}$ and $\underline{H}^*$, and let ∇ and ∇^* be the connections on $\mathcal{H}$ and $\mathcal{H}^*$ defined by the above local systems. The connection ∇^* is called the dual of ∇.

Denote by

$$\langle\ ,\ \rangle \colon \underline{H} \times \underline{H}^* \to \mathbb{C}_{S'}$$

the associated non-degenerate pairing which extends to a pairing $\mathcal{H} \times \mathcal{H}^* \to \mathcal{O}_{S'}$. Denote by T and M the monodromy transformations on $\underline{H}$ and $\underline{H}^*$, respectively. Obviously, these are related by the formula

$$T = (M^*)^{-1},$$

where M^* denotes linear transformation on the dual space.

(2.6.2) Let us see how the operators of covariant differentiation

$$D = \nabla_{d/dt}\colon \mathscr{H} \to \mathscr{H} \quad \text{and} \quad D^* = \nabla^*_{d/dt}\colon \mathscr{H}^* \to \mathscr{H}^*$$

are related to each other.

Proposition 1 If ω is a local section of the sheaf $\mathscr{H}$ and γ is a local section of $\underline{H}^*$, i.e. a horizontal section of $\mathscr{H}^*$, then the derivative of the function $I(t) = \langle \omega, \gamma \rangle$ is

(2.6.3) $$\langle \omega, \gamma \rangle' = \langle D\omega, \gamma \rangle.$$

Proof If the formula is true for ω_1 and ω_2, then it is true for $\omega = \omega_1 + \omega_2$ by virtue of the additivity of both sides of the formula wrt ω. Let us show that if (2.6.3) is true for ω, then it is true for $g\omega$, where $g \in \mathscr{O}_{S'}$:

left-hand side $= \langle g\omega, \gamma \rangle' = (g\langle \omega, \gamma \rangle)' = g'\langle \omega, \gamma \rangle + g\langle \omega, \gamma \rangle'$,
right-hand side $= \langle Dg\omega, \gamma \rangle = \langle g'\omega + gD\omega, \gamma \rangle = g'\langle \omega, \gamma \rangle + g\langle D\omega, \gamma \rangle$.

The formula is true for horizontal sections ω, because then $\langle \omega, \gamma \rangle = \text{const}$ and $\langle \omega, \gamma \rangle' = 0$, and on the other hand $D\omega = 0$ and $\langle D\omega, \gamma \rangle = 0$. The formula is true for any section ω because horizontal sections generate $\mathscr{H}$ over $\mathscr{O}_{S'}$. ■

Proposition 2 If ω and γ are any sections of $\mathscr{H}$ and $\mathscr{H}^*$, then

(2.6.4) $$\langle \omega, \gamma \rangle' = \langle D\omega, \gamma \rangle + \langle \omega, D^*\gamma \rangle.$$

Proof Arguing as above (the formula is true for horizontal sections; it is additive in γ; being true for γ, it is also true for $g\gamma$, $g \in \mathscr{O}_{S'}$), we obtain the proof. ■

(2.6.5) Now let $\omega_1, \ldots, \omega_\mu$ be a local basis of sections of $\mathscr{H}$ and let $\gamma_1 = \omega_1^*, \ldots, \gamma_\mu = \omega_\mu^*$ be the dual basis of the sheaf $\mathscr{H}^*$, $\langle \omega_i, \gamma_j \rangle = \delta_{ij}$.

Proposition 3 If $\Gamma = (\Gamma_{ij})$ is the matrix of coefficients of a connection ∇ in a basis $\omega_1, \ldots, \omega_\mu$ and $\Gamma^* = (\Gamma^*_{ij})$ is a matrix of coefficients of ∇^* in the dual basis $\gamma_1, \ldots, \gamma_\mu$, then

$$\Gamma^* = -\Gamma^t.$$

Proof Recall that Γ and Γ^* are matrices of operators D and D^* in the bases ω and γ:

$$D\omega_j = \sum_{i=1}^{\mu} \Gamma_{ij}\omega_i, \quad D^*\gamma_i = \sum_{j=1}^{\mu} \Gamma^*_{ji}\gamma_j.$$

We have $\langle \omega_j, \gamma_i \rangle' = 0 = \langle D\omega_j, \gamma_i \rangle + \langle \omega_j, D^* \gamma_i \rangle = \Gamma_{ij} + \Gamma^*_{ji}$. ∎

(2.6.6) Let $\omega_1, \ldots, \omega_\mu$ be a basis of local sections of the sheaf $\mathscr{H}$, and let γ be a section of $\mathscr{H}^*$. Then the functions

$$I_1(t) = \langle \omega_1, \gamma \rangle, \ldots, I_\mu(t) = \langle \omega_\mu, \gamma \rangle$$

can be considered as coordinates of γ in the basis dual to $\{\omega_i\}$:

$$\gamma = I_1(t)\gamma_1 + \ldots + I_\mu(t)\gamma_\mu.$$

In particular, if γ is a section of $\underline{H}^*$, i.e. a horizontal section of $\mathscr{H}^*$, then $\nabla^* \gamma = 0$ and the coordinates of γ satisfy the system (2.3.10) of differential equations of the connection ∇^*. Thus if γ is a section of $\underline{H}^*$, then the functions $I_1(t), \ldots, I_\mu(t)$ satisfy a system of differential equations

$$\begin{pmatrix} I'_1 \\ \vdots \\ I'_\mu \end{pmatrix} = \Gamma^t(t) \begin{pmatrix} I_1 \\ \vdots \\ I_\mu \end{pmatrix}, \tag{2.6.7}$$

where $\Gamma = (\Gamma_{ij})$ is the matrix of coefficients of ∇ in basis $\omega_1, \ldots, \omega_\mu$.

If $\gamma_1, \ldots, \gamma_\mu$ form a basis of sections of $\underline{H}^*$, then the columns $\Omega_j = (\Omega_{1j}(t), \ldots, \Omega_{\mu j}(t))^t$, $\Omega_{ij} = \langle \omega_i, \gamma_j \rangle$, form a basis of solutions, and the matrix

$$\Omega(t) = (\Omega_{ij}(t)) \tag{2.6.8}$$

is the fundamental matrix of solutions of the system $I' = \Gamma^t(t) I$.

3 De Rham cohomology

3.1 The Poincaré lemma

We return now to our main subject of investigation – the Milnor fibration $f\colon X \to S$ of a singularity of holomorphic function f and the connection on the locally free sheaf $\mathscr{H} = R^p f_* \mathbb{C}_{X'} \otimes \mathscr{O}_{S'}$ on S', defined (topologically) by the local system $R^p f_* \mathbb{C}_{X'}$. In order to pass to differential forms we have to use the Poincaré lemma.

(3.1.1) The Poincaré lemma (in its holomorphic version) asserts that if X is a complex manifold, then the De Rham complex $(\Omega^{\cdot}_X, \mathrm{d})$ is a resolution of the constant sheaf $\mathbb{C}_X$, i.e. the sequence $0 \to \mathbb{C}_X \to \mathscr{O}_X \xrightarrow{\mathrm{d}} \Omega^1_X \xrightarrow{\mathrm{d}} \ldots$ is exact.

This implies that for a complex manifold X we have an isomorphism $H^p(X, \mathbb{C}) \simeq \mathbb{H}^p(\Omega^{\cdot}_X)$, where $\mathbb{H}^p(\Omega^{\cdot}_X)$ is the hypercohomology of the De Rham complex. If, moreover, X is a Stein manifold, then (see below for the relative case) the cohomology $H^p(X, \mathbb{C})$ is isomorphic to the De Rham cohomology, i.e. to the cohomology of the complex $(H^0(X, \Omega^{\cdot}_X, \mathrm{d})$ (closed forms/exact forms on X). Turning to the relative case, we have the following generalization:

3.2 Relative De Rham cohomology

(3.2.1) *Definition* The sheaves $\mathscr{H}^p_{\mathrm{DR}}(X/S)$ of hypercohomology of the De Rham complex of relative differential forms of a morphism $f\colon X \to S$

(3.2.2) $$\Omega^{\cdot}_{X/S}\colon 0 \to \mathscr{O}_X \xrightarrow{\mathrm{d}} \Omega^1_{X/S} \xrightarrow{\mathrm{d}} \Omega^2_{X/S} \to \dots$$

wrt the functor of direct image f_* are called the *relative De Rham cohomology*

(3.2.3) $$\mathscr{H}^p_{\mathrm{DR}}(X/S) \underset{Df}{=} \mathbb{R}^p f_*(\Omega^{\cdot}_{X/S}).$$

Here $\Omega^p_{X/S} = \Omega^p_f = \Omega^p_X / f^*\Omega^1_S \wedge \Omega^{p-1}_X$.

The main tools for working with hypercohomology are the first and the second spectral sequences of hypercohomology, which for the complex $\Omega^{\cdot}_{X/S}$ are of the form

$$'E^{p,q}_1 = R^q f_*(\Omega^p_{X/S}) \Rightarrow \mathbb{R}^{p+q} f_*(\Omega^{\cdot}_{X/S}),$$

(3.2.4′) $$'E^{p,q}_2 = \mathscr{H}^p(R^q f_* \Omega^{\cdot}_{X/S}) \Rightarrow \mathbb{R}^{p+q} f_*(\Omega^{\cdot}_{X/S}),$$

(3.2.4″) $$''E^{p,q}_2 = R^p f_*(\mathscr{H}^q(\Omega^{\cdot}_{X/S})) \Rightarrow \mathbb{R}^{p+q} f_*(\Omega^{\cdot}_{X/S}).$$

Here $\mathscr{H}^p(\cdot)$ denotes the cohomology sheaves of a complex of sheaves, and $R^q f_* \Omega^{\cdot}_{X/S}$ is the complex $\dots \to R^q f_* \Omega^{p-1}_{X/S} \to R^q f_* \Omega^p_{X/S} \to R^q f_* \Omega^{p+1}_{X/S} \to \dots$.

3.3 De Rham cohomology for smooth Stein morphisms

(3.3.1) A connection of the relative De Rham cohomology with topology is established by the relative Poincaré lemma:

Proposition If a morphism $f\colon X \to S$ is smooth, then

(3.3.2) $$\mathscr{H}^p_{\mathrm{DR}}(X/S) = \mathbb{R}^p f_*(\Omega_{\cdot X/S}) \simeq R^p f_* \mathbb{C}_X \underset{\mathbb{C}_S}{\otimes} \mathscr{O}_S.$$

Proof The relative Poincaré lemma asserts that if f is smooth, then there is an exact sequence $0 \to f^{-1}\mathscr{O}_S \to \Omega^{\cdot}_{X/S}$, i.e. the De Rham complex $\Omega^{\cdot}_{X/S}$ is a resolution of the sheaf $f^{-1}\mathscr{O}_S$. Hence $\mathbb{R}^p f_*(\Omega^{\cdot}_{X/S}) = R^p f_*(f^{-1}\mathscr{O}_S)$. It can be shown that the canonical homomorphism

$$R^p f_*(\mathbb{C}_X) \underset{\mathbb{C}_S}{\otimes} \mathscr{O}_S \to R^p f_*(f^{-1}\mathscr{O}_S)$$

is an isomorphism. ■

(3.3.3) *Corollary* If $f\colon X \to S$ is the Milnor fibration, then the sheaf of relative De Rham cohomology $\mathscr{H}_{\mathrm{DR}}(X/S)$ is a natural extension to the whole S of the sheaf $\mathscr{H} = R^p f_* \mathbb{C}_{X'} \otimes_{\mathbb{C}_{S'}} \mathscr{O}_{S'}$ on S',

$$\mathscr{H}^p_{\mathrm{DR}}(X/S)|_{S'} = \mathscr{H}^p_{\mathrm{DR}}(X'/S') \simeq R^p f_* \mathbb{C}_{X'} \otimes \mathscr{O}_{S'}.$$

This follows from the smoothness of the morphism $f'\colon X' \to S'$.

(3.3.4) *Proposition* For the Milnor fibration $f\colon X \to S$ we have

$$\mathscr{H}^p_{\mathrm{DR}}(X/S) \simeq \mathscr{H}^p(f_* \Omega^{\cdot}_{X/S}),$$

i.e. the sheaves of De Rham cohomology are the cohomology sheaves of the direct image of the relative De Rham complex.

Proof The morphism f is Stein and hence the first spectral sequence (3.2.4′) degenerates $'E_1^{p,q} = R^q f_*(\Omega^p_{X/S}) = 0$ for $q \geqslant 1$ and hence $'E_2^{p,0} = \mathscr{H}^p(f_* \Omega^p_{X/S}) \simeq \mathbb{R}^p f_*(\Omega^{\cdot}_{X/S})$. ■

3.4 Coherence theorem

This is one of the fundamental technical results of the theory.

Brieskorn [Br] proved that if $f\colon X \to S$ is the Milnor fibration of an isolated singularity, then the sheaves $\mathscr{H}^p_{\mathrm{DR}}(X/S)$ are coherent sheaves of $\mathscr{O}_S$-modules.

For proper morphisms coherence follows from Grauert's famous direct image theorem. Brieskorn obtained the proof of the coherence by embedding the morphism $f\colon X \to S$ in a projective morphism $f\colon Y \to S$. Later generalizations and other proofs of the coherentness theorem were obtained by Hamm [H], Buchweitz-Greuel [B-G] and van Straten [Sr].

3.5 On the absence of torsion in the De Rham cohomology sheaves

(3.5.1) The next important question is one about the absence of torsion in the sheaves $\mathscr{H}^p_{\mathrm{DR}}(X/S)$. We'll discuss this in §6 after introducing the Gauss–Manin connection on these sheaves. For an isolated singularity $f\colon (\mathbb{C}^{n+1}, 0) \to (\mathbb{C}, 0)$, the absence of torsion in $\mathscr{H}^p_{\mathrm{DR}}(X/S)$ for $p = n$ was proved by Sebastiani [Se]. For $p < n$, torsion is absent by virtue of the presence of a connection on these sheaves. From this, (3.3.3) and (1.4.5) it follows that for an isolated singularity we have

$$\mathscr{H}^p_{\mathrm{DR}}(X/S) = \begin{cases} \mathscr{O}_S & \text{for } p = 0, \\ 0 & \text{for } 1 \leqslant p < n, \\ \mathscr{O}^\mu_S & \text{for } p = n. \end{cases}$$

The absence of torsion in the sheaves $\mathscr{H}^i_{\mathrm{DR}}(X/S)$ is also important because in such cases the Milnor number, $\mu^{(i)} = \dim H^i(X_t, \mathbb{C}) = b_i$ (the ith Betti number of fibre X_t), is given by the formula

$$\mu^{(i)} = \mathrm{rk}\, \mathscr{H}^i_{\mathrm{DR}}(X/S) = \dim \mathscr{H}^i_{\mathrm{DR}}(X/S)_0 / t\mathscr{H}^i_{\mathrm{DR}}(X/S)_0,$$

and the germ of the De Rham cohomology sheaf $\mathscr{H}^i_{\mathrm{DR}}(X/S)_0$ can be calculated 'upstairs' as shown in the following proposition.

3.6 Relation between $\mathscr{H}^p(f_*\Omega^\cdot_f)$ and $f_*\mathscr{H}^p(\Omega^\cdot_f)$

(3.6.1) *Proposition* If $f\colon X \to S$ is the Milnor fibration of an isolated singularity f (or, in the more general case, of a concentrated non-isolated singularity, see §10), then the De Rham cohomology sheaf $\mathscr{H}^p_{\mathrm{DR}}(X/S) = \mathscr{H}^p(f_*\Omega^\cdot_f)$ enters into the exact sequence

$$(3.6.2) \qquad 0 \to R^p f_*\mathbb{C}_X \underset{\mathbb{C}_S}{\otimes} \mathscr{O}_S \to \mathscr{H}^p(f_*\Omega^\cdot_f) \to f_*\mathscr{H}^p(\Omega^\cdot_f) \to 0,$$

where the last sheaf is concentrated at the point $t_0 = 0 \in S$, and the two first sheaves are locally free of rank $\mu^{(p)} = b_p(X_t)$ on S', and at the point $t_0 = 0$ the first of them has a zero fibre and

$$(3.6.3) \qquad \mathscr{H}^p(f_*\Omega^\cdot_f)_{t_0} = f_*\mathscr{H}^p(\Omega^\cdot_f)_{t_0} = \mathscr{H}^p(\Omega^\cdot_f)_{x_0} = H^p(\Omega^\cdot_{f,x_0}),$$

where $(X, x_0) = (\mathbb{C}^{n+1}, 0)$.

Proof Consider the second spectral sequence of hypercohomology (3.2.4″)

$$''E_2^{p,q} = R^p f_*(\mathscr{H}^q(\Omega^\cdot_f)) \Rightarrow \mathbb{R}^{p+q} f_*(\Omega^\cdot_f) = \mathscr{H}^{p+q}(f_*\Omega^\cdot_f).$$

As f' is smooth, the sheaves $\mathscr{H}^q(\Omega^\cdot_f)$ are concentrated on $X_0 = f^{-1}(0)$ for $q \geqslant 1$ and hence $''E_2^{p,q} = 0$ for $q \geqslant 1$ and $p \geqslant 1$. For $q = 0$ we have $\mathscr{H}^0(\Omega^\cdot_f) = f^{-1}\mathscr{O}_S$ and $''E_2^{p,0} = R^p f_*(f^{-1}\mathscr{O}_S) = R^p f_*\mathbb{C}_X \otimes \mathscr{O}_S$

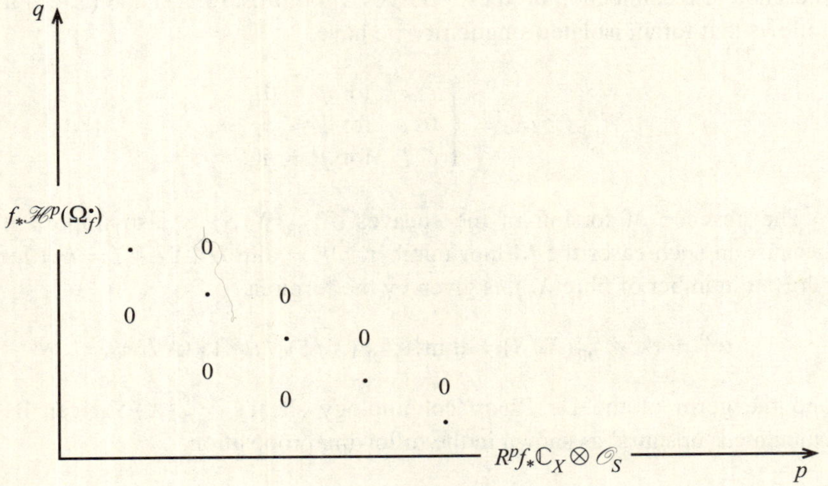

This involves the exact sequence (3.7.2). The germ $R^p f_*(f^{-1}\mathscr{O}_S)_{t_0} = H^p(X_0, \mathscr{O}_{S,t_0}) = 0$ is zero because X_0 is homeomorphic to a cone over the point x_0 and hence is contractible. ∎

Thus we obtain that for an isolated singularity f the Milnor number is equal

$$(3.6.4) \qquad \mu = \dim_{\mathbb{C}} H^n(\Omega^\cdot_{f,x_0})/tH^n(\Omega^\cdot_{f,x_0})$$

(because $\mathscr{H}^n_{\mathrm{DR}}(X/S)$ by the result of Sebastiani has no torsion).

4 Gauss–Manin connection on relative De Rham cohomology

Let $f\colon X \to S$ be the Milnor fibration of a singularity $f\colon (\mathbb{C}^{n+1}, \mathbb{C}) \to (\mathbb{C}, 0)$. We have local systems – cohomological fibrations $\underline{H} = R^p f'_* \mathbb{C}_{X'} = \bigcup_{t\in S'} H^p(X_t, \mathbb{C})$ on $S' = S\backslash\{0\}$, and hence there are corresponding connections on $\mathscr{H} = \underline{H} \otimes_{\mathbb{C}_{S'}} \mathscr{O}_{S'}$ which we call topological.

On the other hand we have an identification of $\mathscr{H}$ with the De Rham cohomology (§3), $\mathscr{H} \simeq \mathscr{H}^p_{\mathrm{DR}}(X'/S')$ and a natural extension of this sheaf over the whole S, $\mathscr{H}^p_{\mathrm{DR}}(X/S) \simeq \mathscr{H}^p(f_*\Omega^\cdot_{X/S})$. We want to calculate the connection on $\mathscr{H}^p_{\mathrm{DR}}(X'/S')$ corresponding to the topological connection on $\mathscr{H}$ and to extend it to $\mathscr{H}^p_{\mathrm{DR}}(X/S)$.

4.1 Identification of sheaves of sections of cohomological fibration and of relative De Rham cohomology

(4.1.1) The identification of $\mathscr{H}$ with $\mathscr{H}^p_{\mathrm{DR}}(X'/S')$ is obtained by means of integration of differential forms (De Rham's theorem).

Let $\underline{H}^* = \bigcup_{t\in S'} H_p(X'_t, \mathbb{C}) = \mathrm{Hom}_{\mathbb{C}_{S'}}(\underline{H}, \mathbb{C}_{S'})$ be the homological fibration on S'. This is a local system dual to $\underline{H}$. We have a non-degenerate pairing

$$\mathscr{H} \times \mathscr{H}^* \to \mathscr{O}_{S'}, \text{ where } \mathscr{H}^* = \underline{H}^* \underset{\mathbb{C}_{S'}}{\otimes} \mathscr{O}_{S'}.$$

Consider a local section σ of the fibration $\underline{H}^*$. We can think of σ as a continuous family of cycles $\sigma(t) \in H_p(X_t, \mathbb{C})$ on hypersurfaces X_t, which is obtained geometrically by continuous deformation of a cycle $\sigma(t_0)$ on one of the fibres X_{t_0}. The homology classes of cycles $\sigma(t)$ do not depend on the deformation (if σ is a section over an open set lying in some sector with the center at the origin $0 \in S$).

Now let ω be a local section of the sheaf $\mathscr{H}^p_{\mathrm{DR}}(X'/S')$. The cohomology class ω is presented by a cocycle, i.e. a p-form ω such that $\mathrm{d}\omega = \mathrm{d}f \wedge \eta$. Consider the function $I(t)$ which is the integral of the form ω over the cycles of the family σ, i.e. the value of $I(t)$ at a point t is equal to the integral of the p-form $\omega_{|X_t}$ over the p-cycle $\sigma(t) \subset X_t$

(4.1.2) $$I(t) = \int_{\sigma(t)} \omega.$$

We obtain a non-degenerative pairing

(4.1.3) $$\mathscr{H}^p_{\mathrm{DR}}(X'/S') \times \mathscr{H}^* \to \mathscr{O}_{S'},\ (\omega, \sigma) \to I(t) = I_{\omega,\sigma}(t),$$

which identifies $\mathscr{H}^p_{\mathrm{DR}}(X'/S')$ with the sheaf dual to $\mathscr{H}^*$, i.e. with the sheaf $\mathscr{H}$.

(4.1.4) To verify that $I(t)$ is, in fact, a section of the sheaf $\mathscr{O}_{S'}$, i.e. that $I(t)$ is a holomorphic function, and also to calculate $I'(t)$, we need Leray's residue theorem. This theorem and its application are natural generalizations of the integral Cauchy formula $g(z_0) = (1/2\pi \mathrm{i}) \int g(z)\,\mathrm{d}z/(z - z_0)$ and its application for the proof of the existence of $g'(z)$ and decomposition of a holomorphic function into series in the standard course of complex analysis.

A fibre $X_t \subset X$ is a submanifold of complex codimension 1 and hence of real codimension 2. There is defined the coboundary Leray operator

(4.1.5) $$\delta\colon H_p(X_t) \to H_{p+1}(X\backslash X_t).$$

One needs to consider a tube neighbourhood N of the submanifold X_t in X

(fibered on disks). Then the boundary $\partial N \subset X\backslash X_t$ is a fibration on circles over X_t and ∂ assigns to a cycle in $H_p(X_t)$ its pre-image on ∂N wrt the projection $\partial N \to X_t$.

(4.1.6) *Leray's residue theorem* asserts that

$$I(t) = \int_{\sigma(t)} \omega = \frac{1}{2\pi \mathrm{i}} \int_{\delta\sigma(t)} \frac{\mathrm{d}f \wedge \omega}{f - t}.$$

This theorem is called the residue theorem because $\omega_{|X_t}$ is the Poincaré residue of the form $\mathrm{d}f \wedge \omega/(f - t)$, $\omega_{|X_t} = \mathrm{Res}_{X_t}[\mathrm{d}f \wedge \omega/(f - t)]$. Indeed, we can consider the equation $y = f(x) - t = 0$, of a smooth fibre X_t as a local coordinate on X. Then $dy = df$ $(t = \text{const})$. However, the Poincaré residue is defined in the following way: we 'chip' a multiplier $dy/y = df/f - t$ and restrict it to the subvariety X_t: $y = 0$.

(4.1.7) *Theorem* The function $I(t) = \int_{\sigma(t)} \omega$ is holomorphic on S'.

To prove this, we apply the integral presentation of $I(t)$ in (4.1.6). If t varies in a small neighbourhood of a fixed point t_0, we can assume that $\delta\sigma(t) = \delta\sigma$ are the same for different t. The form $\mathrm{d}f \wedge \omega/[f(x) - t]$ depends holomorphically on t and we can apply the standard theorem of analysis asserting that an integral of a form depending holomorphically on a parameter t over a fixed chain also depends holomorphically on t. The same presentation enables us to calculate $I'(t)$ as the derivation of an integral depending on a parameter.

4.2 Calculation of the connection on a relative De Rham cohomology sheaf

(4.2.1) We proceed to the calculation of the connection on $\mathscr{H}^p_{\mathrm{DR}}(X'/S')$. Both this connection and the topological connection on $\mathscr{H}$ are dual to the connection on $\mathscr{H}^* = \underline{H}^* \otimes_{\mathbb{C}_{S'}} \mathscr{O}_{S'}$, defined by the homological fibration $\underline{H}^*$. By (2.6.3), if $D = \nabla_{\mathrm{d}/\mathrm{d}t}$ is the covariant derivative of the connection on $\mathscr{H}^p_{\mathrm{DR}}(X'/S')$ and $\sigma(t)$ is a local section of $\underline{H}^*$, then $\langle \omega, \sigma \rangle' = \langle D\omega, \sigma \rangle$. We will calculate the derivation $\langle \omega, \sigma \rangle'$ of the function $I(t) = I_{\omega,\sigma}(t) = \langle \omega, \sigma \rangle$. We use the integral presentation (4.1.6) and differentiate the integral wrt the parameter t

$$I'(t) = \frac{\mathrm{d}}{\mathrm{d}t}\int_{\sigma(t)} \omega = \frac{\mathrm{d}}{\mathrm{d}t}\left(\frac{1}{2\pi \mathrm{i}}\int_{\delta\sigma(t)} \frac{\mathrm{d}f \wedge \omega}{f(x)-t}\right) = \frac{1}{2\pi \mathrm{i}}\int_{\delta\sigma(t)} \frac{\mathrm{d}f \wedge \omega}{[f(x)-t]^2}$$

$$= \frac{1}{2\pi \mathrm{i}}\int_{\delta\sigma(t)} \left\{\frac{\mathrm{d}\omega}{f(x)-t} - \mathrm{d}\left(\frac{\omega}{f(x)-t}\right)\right\} = \frac{1}{2\pi \mathrm{i}}\int_{\delta\sigma(t)} \frac{\mathrm{d}\omega}{f(x)-t}$$

$$= \frac{1}{2\pi \mathrm{i}}\int_{\delta\sigma(t)} \frac{\mathrm{d}f \wedge \eta}{f(x)-t} = \int_{\sigma(t)} \eta = \langle \eta, \sigma\rangle .$$

Thus $\langle D\omega, \sigma\rangle = \langle \eta, \sigma\rangle$ for any section σ and hence $D\omega = \eta$.

(4.2.2) We obtain the following rule for calculation of the connection on $\mathscr{H}^p_{\mathrm{DR}}(X'/S')$, which one can (try to) carry word for word to the De Rham cohomology sheaves $\mathscr{H}^p_{\mathrm{DR}}(X/S)$ to obtain the Gauss–Manin connection defined by Brieskorn [Br]:

$$\nabla\colon \mathscr{H}^p_{\mathrm{DR}}(X/S) \to \Omega^1_S \underset{\mathscr{O}_S}{\otimes} \mathscr{H}^p_{\mathrm{DR}}(X/S)$$

or

$$\nabla_{d/dt} = D\colon \mathscr{H}^p_{\mathrm{DR}}(X/S) \to \mathscr{H}^p_{\mathrm{DR}}(X/S):$$

Take a p-form ω representing a cohomology class in $\mathscr{H}^p_{\mathrm{DR}}(X/S)$. As this is a cycle in the complex $\Omega^{\cdot}_{X/S}$, we have $\mathrm{d}\omega = 0$ in $\Omega^{p+1}_{X/S}$, i.e. $\mathrm{d}\omega = \mathrm{d}f \wedge \eta$, where $\eta \in \Omega^p_X$. Then

(4.2.3) $$D(\omega) = \text{a class of the form } \eta \text{ or}$$
$$\nabla(\omega) = \mathrm{d}f \wedge D(\omega) = \mathrm{d}f \wedge \eta.$$

4.3 The division lemma. The connections on the sheaves $\mathscr{H}^p_{\mathrm{DR}}(X/S)$ for $p \leqslant n-1$

We have not yet obtained a connection on the sheaf $\mathscr{H}^p_{\mathrm{DR}}(X/S)$ because we have not yet varified that the form η represents a cohomology class in $\mathscr{H}^p_{\mathrm{DR}}(X/S)$, i.e. that it is a cocycle in $f_*\Omega^{\cdot}_{X/S}$, i.e. that $d\eta = df \wedge \xi$ is divisible by df, i.e. $d\eta$ is a coboundary in the Koszul complex (see (10.5.1)). It is only clear that $d\eta$ is a cocycle in the Koszul complex, i.e. that $\mathrm{d}f \wedge d\eta = 0$. Indeed, we have $\mathrm{d}\omega = \mathrm{d}f \wedge \eta$, hence $\mathrm{d}(\mathrm{d}\omega) = -\mathrm{d}f \wedge d\eta = 0$.

(4.3.1) The so-called division lemma is the assertion about the vanishing of the cohomology of the Koszul complex. Recall the definition.

Let A be a commutative ring and $g_1, \ldots, g_m \in A$. The *cohomological Koszul complex* $K^{\cdot}(g)$ is the complex

$K^\cdot(g)$:

$$0 \to A \to A^m \to \Lambda^2 A^m \to \ldots \to \Lambda^p A^m \xrightarrow{g\wedge\cdot} \Lambda^{p+1} A^m \to \ldots \to \Lambda^m A^m \to$$

in which the differentials are homomorphisms, $\Lambda^p A^m \xrightarrow{g\wedge\cdot} \Lambda^{p+1} A^m$, which send ϕ to $g \wedge \varphi$, $g = g_1 e_1 + \ldots + g_m e_m \in A^m = \oplus_{i=1}^m A e_i$.

In our context the Koszul complex appears in the form of the complex of differential forms. Let X be a complex manifold of dimension $n+1$, and $f(x)$ be a holomorphic function on X. Consider the complex $(\Omega^\cdot_{X/S}, \mathrm{d}f)$:

(4.3.2)

$$0 \to \mathscr{O}_X \xrightarrow{\mathrm{d}f\wedge} \Omega^1_X \to \ldots \to \Omega^p_X \xrightarrow{\mathrm{d}f\wedge} \Omega^{p+1}_X \to \ldots \to \Omega^n_X \xrightarrow{\mathrm{d}f\wedge} \Omega^{n+1}_X \to 0$$

in which the differential is the exterior product by the 1-form df. The cohomology of this complex

$$\mathscr{K}^p(f) = \mathrm{Ker}\,(\Omega^p \xrightarrow{\mathrm{d}f} \Omega^{p+1})/\mathrm{Im}\,(\Omega^{p-1} \xrightarrow{\mathrm{d}f} \Omega^p)$$

is called the *Koszul cohomology*.

It is obvious that for a point $x \in X$ the complex $\Omega^\cdot_{X,x}$ is the Koszul complex, in which $A = \mathscr{O}_{X,x} = \mathscr{O}_{\mathbb{C}^{n+1},0}$, $A^m = \Omega^1_{X,x} = \oplus_{i=0}^n \mathscr{O}_{X,x}\,\mathrm{d}x_i$, and the sequence g is the sequence of partial derivatives $f'_{x_0}, \ldots, f'_{x_n}$, $\mathrm{d}f = f'_{x_0}\,\mathrm{d}x_0 + \ldots + f'_{x_n}\,\mathrm{d}x_n$. There is a well-known lemma:

(4.3.3) *Division lemma (De Rham)* If $g_1, \ldots, g_m$ is a regular sequence, then the cohomologies of the Koszul complex are

$$H^p(K^\cdot(g)) = \begin{cases} 0, & p < m, \\ A/(g_1, \ldots, g_m), & p = m, \end{cases}$$

i.e. if $\omega \in \Lambda^p A^m$ and $p < m$, then

$$g \wedge \omega = 0 \Leftrightarrow \omega = g \wedge \eta, \text{ for some } \eta \in \Lambda^{p-1} A^m. \qquad \blacksquare$$

(4.3.4) If $0 \in X$ is an isolated critical point of a function f, i.e. 0 is a unique solution of the system $f'_{x_0} = \ldots = f'_{x_n} = 0$, then $f'_{x_0}, \ldots, f'_{x_n}$ is a regular sequence, and the division lemma asserts that $\mathscr{K}^p(f) = 0$ for $p \leqslant n$, and the sheaf $\mathscr{K}^{n+1}(f) = \Omega^{n+1}_X / df \wedge \Omega^n_X = \Omega^{n+1}_{X/S}$ is concentrated at the point 0 and $\Omega^{n+1}_{X/S} \simeq \mathscr{O}_{X,0}/J_f$, where $J_f = (f'_{x_0}, \ldots, f'_{x_n})$ is the Jacobi ideal of f. The algebra

(4.3.5) $$Q_f = \mathscr{O}_{X,0}/J_f$$

is called the local ring of the critical point 0 of f [AGV] or the *Milnor algebra* or the *Jacobi algebra* of the singularity f. The Q_f-module $\Omega^{n+1}_{X/S} \simeq Q_f$ is often denoted by

(4.3.6) $$\Omega^{n+1}_{X/S} = \Omega_f \text{ or } Q^f.$$

Thus if f is an isolated singularity, then we have the exact sequence

$$0 \to \mathscr{O}_X \to \Omega^1_X \overset{df\wedge}{\to} \ldots \to \Omega^n_X \overset{df\wedge}{\to} \Omega^{n+1}_X \to \Omega_f \to 0;$$

and if $\omega \in \Omega^p_X$ and $p \leqslant n$, then $df \wedge \omega = 0 \Leftrightarrow \omega = df \wedge \eta$.

(4.3.7) We now return to the definition of the connection on $\mathscr{H}^p_{\mathrm{DR}}(X/S)$. We saw that the $(p+1)$-form $\mathrm{d}\eta$ is a cocycle, $\mathrm{d}f \wedge d\eta = 0$. However, if f has an isolated singularity and $p+1 \leqslant n$, i.e. $p \leqslant n-1$, then $d\eta$ is a coboundary as well, $\mathrm{d}\eta = \mathrm{d}f \wedge \xi$. This means that the form η defines a cocycle in the complex $f_*\Omega^\cdot_{X/S}$ and we indeed obtain a connection $\nabla_{\mathrm{d}/\mathrm{d}t}\colon \mathscr{H}^p_{\mathrm{DR}}(X/S) \to \mathscr{H}^p_{\mathrm{DR}}(X/S)$.

(4.3.8) It is easy to verify the correctness of the definition of $D = \nabla_{d/dt}$, i.e. that the class $\eta = D(\omega)$ is independent of the choice of the form ω in a class of $\mathscr{H}^p_{\mathrm{DR}}(X/S)$. Moreover, D really is a covariant derivation, i.e. it satisfies the Leibniz identity. We have: $\mathrm{d}\omega = \mathrm{d}f \wedge \eta$, $\eta = D(\omega)$; $\mathrm{d}(g\omega) = \mathrm{d}g \wedge \omega + g{\cdot}\mathrm{d}\omega = \mathrm{d}f \wedge (\mathrm{d}g/\mathrm{d}f)\omega + \mathrm{d}f \wedge g\eta$, i.e. $D(g\omega) = g'\omega + gD(\omega)$. (Here f is a local coordinate on S and $\mathrm{d}g = (\mathrm{d}g/\mathrm{d}f)\,\mathrm{d}f$.)

4.4 The sheaf $'\mathscr{H} = f_*\Omega^n_{X/S}/\mathrm{d}(f_*\Omega^{n-1}_{X/S})$

The previous construction does not work for $p = n$, because in this case the $(n+1)$-form $\mathrm{d}\eta$ need not be a coboundary, i.e. it may not be divisible by $\mathrm{d}f$. We obtain that the operator $D = \nabla_{d/dt} = \partial_\cdot$ acts from $\mathscr{H}^p_{\mathrm{DR}}(X/S)$ not to itself but to a greater module,

(4.4.1) $$\nabla\colon \mathscr{H}^n_{\mathrm{DR}}(X/S) \to \Omega^1_S \otimes {'\mathscr{H}}, \quad {'\mathscr{H}} = f_*\Omega^n_{X/S}/\mathrm{d}(f_*\Omega^{n-1}_{X/S}).$$

The sheaves $\mathscr{H}^n_{\mathrm{DR}}(X/S)$ and $'\mathscr{H}$ are defined from the complex $f_*\Omega^\cdot_{X/S}$:

$f_*\Omega^\cdot_{X/S}$:

$$0 \to f_*\mathscr{O}_X \overset{\mathrm{d}}{\to} f_*\Omega^1_{X/S} \to \ldots \to f_*\Omega^{n-1}_{X/S} \overset{\mathrm{d}_{n-1}}{\to} f_*\Omega^n_{X/S} \overset{\mathrm{d}_n}{\to} f_*\Omega^{n+1}_{X/S} \to 0,$$

$$\mathscr{H}^n_{\mathrm{DR}}(X/S) = \operatorname{Ker} \mathrm{d}_n/\operatorname{Im} \mathrm{d}_{n-1}, \quad {'\mathscr{H}} = \operatorname{Coker} \mathrm{d}_{n-1} = f_*\Omega^n_{X/S}/\operatorname{Im} \mathrm{d}_{n-1}.$$

Hence there is the exact sequence

(4.4.2) $$0 \to \mathscr{H}^n_{\mathrm{DR}}(X/S) \to {'\mathscr{H}} \overset{\mathrm{d}}{\to} f_*\Omega^{n-1}_{X/S} \to 0,$$

and since the sheaf $f_*\Omega^{n+1}_{X/S} = f_*(\Omega_f)$ is concentrated at the point $0 \in S$, the restrictions of $\mathscr{H}^n_{\mathrm{DR}}(X/S)$ and $'\mathscr{H}$ coincide on S'.

(4.4.3) We obtained a connection in some more general sense [M]: let $E \subset F$ be two $\mathscr{O} = \mathscr{O}_{S,0}$-modules; then a $\mathbb{C}$-linear map $D\colon E \to F$, satisfying the Leibniz rule $D(ge) = (dg/dt)e + gD(e)$, $e \in E$, $g \in \mathcal{O}$, is called a *connection in the pair* (E, F). Using localization, we obtain the connection in the old sense.

4.5 Meromorphic connections

Let $\mathcal{O} = \mathcal{O}_{S,0} = \mathbb{C}\{t\}$ and let $\mathcal{K} = \mathcal{O}_{(t)} = \mathcal{O}[t^{-1}]$ be a field of germs of functions meromorphic at the point $0 \in S$. Here $\mathcal{K}$ is a field of fractions of the ring $\mathcal{O}$. Let $M = \mathcal{K}e_1 \oplus \ldots \oplus \mathcal{K}e_\mu$ be a vector space over $\mathcal{K}$, $\dim_{\mathcal{K}} M = \mu$. The definition of a connection transfers word for word to this situation.

(4.5.1) *Definition* A *meromorphic connection on* M is a $\mathbb{C}$-linear map $\nabla\colon M \to \Omega_{S,0} \otimes_{\mathcal{O}} M$ or a $\mathbb{C}$-linear map

$$D = \nabla_{d/dt}\colon M \to M,$$

satisfying the Leibniz rule: $D(gm) = g'm + gD(m)$.

(4.5.2) As in §2 a meromorphic connection D is defined by the matrix $\Gamma(t) = (\Gamma_{ij}(t))$ of connection form coefficients in a basis $e_1, \ldots, e_\mu$ of a vector space M, $D(e_j) = \sum_{i=1}^{\mu} \Gamma_{ij}(t)e_i$. The condition of horizontality $s = ey \in \operatorname{Ker} D$ is given by a system of linear differential equations $y' = -\Gamma(t)y$ with a singularity at the point $0 \in S$, in which coefficients $\Gamma_{ij}(t)$ are meromorphic functions.

(4.5.3) Now if ∇ is a connection on an $\mathcal{O}$-module E, then it is naturally extended to a connection on the vector space $M = E_{(t)} = E \otimes_{\mathcal{O}} \mathcal{K}$ by the 'rule of differentiation of a product (fraction)'

$$D\left(s \otimes \frac{1}{t^k}\right) = D(s) \otimes \frac{1}{t^k} - s \otimes \frac{k}{t^{k+1}}, \; s \in E.$$

(4.5.4) In the same way we can localize the situation in the case of a connection D on a pair (E, F), where $E \subset F$ and F/E is torsion, i.e. $\dim_{\mathbb{C}} F/E < \infty$. In this case $E \otimes_{\mathcal{O}} \mathcal{K} = F \otimes_{\mathcal{O}} \mathcal{K} = M$ and the connection $D\colon E \to F$ defines a meromorphic connection on M.

(4.5.5) *Definition* A *lattice* in a vector space M, $\dim_{\mathcal{K}} M = \mu$, is an $\mathcal{O}$-submodule $E \subset M$ of rank μ, i.e. a finitely generated submodule E such that $\mathcal{K}E = M$.

(4.5.6) To return to our discussion: let $i: S' \subset S$ be an inclusion and $\mathscr{H} = \underline{H} \otimes_{\mathbb{C}_{S'}} \mathscr{O}_{S'} \simeq \mathscr{H}^n_{\mathrm{DR}}(X'/S')$ be a locally free sheaf on S'. Denote by $\mathscr{H}[0] \subset i_*(\mathscr{H})$ a subsheaf of the direct image of $\mathscr{H}$, in which the germ at $0 \in S$ coincides with the localization of the module $\mathscr{H}^n_{\mathrm{DR}}(X/S)_0$ as well as $'\mathscr{H}_0$. We obtain a meromorphic Gauss–Manin connection ∇ on the sheaf $\mathscr{H}[0]$, in which the locally free submodules $\mathscr{H}^n_{\mathrm{DR}}(X/S) \subset {}'\mathscr{H} \subset \mathscr{H}[0]$ are lattices (we will prove in §6 that there is no torsion in the sheaves $\mathscr{H}^n_{\mathrm{DR}}(X/S)$ and $'\mathscr{H}$ and hence they are locally free).

4.6 The Gauss–Manin connection as a connecting homomorphism

(4.6.1) We can give a more conceptual definition of the Gauss–Manin connection $\nabla: \mathscr{H}^n_{\mathrm{DR}}(X/S) \to \Omega^1_X \otimes_{\mathscr{O}_S} {}'\mathscr{H}$. Denote by $\overline{\Omega}^{\cdot}_{X/S}$ a 'shortened' complex $\Omega^{\cdot}_{X/S}$, in which the term $\Omega^{n+1}_{X/S}$ is changed by zero:

$$\overline{\Omega}^{\cdot}_{X/S}\colon 0 \to \mathscr{O}_X \to \Omega^1_{X/S} \to \dots \xrightarrow{d} \Omega^{n-1}_{X/S} \xrightarrow{d} \Omega^n_{X/S} \to 0.$$

Then $\mathscr{H}^p(f_*\overline{\Omega}^{\cdot}_{X/S}) = \mathscr{H}^p(f_*\Omega^{\cdot}_{X/S}) = \mathscr{H}^p_{\mathrm{DR}}(X/S)$ for $p \leqslant n-1$, and $\mathscr{H}^n(f_*\overline{\Omega}^{\cdot}_{X/S}) = {}'\mathscr{H}$.

By the definition of the sheaves of relative differentials there is an exact sequence of complexes

$$f^*\Omega^1_S \underset{\mathscr{O}_X}{\otimes} \overline{\Omega}^{\cdot-1}_{X/S} \xrightarrow{\varphi} \Omega^{\cdot}_X \to \Omega^{\cdot}_{X/S} \to 0, \text{ where } \varphi\colon \mathrm{d}t \otimes \omega \mapsto \mathrm{d}f \wedge \omega.$$

If f has an isolated singularity, then the division lemma means that φ is a monomorphism, i.e. the sequences

$$0 \to f^*\Omega^1_S \otimes \Omega^{p-1}_{X/S} \xrightarrow{\varphi} \Omega^p_X \to \Omega^p_{X/S} \to 0 \tag{4.6.2}$$

are exact. Indeed, if $\varphi(\mathrm{d}t \otimes \omega) = \mathrm{d}f \wedge \omega = 0$, then $\omega = \mathrm{d}f \wedge \eta$, i.e. $\omega = 0$ in $\Omega^{p-1}_{X/S}$, i.e. $\mathrm{Ker}\,\varphi = 0$.

Applying the hypercohomology functor $\mathbb{R}f_*$ to the exact sequence of complexes, or in our situation of Stein morphism f, applying functor f_* to the sequences (4.6.2), we again obtain exact sequences and hence an exact sequence of complexes

$$0 \to \Omega^1_X \underset{\mathscr{O}_S}{\otimes} f_*\overline{\Omega}^{\cdot-1}_{X/S} \to f_*\Omega^{\cdot}_X \to f_*\Omega^{\cdot}_{X/S} \to 0, \tag{4.6.3}$$

or in more detail:

$$\begin{array}{ccccccccccc}
 & & 0 & & 0 & & 0 & & 0 & & \\
 & & \downarrow & & \downarrow & & \downarrow & & \downarrow & & \\
0 & \to & \Omega^1_S \otimes f_*\mathcal{O}_X & \to\dots\to & \Omega^1_S \otimes f_*\Omega^{p-1}_{X/S} & \to\dots\to & \Omega^1_S \otimes f_*\Omega^{n-1}_{X/S} & \to & \Omega^1_S \otimes f_*\Omega^{n}_{X/S} & \to & 0\\
\downarrow & & \downarrow & & \downarrow & & \downarrow & & \downarrow & & \\
0 \to f_*\mathcal{O}_X \to & & f_*\Omega^1_X & \to\dots\to & f_*\Omega^p_X & \xrightarrow{d}\dots\to & f_*\Omega^n_X & \to & f_*\Omega^{n+1}_X & \to & 0\\
\| & & \downarrow & & \downarrow & & \downarrow & & \downarrow & & \\
0 \to f_*\mathcal{O}_X \to & & f_*\Omega^1_{X/S} & \to\dots\to & f_*\Omega^p_{X/S} & \to\dots\to & f_*\Omega^n_{X/S} & \to & f_*\Omega^{n+1}_{X/S} & \to & 0\\
\downarrow & & \downarrow & & \downarrow & & \downarrow & & \downarrow & & \\
0 & & 0 & & 0 & & 0 & & 0. & &
\end{array}$$

Consider the connecting homomorphism ∂ in the exact cohomology sequence of this exact sequence of complexes

(4.6.4)

$$\begin{array}{ccccc}
\dots\to \mathcal{H}^p(f_*\Omega^\cdot_X) \to \mathcal{H}^p(f_*\Omega^\cdot_{X/S}) & \xrightarrow{\partial} & \Omega^1_S \underset{\mathcal{O}_S}{\otimes} \mathcal{H}^p(f_*\Omega^\cdot_{X/S}) & \to \mathcal{H}^{p+1}(f_*\Omega^\cdot_X) \to\dots \\
\| & & \| & \\
\mathcal{H}^p_{\mathrm{DR}}(X/S) & \xrightarrow{\partial} & \Omega^1_S \otimes \mathcal{H}^p_{\mathrm{DR}}(X/S). &
\end{array}$$

Remembering the rule of calculation of the connecting homomorphism ∂ in an exact cohomology sequence, we obtain that

$$\partial\colon \mathcal{H}^p_{\mathrm{DR}}(X/S) \to \Omega^1_S \otimes \mathcal{H}^p_{\mathrm{DR}}(X/S) \quad \text{or} \quad \partial\colon \mathcal{H}^p_{\mathrm{DR}}(X/S) \to \mathcal{H}^p_{\mathrm{DR}}(X/S)$$

(if we identify $\Omega^1_S = \mathcal{O}_S\, \mathrm{d}t$ with $\mathcal{O}_S$) exactly coincides with the Brieskorn definition of the Gauss–Manin connection ∇, i.e. it coincides with the operator $\partial_t = \nabla_{d/dt}\colon \mathcal{H}^p_{\mathrm{DR}}(X/S) \to \mathcal{H}^p_{\mathrm{DR}}(X/S)$. For $p = n$ we obtain accordingly the Gauss–Manin connection $\partial = \nabla$,

$$\partial\colon \mathcal{H}^p_{\mathrm{DR}}(X/S) \to \Omega^1_S \underset{\mathcal{O}_S}{\otimes} {}'\mathcal{H}.$$

Now note that X is smooth and by the Poincaré lemma the complex $\Omega^\cdot_X$ is a resolution of the sheaf $\mathbb{C}_X$ and hence $\mathcal{H}^p(f_*\Omega^\cdot_X) = \mathbb{R}^p f_*(\Omega^\cdot_X) = R^p f_*\mathbb{C}_X$. However, this sheaf has a zero germ at the point $0 \in S$ by (3.7.1). From (4.6.4) we obtain:

(4.6.5) *Corollary* The Gauss–Manin connection on the De Rham cohomology gives the isomorphisms

$$\partial_t\colon \mathcal{H}^p_{\mathrm{DR}}(X/S)_0 \underset{\simeq}{\to} \mathcal{H}^p_{\mathrm{DR}}(X/S)_0, \quad 1 \leqslant p \leqslant n-1$$

$$\partial_t\colon \mathcal{H}^n_{\mathrm{DR}}(X/S)_0 \underset{\simeq}{\to} {}'\mathcal{H}_0, \quad p = n$$

$$(\text{and } 0 \to \mathbb{C} \to \mathcal{H}^0_{\mathrm{DR}}(X/S)_0 \xrightarrow{\partial_t} \mathcal{H}^0_{\mathrm{DR}}(X/S)_0 \to 0).$$

(4.6.6) *Remark* As noted by Malgrange [M], this enables us to calculate the Betti numbers $b_p = \dim H^p(X_t, \mathbb{C})$ of the Milnor fibre (i.e. the Milnor numbers) purely analytically without appealing to a result of Milnor about the homotopy equivalence of X_t to a bouquet $S^n \vee \ldots \vee S^n$ of spheres. The presence of a connection on $\mathscr{H}^p_{\mathrm{DR}}(X/S)$ for $1 \leqslant p \leqslant n-1$ involves (see (6.1.1)) that $\mathscr{H}^p_{\mathrm{DR}}(X/S)$ are locally free sheaves of rank b_p. Moreover, $\operatorname{Ker} \partial_t$ is a local system of rank b_p (by the Cauchy theorem on solutions of a system of linear differential equations). But ∂_t is an isomorphism and hence $\operatorname{Ker} \partial_t = 0$ and $b_p = 0$ for $1 \leqslant p \leqslant n-1$.

(4.6.7) We give one more proof of the coincidence of the Gauss–Manin connection on $\mathscr{H}^n_{\mathrm{DR}}(X'/S')$ (defined by Brieskorn) with the topological connection on $\mathscr{H} = R^n f_* \mathbb{C}_{X'} \otimes \mathscr{O}_{S'}$.

To prove that under an isomorphism of sheaves the connections correspond one to another, it is sufficient to show that horizontal sections of each of them correspond one to another. Again consider the exact hypercohomology sequence (4.6.4) corresponding to the sequence of complexes (4.6.3) on S', i.e. for $f'\colon X' \to S'$,

$$\begin{array}{ccccc} \ldots \to \mathbb{R}^n f'_*(\Omega^\cdot_{X'}) \xrightarrow{p} & \mathbb{R}^n f'_*(\Omega^\cdot_{X'/S'}) & \xrightarrow{\partial} & \mathbb{R}^{n+1} f'_*(f^*\Omega^1_{S'} \otimes \Omega^{\cdot-1}_{X'/S'}) & \to \ldots \\ & \| & & \| & \\ & \mathscr{H}^n_{\mathrm{DR}}(X'/S') & \xrightarrow{\nabla} & \Omega^1_S \otimes \mathscr{H}^n_{\mathrm{DR}}(X'/S'). & \end{array}$$

Hence $\operatorname{Ker} \nabla = \operatorname{Ker} \partial = \operatorname{Im}(\mathbb{R}^n f'_*(\Omega^\cdot_{X'}) \to \mathbb{R}^n f'_*(\Omega^\cdot_{X'/S'})$. Complexes $\Omega^\cdot_{X'}$ and $\Omega^\cdot_{X'/S'}$ are resolutions of the sheaves $\mathbb{C}_{X'}$ and $f^{-1}\mathscr{O}_{S'}$, respectively. Hence $\operatorname{Ker} \nabla$ corresponds to

$$\operatorname{Im}(R^n f'_* \mathbb{C}_{X'} \to R^n f_*(f^{-1}\mathscr{O}_{S'})) = \operatorname{Im}\left(R^n f'_* \mathbb{C}_{X'} \to R^n f'_* \mathbb{C}_{X'} \underset{\mathbb{C}_S}{\otimes} \mathscr{O}_{S'}\right),$$

i.e. to horizontal sections of the topological connection.

5 Brieskorn lattices

First we summarise briefly the results of §4. Let $f\colon X \to S$ be the Milnor fibration of an isolated singularity and $\underline{H} = R^n f'_* \mathbb{C}_{X'} = \bigcup_{t \in X'} H^n(X_t, \mathbb{C})$ be the cohomological fibration on S'. The locally free sheaf $\mathscr{H} = \underline{H} \otimes \mathscr{O}_{S'}$ on S' is identified with the De Rham cohomology sheaf $\mathscr{H}^n_{\mathrm{DR}}(X'/S')$, to which we transfer the topological connection on $\mathscr{H}$. The sheaf $\mathscr{H}^n_{\mathrm{DR}}(X'/S')$ naturally extends to the locally free sheaf $\mathscr{H}^n_{\mathrm{DR}}(X/S)$ on the whole of S. The attempt to extend the connection ∇ onto $\mathscr{H}^n_{\mathrm{DR}}(X/S)$ leads us to a locally free sheaf $'\mathscr{H} \supset \mathscr{H}^n_{\mathrm{DR}}(X/S)$ coinciding with $\mathscr{H}^n_{\mathrm{DR}}(X/S)$ on S'. The covariant differentiation gives an isomorphism $\partial_t = \nabla_{d/dt}\colon \mathscr{H}^n_{\mathrm{DR}}(X/S) \xrightarrow{\simeq} {}'\mathscr{H}$. The identification of the sheaf $'\mathscr{H} = f_* \Omega^n_{X/S} / d(f_* \Omega^{n-1}_{X/S})$ as well as of the sheaf $\mathscr{H}^n_{\mathrm{DR}}(X/S)$ with the sheaf $\mathscr{H}$ on S' is realized by means of the De Rham theorem: a local section of $'\mathscr{H}$ is represented by a class of an n-form ω on X, the restrictions $\omega_{|X_t}$ (these are forms of the largest degree on X_t) give a family of cohomology classes $w(t) \in H^n(X_t, \mathbb{C})$, i.e. a section $w = s[\omega]$ of the sheaf $\mathscr{H} = R^n f_* \mathbb{C}_{S'} \otimes \mathscr{O}_{S'}$. The germs of the sheaves $\mathscr{H}^n_{\mathrm{DR}}(X/S)_0$ and $'\mathscr{H}_0$ at the point $0 \in S$ have the same localization wrt t, which is a germ of the sheaf $\mathscr{H}[0] \subset i_*(\mathscr{H})$. The connection ∇ defines a meromorphic connection on $\mathscr{H}[0]$, and $\mathscr{H}^n_{\mathrm{DR}}(X/S) \subset {}'\mathscr{H} \subset \mathscr{H}[0]$ are two lattices in $\mathscr{H}[0]$.

5.1 Brieskorn lattice $''\mathscr{H}$

(5.1.1) Brieskorn considered one more natural lattice, $''\mathscr{H}$ in $\mathscr{H}[0]$, associated with $(n+1)$-forms for which the identification with $\mathscr{H}$ on S' is realized by means of a Poincaré residue.

The division lemma implies the exact sequence

$$0 \to \Omega^n_{X/S} \xrightarrow{\mathrm{d}f\wedge} \Omega^{n+1}_X \to \Omega^{n+1}_{X/S} \to 0.$$

Taking the quotient by the subsheaf $d\Omega^{n-1}_{X/S} \subset \Omega^n_{X/S}$ and applying f_*, we obtain an exact sequence

$$0 \to {}'\mathscr{H} \xrightarrow{\mathrm{d}f\wedge} {}''\mathscr{H} \xrightarrow{r} f_*\Omega_f \to 0, \tag{5.1.2}$$

where the sheaf

$$''\mathscr{H} = f_*\Omega^{n+1}_X / \mathrm{d}f \wedge \mathrm{d}(f_*\Omega^{n-1}_X) \tag{5.1.3}$$

is called the *Brieskorn lattice*.

(5.1.4) The sheaf $f_*\Omega_f = f_*\Omega^{n+1}_{X/S}$ is concentrated at the point $0 \in S$ and so the sheaves $'\mathscr{H} \subset^{\mathrm{d}f\wedge} {}''\mathscr{H}$ coincide on S'. The sheaf $''\mathscr{H}$ is free of torsion (§6) and so $''\mathscr{H} \subset \mathscr{H}[0]$.

(5.1.5) The identification of $''\mathscr{H}_{|S'}$ with $\mathscr{H}$ or with $'\mathscr{H}_{|S'}$ is realized in the following way. The inclusion $'\mathscr{H} \subset {}''\mathscr{H}$ in (5.1.2) is given by multiplication by $\mathrm{d}f$, hence the inverse mapping must consist of division by $\mathrm{d}f$. This is possible on X' because the Koszul complex is exact there. Namely, let ω be a $(n+1)$-form on X representing a section of $''\mathscr{H}$. The fibres X_t are hypersurfaces defined by equations $f(x) - t = 0$, $t = \mathrm{const}$. For $t \neq 0$ the fibres X_t are smooth and we can consider the function $f(x) - t$ as a local parameter $y = f(x) - t$; $\mathrm{d}y = \mathrm{d}f$ because $t = \mathrm{const}$. The form ω of the largest degree can be represented in the form $\omega = \mathrm{d}f \wedge \psi$. Then the restriction $\psi_{|X_t}$ does not depend on the form ψ and it is nothing but the Poincaré residue of the form $\omega/[f(x) - t]$,

$$\psi_{|X_t} = \mathrm{Res}_{X_t} \frac{\omega}{f - t}.$$

We obtain a family of forms on fibres X_t, i.e. a relative n-form (more exactly a section of $\Omega^n_{X'/S'}/d\Omega^{n-1}_{X'/S'}$) which is denoted by

$$\frac{\omega}{df} = \mathrm{Res}\,\frac{\omega}{f - t} \tag{5.1.6}$$

and is called the *Gelfand–Leray form* of the form ω. Hence we obtain a section $w = s[\omega]$ of the sheaf $\mathscr{H}$, where $w(t) \in H^n(X_t, \mathbb{C})$ is represented by the form $\psi_{|X_t} = \mathrm{Res}_{X_t}\,\omega/[f(x) - t]$. The section $s[\omega]$ Varchenko [AGV] calls a *geometric section* of the sheaf $\mathscr{H}$. Thus we obtained the identification $''\mathscr{H}_{|S'} = \mathscr{H}$.

5.2 Calculation of the Gauss–Manin connection ∇ on $'\mathscr{H}$

(5.2.1) We defined the covariant derivative $D = \partial_t = \nabla_{d/dt}$ of the Gauss–Manin connection on $\mathscr{H}^n_{\mathrm{DR}}(X/S)$, ∂_t: $\mathscr{H}^n_{\mathrm{DR}}(X/S) \to {}'\mathscr{H}$, which to an n-form ω representing a section of $\mathscr{H}^n_{\mathrm{DR}}(X/S)$ assigns an n-form $\partial_t(\omega)$ defined by the rule

$$d\omega = df \wedge \partial_t(\omega).$$

Taking into account the inclusion $\mathscr{H}^n_{\mathrm{DR}}(X/S) \overset{df\wedge}{\subset} {}''\mathscr{H}$, we see that the covariant derivative ∂_t: $\mathscr{H}^n_{\mathrm{DR}}(X/S) \to {}''\mathscr{H}$ reduces simply to the differential d, $\partial_t(\omega) = d\omega$. The same formula defines the extension of the connection on $\mathscr{H}^n_{\mathrm{DR}}(X/S)$ to $'\mathscr{H}$,

$$\partial_t\colon {}'\mathscr{H} \to {}''\mathscr{H},\ \partial_t(\omega) = d\omega. \tag{5.2.2}$$

5.3 Increasing filtration on $\mathscr{H}^{(0)}$

(5.3.1) We join together the exact sequences (4.4.2) and (5.1.2) connecting the sheaves $\mathscr{H}^n_{\mathrm{DR}}(X/S)$, $'\mathscr{H}$, $''\mathscr{H}$ and operators ∂_t. However, before this we introduce new notation for these sheaves following that used by K. Saito [SK] (the reason for this new notation will become clear below). These sheaves are first in a series of sheaves forming an increasing filtration.

Put

$$(5.3.2)\qquad \begin{aligned} \mathscr{H}^{(-2)} &= \mathscr{H}^n_{\mathrm{DR}}(X/S) = \mathscr{H}^n(f_*\Omega^{\cdot}_{X/S}) \\ \mathscr{H}^{(-1)} &= {}'\mathscr{H} = f_*\Omega^n_{X/S}/\mathrm{d}(f_*\Omega^{n-1}_{X/S}), \\ \mathscr{H}^{(0)} &= {}''\mathscr{H} = f_*\Omega^{n+1}_X/\mathrm{d}f \wedge d(f_*\Omega^{n-1}_X). \end{aligned}$$

All these sheaves naturally coincide (are identified) with the sheaf $\mathscr{H}$ on S', and their germs at the point $0 \in S$ can be calculated on X by means of complexes of germs at the point $x = 0 \in \mathbb{C}^{n+1}$:

$$(5.3.3)\qquad \begin{aligned} &\mathscr{H}^{(-2)}_0 = H^n(\Omega^{\cdot}_{X/S,x}),\ \mathscr{H}^{(-1)}_0 = \Omega^n_{X,x}/\mathrm{d}f \wedge \Omega^{n-1}_{X,x} + \mathrm{d}\Omega^{n-1}_{X,x} \\ &\mathscr{H}^{(0)}_0 = \Omega^{n+1}_{X,x}/\mathrm{d}f \wedge \mathrm{d}\Omega^{n-1}_{X,x}. \end{aligned}$$

(5.3.4) *Proposition* There is a commutative diagram

$$\begin{array}{ccccccc} 0 \to & \mathscr{H}^{(-1)} & \xrightarrow{\mathrm{d}f\wedge} & \mathscr{H}^{(0)} & \xrightarrow{r} & f_*\Omega_f & \to 0 \\ & \wr\uparrow \partial_t = \mathrm{d} & & \wr\uparrow \partial_t = \mathrm{d} & & \| & \\ 0 \to & \mathscr{H}^{(-2)} & \to & \mathscr{H}^{(-1)} & \xrightarrow{d} & f_*\Omega_f & \to 0 \end{array}$$

where $\Omega_f = \Omega^{n+1}_{X/S}$, and r is a mapping to the quotient. Both operators ∂_t in the diagram are isomorphisms.

Proof The commutativity of the diagram is obvious. Let us verify that $\partial_t\colon \mathscr{H}^{(-1)} \to \mathscr{H}^{(0)}$ is an isomorphism. Analogically we can verify that the second operator $\partial_t\colon \mathscr{H}^{(-2)} \to \mathscr{H}^{(-1)}$ is an isomorphism and, moreover, besides this was already verified in (4.6.5). It is useful to carry out such reasoning by means of the following commutative diagram

$$(5.3.5)\qquad \begin{array}{ccccccc} & & 0 & & & & \\ & & \uparrow & & & & \\ & & f_*\Omega^{n-1}_X & & & & \\ & & \uparrow -\mathrm{d}f & & & & \\ \ldots \to f_*\Omega^{n-1}_X & \xrightarrow{\mathrm{d}} & f_*\Omega^n_X & \xrightarrow{\mathrm{d}} & f_*\Omega^{n+1}_X & \to 0 & \\ & & \uparrow -\mathrm{d}f & & \uparrow -\mathrm{d}f & & \\ & \ldots \to & f_*\Omega^{n-1}_X & \xrightarrow{\mathrm{d}} & f_*\Omega^n_X & \xrightarrow{\mathrm{d}} & f_*\Omega^{n+1}_X \to 0, \\ & & \uparrow & & \uparrow & & \uparrow \\ & & \vdots & & \vdots & & \vdots \end{array}$$

which we will call a ‘fragment’ because it is a part of a double complex, in which the complexes $(f_*\Omega^{\cdot}_X, \mathrm{d})$, and $(f_*\Omega^{\cdot}_X, -\mathrm{d}f\wedge)$ are bound and whose natural origin will be explained in (II.5.3.10).

Obviously, $\partial_t\colon \mathscr{H}^{(-1)} \to \mathscr{H}^{(0)}$ is an epimorphism because an element of $\mathscr{H}^{(0)}$ is represented by an $(n+1)$-form ω on X and it can be represented in the form $\omega = \mathrm{d}\eta$ because the complex $(\Omega^{\cdot}_X, \mathrm{d})$ is exact in dimensions

$\geqslant 1$. Let us show that ∂_t is a monomorphism. Let $\omega \in f_*\Omega_X^n$ represent an element of $\mathscr{H}^{(-1)}$ and let $\partial_t([\omega]) = 0$ in $\mathscr{H}^{(0)}$, i.e. $\mathrm{d}\omega = \mathrm{d}f \wedge \mathrm{d}\eta$, $\eta \in f_*\Omega_X^{n-1}$. We have to prove that $[\omega] = 0$ in $\mathscr{H}^{(-1)}$, i.e. $\omega = \mathrm{d}f \wedge \xi + \mathrm{d}\xi_1$, ξ and $\xi_1 \in f_*\Omega_X^{n-1}$. We have $\mathrm{d}\omega = \mathrm{d}f \wedge d\eta = -d(\mathrm{d}f \wedge \eta)$, i.e. $\mathrm{d}(\omega + \mathrm{d}f \wedge \eta) = 0$. Again from the exactness of $(\Omega_X^{\cdot}, d)$ we have that $\omega + \mathrm{d}f \wedge \eta = \mathrm{d}\xi_1$, i.e. $\omega = -\mathrm{d}f \wedge \eta + \mathrm{d}\xi_1$. ∎

It follows from Proposition (5.3.4) that we can restore $\mathscr{H}^{(-2)}$ by means of $df\wedge\colon \mathscr{H}^{(-1)} \subset \mathscr{H}^{(0)}$ and $\partial_t\colon \mathscr{H}^{(-1)} \overset{\sim}{\to} \mathscr{H}^{(0)}$,

(5.3.6) $$\mathscr{H}^{(-2)} = \{\omega \in \mathscr{H}^{(-1)}\colon \partial_t(\omega) \in \mathscr{H}^{(-1)}\}.$$

This enables us to define an increasing filtration on $\mathscr{H}^{(0)}$,

(5.3.7) $$\begin{aligned}\mathscr{H}^{(-k)} &= \{\omega \in \mathscr{H}^{(-k+1)}\colon \partial_t(\omega) \in \mathscr{H}^{(-k+1)}\}\\ &= \{\omega \in \mathscr{H}^{(-1)}\colon \partial_t^{k-1}(\omega) \in \mathscr{H}^{(-1)}\}.\end{aligned}$$

We obtain a commutative diagram

(5.3.8) $$\begin{array}{ccccccccccccc} \ldots \subset & \mathscr{H}^{(-k)} & \subset \ldots \subset & \mathscr{H}^{(-2)} & \subset & \mathscr{H}^{(-1)} & \overset{df\wedge}{\subset} & \mathscr{H}^{(0)} & \to & f_*\Omega_f & \to 0 \\ & \wr\uparrow \partial_t & & \wr\uparrow \partial_t & & \wr\uparrow \partial_t & & \wr\uparrow \partial_t = \mathrm{d} & & \| & \\ \ldots \subset & \mathscr{H}^{(-k-1)} & \subset \ldots \subset & \mathscr{H}^{(-3)} & \subset & \mathscr{H}^{(-2)} & \subset & \mathscr{H}^{(-1)} & \overset{d}{\to} & f_*\Omega_f & \to 0, \end{array}$$

where, as it is easy to see, all inclusions have the same quotients,

(5.3.9) $$0 \to \mathscr{H}^{(-k-1)} \subset \mathscr{H}^{(-k)} \overset{r(k)}{\to} f_*\Omega_f \to 0.$$

5.4 A practical method of calculation of the Gauss–Manin connection

(5.4.1) Let $\nabla_{d/dt} = \partial_t\colon \mathscr{M} \to \mathscr{M}$ be a meromorphic Gauss–Manin connection on a vector space $\mathscr{M} = \mathscr{H}^{(0)} \otimes_{\mathscr{O}_{S,0}} \mathscr{K}$ over the field $\mathscr{K} = \mathscr{O}_{S,0}[t^{-1}]$ (4.5.6). To calculate the connection ∂_t practically it is convenient to use the Brieskorn lattice $\mathscr{H}_0^{(0)} = \Omega_{X,0}^{n+1}/\mathrm{d}f \wedge \mathrm{d}\Omega_{X,0}^{n-1}$ because $\Omega_{X,0}^{n+1}$ is a locally free module of rank 1 and differential forms $\omega = g(x)\,\mathrm{d}x$, $\mathrm{d}x = \mathrm{d}x_0 \wedge \ldots \wedge \mathrm{d}x_n$, are identified with functions,

$$\Omega_{X,0}^{n+1} \simeq \mathscr{O}_{X,0},\ \omega = g(x)\,\mathrm{d}x \leftrightarrow g(x).$$

Consider the exact sequence (5.1.2)

$$\begin{array}{ccccccccc} 0 \to & \mathscr{H}^{(-1)} & \overset{df\wedge}{\subset} & \mathscr{H}^{(0)} & \overset{r}{\to} & f_*\Omega_f & \to 0 \\ & \twoheaduparrow & & \twoheaduparrow & & \wr\| & \\ 0 \to & \mathscr{O}_X^{n+1} & \to & \mathscr{O}_X & \to & Q_f & \to 0. \end{array}$$

Under these epimorphisms the module $Q^f = f_*\Omega_f$ (4.4.6) is identified with the local Artinian algebra $Q_f = \mathscr{O}_{X,0}/J_f$.

As we know (5.2.2), the connection ∂_t on the lattice $\mathscr{H}^{(-1)} \overset{\mathrm{d}f}{\subset} \mathscr{H}^{(0)}$ is calculated by the formula:

(5.4.2) for $\mathrm{d}f \wedge \xi \in \mathrm{d}f \wedge \mathscr{H}^{(-1)}$ we have $\partial_t(\mathrm{d}f \wedge \xi) = \mathrm{d}\xi$.

On the other hand, the $\mathscr{O}_{S,0}$-module $f_*\Omega_f$ is concentrated at the point $0 \in S$. Hence there exist such k that $f^k\mathscr{H}^{(0)} \subset \mathscr{H}^{(-1)}$. Hence to calculate ∂_t for an element $\omega \in \mathscr{H}^{(0)}$, we need first to 'drive' ω into $\mathscr{H}^{(-1)}$ and then to use the previous rule: if $f^k\omega = \mathrm{d}f \wedge \xi$, then $\partial_t(f^k\omega) = \mathrm{d}\xi$; but $\partial_t(f^k\omega) = kf^{k-1}\omega + f^k\partial_t\omega$ and hence we have in $\mathscr{M}$

(5.4.3) $$\partial_t\omega = -\frac{k}{f}\omega + \frac{\mathrm{d}\xi}{f^k}.$$

The $\mathscr{O}_{S,0}$-module $\mathscr{H}^{(0)}$ is generated by a class of the form dx. So to 'drive' $\mathscr{H}^{(0)}$ into $\mathscr{H}^{(-1)}$, it is sufficient to find such k that

(5.4.4) $$f^k\,\mathrm{d}x = \mathrm{d}f \wedge \xi,$$

where $\xi = \sum_{i=0}^n h_i(x)\,\mathrm{d}x_0 \wedge \ldots \wedge \mathrm{d}\hat{x}_i \wedge \ldots \wedge \mathrm{d}x_n \in \Omega^n_{X,0}$ is an n-form. However, $\mathrm{d}f \wedge \xi = \sum_{i=0}^n (-1)^i h_i(x) f'_{x_i}\,\mathrm{d}x$ and so the calculation of the connection of $\mathscr{H}^{(0)}$ is reduced to the question of finding k such that

(5.4.5) $$f^k \in (f'_{x_0}, \ldots, f'_{x_n}) = J_f$$

and explicit expression of f^k by the derivatives f'_{x_i} (the calculation of the form ξ). The identification of $\Omega^{n+1}_{X,0}$ with $\mathscr{O}_{X,0}$, $\omega = g(x)\,dx \leftrightarrow g(x)$, enables us to rewrite the calculation of the connection ∂_t in terms of functions $g(x) \in \mathscr{O}_{X,0}$.

Let us consider an important example.

5.5 Calculation of the Gauss–Manin connection of quasihomogeneous isolated singularities

(5.5.1) Let $f(x) = \sum_m c_m x^m \in \mathbb{C}[x_0, \ldots, x_n]$ be a quasihomogeneous singularity of degree $d = 1$ with weights $\mathrm{wt}(x_i) = w_i$, where for $m = (m_0, \ldots, m_0)$, x^m denotes $x_0^{m_0} \cdot \cdots \cdot x_n^{m_n}$. In this case the connection ∂_t on $\mathscr{H}^{(0)}$ is calculated explicitly and completely. This arises from two circumstances. Firstly $f \in J$, i.e. $k = 1$, and from the Euler identity $\sum_{i=0}^n w_i x_i f'_{x_i} = f(x)$ for the form ξ such that $f(x)\,\mathrm{d}x = \mathrm{d}f \wedge \xi$ we have an explicit formula

(5.5.2) $$\xi = \sum_{i=0}^n (-1)^i w_i x_i\,\mathrm{d}x_0 \wedge \ldots \wedge \mathrm{d}\hat{x}_i \wedge \ldots \wedge \mathrm{d}x_n.$$

Secondly the image of the inclusion $\mathscr{H}^{(-1)} \overset{\mathrm{d}f}{\subset} \mathscr{H}^{(0)}$ coincides with $f\mathscr{H}^{(0)} = t\mathscr{H}^{(0)}$, i.e. with a submodule of $\mathscr{H}^{(0)}$ generated by the maximal ideal $(t) \subset \mathscr{O}_{S,0}$. So (by Nakayama's lemma) to find a basis $\omega_1, \ldots, \omega_\mu$ of the $\mathscr{O}_{S,0}$-module $\mathscr{H}^{(0)}$, it is sufficient to find a basis in a vector space $f_*\Omega_f = Q^f \simeq Q_f = \mathscr{O}_{X,0}/J_f$.

Let $x^m = x_0^{m_0} \dots x_n^{m_n} \in Q_f$ be a monomial basis in Q_f, $m \in A$, $|A| = \mu$. Then the forms $\omega_m = x^m \,\mathrm{d}x$, $m \in A$, represent a basis of the $\mathscr{O}_{S,0}$-module $\mathscr{H}^{(0)}$. From $f(x)\,\mathrm{d}x = \mathrm{d}f \wedge \xi$ we have $f(x)\omega_m = \mathrm{d}f \wedge x^m\xi$ and this implies that

$$\partial_t\omega_m = \frac{1}{f(x)}[-\omega_m + \mathrm{d}(x^m\xi)] = \frac{1}{f(x)}\left\{-\omega_m + \left[\sum_{i=0}^{n} w_i(m_i+1)\right]\omega_m\right\}.$$

Putting

$$a(m) = \sum_{i=0}^{n} w_i(m_i+1), \tag{5.5.3}$$

we then obtain:

(5.5.4) *Proposition* For a quasihomogeneous singularity $f(x)$, $\deg f(x) = 1$, $\mathrm{wt}(x_i) = w_i$, the monomial basis $\omega_m \in \mathscr{H}^{(0)}$, $m \in A$, $|A| = \mu$, is a basis of eigenvectors of the operator $t\partial_t$,

$$t\partial_t\omega_m = [a(m) - 1]\omega_m.$$

■

Thus in the case of a quasihomogeneous singularity the Brieskorn lattice $\mathscr{H}^{(0)}$ is saturated, and the residue of the connection $\mathrm{Res}_{\mathscr{H}^{(0)}}\nabla$ is a semisimple operator (see §8).

(5.5.5) *Corollary* The monodromy T of a quasihomogeneous singularity is semisimple and its eigenvalues are

$$\lambda_m = \mathrm{e}^{-2\pi\mathrm{i}\alpha_m},\ \alpha_m = a(m),\ m \in A,\ |A| = \mu.$$

(5.5.6) *Remark* In general the Brieskorn lattice $\mathscr{H}^{(0)}$ is not saturated, i.e. in its basis the system of differential equations defining horizontal sections has a pole of order > 1. Scherk [Sc4] developed methods of construction of the saturation $\mathscr{H}^{(0)}$ of the lattice $\overline{\mathscr{H}}^{(0)}$, the choice of a convenient basis in $\mathscr{H}^{(0)}$ and explicit calculation of the connection ∂_t. The calculations are rather non-trivial. Scherk carries out his method in detail for the singularity $x^5 + y^5 + x^2y^2$ in [Sc1]. This is a singularity of type $T_{2,5,5}$ in the classification of Arnold [AGV]. It is equivalent to the singularity $(x^3 + y^2)(x^2 + y^3)$. This was the first singularity whose monodromy has infinite order to be discovered [A]. For the results of the calculation for the singularities $T_{p,q,r}$ (see (II.7.3.5) and for the singularity $ax^5 + y^6 + x^4y$, $a \in \mathbb{C}$, see [Sc-S]).

6 Absence of torsion in sheaves $\mathscr{H}^{(-i)}$ of isolated singularities

Let $f\colon X \to S$ be the Milnor fibration of an isolated singularity $f\colon (\mathbb{C}^{n+1}, \mathbb{C}) \to (\mathbb{C}, 0)$, and let $\mathscr{H}^{(-i)}$ be the sheaves introduced in (5.3.7). In particular $\mathscr{H}^{(-2)} = \mathscr{H}^{n}_{\mathrm{DR}}(X/S)$ and $\mathscr{H}^{(0)} = f_*\Omega_X^{n+1}/\mathrm{d}f \wedge \mathrm{d}(f_*\Omega_X^{n-1})$ the Brieskorn lattice. These are coherent $\mathscr{O}_S$-modules on a disk S. We'll prove that these sheaves are torsion free (this is a result of Sebastiani – see [Se, M]) and it involves that they are locally free sheaves of rank μ.

6.1 The presence of a connection implies the absence of torsion

(6.1.1) *Lemma* Let E be a finitely generated $\mathscr{O}$-module, $\mathscr{O} = \mathscr{O}_{\mathbb{C}^r,0}$. If there is a connection on E, then E is free.

Proof Assume for simplicity that $r = 1$. Let $e_1, \dots, e_p$ be a minimal system of generators of E. To construct this one can consider a basis $\overline{e}_1, \dots, \overline{e}_p$ of a vector space $E/\mathscr{m}E$ and the pre-images $e_i \in E$ of elements $\overline{e}_i$. Then by Nakayama's lemma $e_1, \dots, e_p$ generate E. We next prove that $e_1, \dots, e_p$ are free generators, i.e. there are no relations between them. The proof is by *reductio ad absurdum*. Assume that $f_1e_1 + \dots + f_pe_p = 0$ is a non-trivial relation, $f_i(t) \in \mathscr{O}$. Then $f_1(0) = \dots = f_p(0) = 0$ because otherwise the system of generators is not minimal. Let k be the minimal order of vanishing of functions $f_i(t)$, which can appear in such relations. Then $k \geqslant 1$. Let us take the relation $\sum f_ie_i = 0$ with minimal k and apply the operator $D = \nabla_{\mathrm{d}/\mathrm{d}t}$ to it. We obtain a relation between the e_i for which at least one of the coefficients has the order of vanishing $\leqslant k - 1$, and this contradicts the minimality of k. Indeed, if $D(e_j) = \sum_{i=1}^{p} \Gamma_{ij}e_i$, then

$$
\begin{aligned}
0 = D\left(\sum_{j=1}^{p} f_je_j\right) &= \sum_{j=1}^{p} f'_{j'}e_j + \sum_{j=1}^{p} f_jD(e_j)\\
&= \sum_{j} f'_je_j + \sum_{j} f_j\left(\sum_{i} \Gamma_{ij}e_j\right)\\
&= \sum_{j=1}^{p}\left(f'_j + \sum_{l=1}^{p} f_1\Gamma_{jl}\right)e_j.
\end{aligned}
$$

If f'_{j_0} has order k at 0, then f'_{j_0} as well as the entire coefficient has order $k - 1$. ■

(6.1.2) Lemma (1.1) implies that the sheaves $\mathcal{H}^p_{\mathrm{DR}}(X/S)$ are locally free for $p \leqslant n-1$, because there are Gauss–Manin connections on them.

Before passing to the sheaves $\mathcal{H}^{(-i)}$ we prove an important result.

6.2 A theorem of Malgrange

(6.2.1) *Theorem* [M1] Let ω be an n-form on X (a section of the sheaf $\mathcal{H}^{(-1)}$), and let γ be a section of the homology fibration $\underline{H}^*$ (a family of n-cycles) in a sector containing the zero ray, $\arg t = 0$. Then

$$\lim_{\substack{t\to 0\\ \arg t=0}} \int_{\gamma(t)} \omega = 0$$

Proof Consider a preimage $Y = f^{-1}([0, t_0]) \subset X$ of a segment $[0, t_0]$ on the axis $\arg t = 0$. Let Γ be an n-cycle on X_{t_0} of a class $\gamma(t_0) \in H^n(X_{t_0}, \mathbb{C})$.

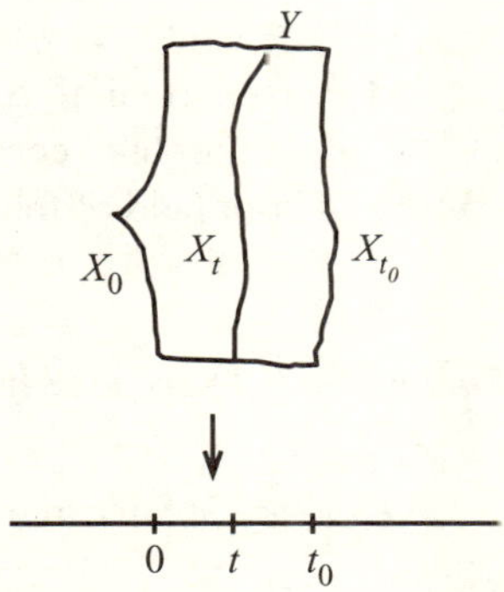

Since X_0 is contractible, Y is also contractible and $H^n(Y, \mathbb{C}) = 0$. Therefore $\Gamma = \partial\Delta$, where Δ is an $(n+1)$-chain on Y. Then by Stokes–Herrera theorem

$$I(t_0) = \int_{\gamma(t_0)} \omega = \int_{\Gamma} \omega = \int_{\Delta} d\omega.$$

Consider an $(n+1)$-chain $\Delta_t = f^{-1}([0, t]) \cap \Delta$, $t \in (0, t_0]$. Then

$$I(t) = \int_{\gamma(t)} \omega = \int_{\Delta_t} d\omega.$$

Indeed, $\Delta = \Delta_t + \Delta'$, where Δ' is a chain on $f^{-1}([t, t_0])$ and $\partial\Delta' = \Gamma - \Gamma_t$. Hence Γ_t is a cycle representing $\gamma(t)$ and $I(t_0) = \int_\Delta d\omega$ $= \int_{\Delta_t} d\omega + \int_{\Delta'} d\omega = \int_{\Delta_t} d\omega + \int_\Gamma \omega - \int_{\Gamma_t} \omega = \int_{\Delta_t} d\omega + I(t_0) - I(t)$. It follows that $I(t) = \int_{\Delta_t} d\omega$. Now by a classical theorem of analysis $\lim_{t\to 0} \int_{\Delta_t} d\omega = \int_{\Delta_0} d\omega$, where $\Delta_0 = X_0 \cap \Delta$ is an n-chain on X_0. However, $X_0\backslash\operatorname{Sing} X_0$ is an n-manifold and $\mathrm{d}\omega$ is an $(n+1)$-coboundary. Hence $\int_{\Delta_0} \mathrm{d}\omega = 0$. ■

6.3 Connection on a pair (E, F)

(6.3.1) We make one more general remark. Let $E \subset F$ be $\mathscr{O}$-modules, $\mathscr{O} = \mathscr{O}_{S,0}$, and let D be a connection on the pair (E, F), i.e. D: $E \to F$ is a $\mathbb{C}$-linear homomorphism satisfying the Leibniz rule. Denote by E^τ and F^τ the torsion submodules of E and F. Then

(6.3.2) $$D(E^\tau) \subset F^\tau.$$

Indeed, assume that $e \in E^\tau$ and $t^k e = 0$, then $0 = D(t^k e) = kt^{k-1}e + t^k D(e)$. Multiplying this equality by t, we obtain $t^{k+1}D(e) = 0$, i.e. $D(e) \in F^\tau$.

(6.3.3) Going to quotients $\overline{E} = E/E^\tau$ and $\overline{F} = F/F^\tau$, we obtain a connection $\overline{D}$ on the pair $(\overline{E}, \overline{F})$. Moreover, if $\dim_{\mathbb{C}} F/E < \infty$, i.e. F/E is a torsion, then D defines a meromorphic connection on the space $E \otimes_{\mathscr{O}} \mathscr{K} = F \otimes_{\mathscr{O}} \mathscr{K}$, where $\mathscr{K}$ is a field of fractions of $\mathscr{O}$.

6.4 Sheaves $\mathscr{H}^{(-p)}$ are locally free

(6.4.1) *Theorem* $\mathscr{H}^{(-2)} = \mathscr{H}_{\mathrm{DR}}(X/S)$, $\mathscr{H}^{(-1)} = f_*\Omega^n_X/d(f_*\Omega^{n-1}_X)$ and $\mathscr{H}^{(0)} = f_*\Omega^{n+1}_X/\mathrm{d}f \wedge \mathrm{d}(f_*\Omega^{n-1}_X)$ are locally free $\mathscr{O}_S$-modules of rank μ.

Proof It is sufficient to prove that these sheaves are torsion free. As $\mathscr{H}^{(-2)} \subset \mathscr{H}^{(-1)} \subset \mathscr{H}^{(0)}$ are submodules it is sufficient to prove that $\mathscr{H}^{(0)}$ is torsion free. Let $\mathscr{H}^{(-1)}_0 = E$, $\mathscr{H}^{(0)}_0 = F$. We then have $E \subset F$ and D: $E \overset{\sim}{\to} F$ is an isomorphism by (5.3.4). Assume that the torsion module $F^\tau \neq 0$. Then $E^\tau \subset F^\tau$ and from (6.1.1) and (6.3.2) it follows that $E^\tau \neq F^\tau$ otherwise there is a connection on $E^\tau = F^\tau$ and hence $F^\tau = 0$. As D: $E \overset{\sim}{\to} F$ is an isomorphism, there exists $\omega \notin E^\tau$ such that $D(\omega) \in F^\tau$. Here ω is represented by a section of $\mathscr{H}^{(-1)}$ such that $\omega_{|S'} \neq 0$ and $D(\omega)_{|S'} = \mathrm{d}\omega_{|S'} = 0$. Let $\gamma(t)$ be a local section of the homological fibration. Then $(d/\mathrm{d}t)\int_{\gamma(t)}\omega = \int_{\gamma(t)} \mathrm{d}\omega = 0$. This means that the function $I(t) = \int_{\gamma(t)}\omega = \text{const.}$ However, by Theorem (2.1) $\lim_{t\to 0} I(t) = 0$. Hence $I(t) \equiv 0$ for any section $\gamma(t)$ of the fibration $\underline{H}^*$. This means that ω is zero as a section of the sheaf $\mathscr{H} = \mathscr{H}^{(-1)}_{|S'}$, i.e. $\omega \in E^\tau$. We therefore obtain a contradiction and hence $F^\tau = 0$. ■

7 Singular points of systems of linear differential equations

7.1 Differential equations of Fuchsian type

We begin by recalling the notion of a regular singular point of a linear differential equation of order μ or of a system of linear differential equations of first order.

(7.1.1) In the 1860s L. Fuchs (1833–1902) investigated linear differential equations

$$u^{(\mu)} + p_1(t)u^{(\mu-1)} + \ldots + p_{\mu-1}(t)u' + p_\mu(t)u = 0,$$

in which $p_i(t)$ are meromorphic functions on the complex plane (or on some of its domain) having a finite number of singular points. We are interested in a local case and so we'll assume that $p_i(t)$ are holomorphic on a punctured disk $S' = \{0 < |t| < \delta\}$ and consider a singular point $t = 0$. By Cauchy's theorem, in a neighbourhood of a point $t_0 \in S'$ the equation has a fundamental system of solutions $u_1(t), \ldots, u_\mu(t)$, which can be continued along a counter-clockwise oriented circle around the origin. Returning to the initial point we get a new fundamental system of solutions which is obtained from the initial system by multiplication by a monodromy matrix T.

(7.1.2) Fuchs proved (see below) that if $\lambda_1, \ldots, \lambda_k$ are roots of the characteristic equation $|T - \lambda E| = 0$ of multiplicities $m_1, \ldots, m_k$, $\alpha_j = (1/2\pi\mathrm{i}) \log \lambda_j$, then we can choose a fundamental system of solutions in the form of linear combinations of functions

$$t^{\alpha_j}\varphi_{j1}(t),\ t^{\alpha_j}\log t{\cdot}\varphi_{j2}(t),\ \ldots,\ t^{\alpha_j}\log^{m_j-1} t{\cdot}\varphi_{jm_j},\ j = 1, \ldots, k,$$

where $\varphi_{j1}(t)$ are holomorphic on S'.

(7.1.3) A singular point $t = 0$ for which functions $\varphi_{j1}(t)$ have only poles (but not essentially singularities) was later called a regular singular point.

(7.1.4) Fuchs proved (1866) that for a singular point $t = 0$ to be regular it is necessary and sufficient to have $a_j(t) = p_j(t)t^j$ holomorphic at 0, i.e. to have $p_j(t)$ of the form $p_j(t) = a_j(t)/t^j$. If all $a_j(t) = a_j = \mathrm{const}$, then we obtain the so-called Euler equation. Linear differential equations in which all singular points are regular, are called equations of Fuchsian type.

7.2 Systems of linear differential equations and connections

Consider a system of linear differential equations

$$(7.2.1)\qquad \begin{cases} y_1' = a_{11}(t)y_1 + \ldots + a_{1\mu}(t)y_\mu \\ \vdots \\ y_\mu' = a_{\mu 1}(t)y_1 + \ldots + a_{\mu\mu}(t)y_\mu, \end{cases}$$

or

$$(7.2.2)\qquad y' = A(t)y,$$

where $y = (y_1, \ldots, y_\mu)^t$ is a column of unknown functions, $A(t) = (a_{ij}(t))$ is a matrix of system, and $a_{ij}(t)$ are functions holomorphic on the punctured disk S'.

By Cauchy's theorem on the existence and uniqueness of solutions in the neighborhood of any point $t_0 \in S'$ there exists a fundamental system (basis) of solutions $\overline{y}_1, \ldots, \overline{y}_\mu$, where $\overline{y}_j = (y_{1j}, \ldots, y_{\mu j})^t$. Writing solutions $\overline{y}_j$ as columns of a matrix $Y(t) = (y_{ij}(t))$, we obtain a fundamental matrix of solutions. Going around the origin $0 \in S$ in a counter-clockwise direction we obtain a linear transformation in the space of solutions of system (7.2.2) in the neighborhood of the point t_0. This is the monodromy operator T. Let T also denote the monodromy matrix wrt basis $\overline{y}_1, \ldots, \overline{y}_\mu$, i.e. columns of T consist of coordinates of the vectors $T\overline{y}_j$. Then after a complete revolution around the point $0 \in S$ the matrix transforms to $Y(t)T = (\overline{y}_1, \ldots, \overline{y}_\mu)T$.

In other words a space of solutions of the system $y' = A(t)y$ defines a local system $\underline{E}$ on S'. The local system $\underline{E}$ defines a connection ∇ on the trivial sheaf $\mathscr{E} = \underline{E} \otimes_{\mathbb{C}_{S'}} \mathscr{O}_{S'} \simeq \mathscr{O}_{S'}^{\mu} = \oplus_{j=1}^{\mu} \mathscr{O}_{S'} e_j$. If $\Gamma(t) = (\Gamma_{ij}(t))$ is the matrix of connection ∇ coefficients wrt the basis e, then we can consider the system (7.2.2) as a system of differential equations defining horizontal sections of the connection if we set

$$(7.2.3)\qquad A(t) = -\Gamma(t).$$

Then the monodromy of system (7.2.2) is nothing but the monodromy of the connection ∇.

(7.2.4) The matrix $A(t)$ of system (7.2.2) (the connection matrix) depends on the choice of a basis in $\mathscr{E}$. If $Q(t)$ is a transition matrix to a new basis and new coordinate functions v are connected with old coordinate functions by $y = Q(t)v$, then a substitution in the equation $y' = A(t)y$ shows that this equation transforms to $v' = B(t)v$, where $B = Q^{-1}AQ - Q^{-1}Q'$.

7.3 Decomposition of a fundamental matrix $Y(t)$

The matrix function $Y(t)$ has a singularity at the point $0 \in S$. As in the case of functions the singularity is characterized by: (i) a revolution around the point 0, i.e. by ramification; and (ii) approaching $0 \in S$. We can separate these effects.

Let T be a monodromy matrix of system (7.2.2) wrt a basis $\overline{y}_1, \ldots, \overline{y}_\mu$. Let us represent the invertible matrix T by the form

(7.3.2) $$T = e^{2\pi i R}.$$

Recall that using series we can define functions $f(A)$ in matrix A. In particular, for a matrix $R \in M(\mu, \mathbb{C})$ we can consider a matrix function

(7.3.3) $$t^R = e^{R \ln t} = \sum_{k=0}^{\infty} \frac{(R \ln t)^k}{k!}.$$

The function t^R has the same monodromy as $Y(t)$: after rounding $0 \in S$ the function t^R is multiplied by $e^{2\pi i R} = T$. Hence we can represent $Y(t)$ by the form

(7.3.4) $$Y(t) = Z(t) \cdot t^R,$$

where $Z(t)$ is a single-valued analytic function on S'.

The matrix T and hence the matrix R depend on the choice of the basis $\overline{y}_1, \ldots, \overline{y}_\mu$, i.e. on the choice of the fundamental matrix $Y(t)$. Changing $Y(t)$ to $Y(t)C$, where $C \in GL(n, \mathbb{C})$ (transition to a new basis), we obtain

$$Y(t)C = Z(t)C(C^{-1}t^R C) = Z(t)C \cdot t^{C^{-1}RC},$$

i.e. the matrix R transforms to a matrix $C^{-1}RC$. Hence changing the basis $\overline{y}_1, \ldots, \overline{y}_\mu$, we can assume that R has a Jordan normal form. In particular, if R is a Jordan block of order k with eigenvalue α,

(7.3.5)
$$R = \begin{pmatrix} \alpha & & & & 0 \\ 1 & \alpha & & & \\ & 1 & \ddots & & \\ & & \ddots & \ddots & \\ 0 & & & 1 & \alpha \end{pmatrix} = \alpha E + N, \text{ where } N = \begin{pmatrix} 0 & & \cdots & 0 \\ 1 & & & \\ \vdots & \ddots & \ddots & \vdots \\ 0 & \cdots & 1 & 0 \end{pmatrix},$$

then

(7.3.6) $$t^R = e^{R \ln t} = t^\alpha \cdot \begin{pmatrix} 1 & & & 0 \\ \ln t & 1 & & \\ \vdots & \ddots & \ddots & \\ \frac{(\ln t)^{n-1}}{(n-1)!} & \cdots & \ln t & 1 \end{pmatrix},$$

$$T = e^{2\pi i R} = \lambda \cdot e^{2\pi i N} = \lambda \cdot T_u,$$

where $\lambda = e^{2\pi i \alpha}$ is an eigenvalue of the monodromy, and T_u is an unipotent part of the monodromy,

(7.3.7) $$N = \frac{1}{2\pi i} \log T_u.$$

7.4 Regular singular points

(7.4.1) *Definition* A singular point $0 \in S$ of system (7.2.2) is called *regular*, if in the representation of the fundamental matrix $Y(t) = Z(t)t^R$ the matrix function $Z(t)$ is meromorphic, i.e. functions $z_{ij}(t)$ have only poles (and haven't essential singularities).

(7.4.2) This definition can be reformulated in terms of the rate of growth of solutions as $t \to 0$. Let $g(t)$ be a multivalued function on S'. Let us make a cut along the real ray $\mathbb{R}^+ \cap S'$. Let S_1 be S' with this cut. One says that a function $g(t)$ has a *moderate rate of growth at* 0, if any branch of $g(t)$ on S_1 grows at most as a degree of $1/|t|$, i.e. $|g(t)| \leqslant A/|t|^n$ for some n. A vector function has a moderate rate of growth, if all its coordinates have a moderate rate of growth.

(7.4.3) It is easy to see that a singular point $0 \in S$ of system (7.2.2) is regular if and only if any of its solutions has a moderate rate of growth at 0.

We have characterized regular singular points in terms of solutions of the equation. Now we'll do it in terms of the equation (7.2.2) itself.

7.5 Simple singular points

(7.5.1) *Definition* The point $t = 0$ is called a *simple singular point* of system (7.2.2), if its matrix $A(t)$ has at the point $t = 0$ a pole of at most first order, i.e. if $A(t)$ can be written in the form $A(t) = A_0/t + A_1(t)$, where A_0 is constant and $A_1(t)$ is a holomorphic matrix. The matrix A_0 is called a *residue* of system (7.2.2) at the point 0.

A system (7.2.2) with a simple singular point is written in the form

(7.5.2) $$t\partial_t y = [A_0 + tA_1(t)]y,$$

where $\partial_t = d/dt$.

(7.5.3) *Proposition* If $t = 0$ is a regular singular point of system (7.2.2), there exists a meromorphic change of unknown functions $y = Q(t)v$ ($Q(t)$ is a meromorphic matrix) which transforms system (7.2.2) to a system with a simple singular point.

Indeed, let $Y(t) = Z(t)t^R$, where $Z(t)$ is a meromorphic matrix. Let us take $Q(t) = Z(t)$, i.e. make the change $y = Z(t)v$. As $Y = Z \cdot t^R$ is a solution of the system $y' = Ay$ (7.2.2), we obtain that this change transforms system (7.2.2) to the system

$$v' = \frac{1}{t} R \cdot v,$$

with a simple singular point. Moreover, we obtain a system of special (Eulerian) type, for which $A_1(t) \equiv 0$. Obviously, these singularities are regular.

(7.5.4) *Lemma* A system of linear differential equations

(7.5.5) $$y' = \frac{R}{t} y,$$

where $R \in M(\mu, \mathbb{C})$ has a fundamental matrix of solutions

$$Y(t) = t^R$$

and, consequently, has a regular singularity at the point $0 \in S$.

The proof of this consists of an explicit solution of system (7.5.5) by 'school' methods. Namely, a system of linear equations $y' = Ry$ with constant coefficients and in variable z has a fundamental matrix of solutions $Y(z) = e^{Rz}$ (it is a 'differential equation with separable variables'). A change of the variable $z = \ln t$ in system (7.5.5) leads to a system $y' = Ry$. Hence system (7.5.5) has a fundamental matrix of solutions $Y = e^{R \ln t} =_{\mathrm{df}} t^R$.

A linear change of unknown functions $y = C\tilde{y}$ in system (7.5.5) leads to a transform of the matrix R, $R \mapsto C^{-1}RC$. By such a change we can transform R into a Jordan normal form. Thus, system (7.5.5) falls apart into several disjoint systems corresponding to Jordan blocks. Denote by $\mathscr{E}^{\alpha,q}$ a system (7.5.5) where R is one Jordan block of order q with eigenvalue α:

($\mathscr{E}^{\alpha,q}$) $$ty' = Ry,$$

where $R = \alpha E + N$ is a triangle matrix in (7.3.5).

7.6 Simple singular points are regular

The classical Sauvage theorem (Sauvage, 1853–1920) asserts:

(7.6.1) *Theorem* (Sauvage, 1886) If $t = 0$ is a simple singular point of system (7.2.2), then it is regular.

This theorem follows from the following theorem, a proof of which can be found in the [C-L]:

(7.6.2) *Key-Lemma* (Lemme-clef) If the residue A_0 of an equation $ty' = [A_0 + tA_1(t)]y$ with a simple singular point has no eigenvalues, which differ by an integer, then there exists a holomorphic change of unknown functions $Q(t)y$, which transforms this equation to an equation (7.5.5) (i.e. reduces it to the Eulerian case).

The condition about eigenvalues in this lemma can be satisfied by means of the Lemma about shift (7.7.6). We'll prove this below.

(7.6.3) Thus, a singular point $0 \in S$ of a system $y' = A(t)y$ is regular if and only if there exists a meromorphic change $y = Q(t)v$, transforming the system to a system with a simple singular point.

(7.6.4) *Remark* Moreover, by the classical Horn theorem (Horn, 1892) such a matrix $Q(t)$ can always be found in the form of a polynomial

$$Q(t) = Q_0 + Q_1 t + \ldots + Q_N t^N, \; Q_i \in M(n, \mathbb{C}).$$

J. Moser and D. Kutz (see [Br]) obtained an estimate for the degree N expressed in the order of pole k of matrix $A(t)$. Namely, let $A(t) = t^{-k}(\sum_{i=0}^{\infty} A_i t^i)$ be a meromorphic matrix with a pole of order $k > 1$ and a nilpotent matrix A_0 of rank r. Then we have:

(7.6.5) *Proposition* A system $y' = A(t)y$ with a regular singular point can be transformed into a system with a simple singular point by means of a change $y = Q(t)v$, where $Q(t)$ is a matrix of polynomials of degree $N \leqslant 2[(k-2)(n-1) + r]$.

A practical method of finding the matrix $Q(t)$ is reduced to a solution of a system of algebraic equations. Unknowns of this system are the elements of matrices $Q_0, \ldots, Q_N$. Equations of this system express the condition that after the change $y = Q(t)v$ we get a system with simple singularity, i.e. the matrix $Q^{-1}AQ - Q^{-1}Q'$ has at most a pole of the first order.

7.7 Connections with regular singularities

Let $\mathscr{K} = \mathscr{O}[t^{-1}]$ be a field of fractions of the ring $\mathscr{O} = \mathscr{O}_{S,0}$, $M = \oplus_{j=1}^{\mu} \mathscr{K} e_j$ be a vector space over $\mathscr{K}$, and $\nabla \colon M \to \Omega^1_{S,0} \otimes M$ be a meromorphic connection on M (4.6.1). If a basis e in M is chosen, and $\Gamma = (\Gamma_{ij}(t))$ is a matrix of coefficients of connection forms in this basis, then the horizontality condition $\nabla(m) = 0$ of a section $m = ey$ is written in the form of a system of linear equations (7.2.2), where $A(t) = -\Gamma(t)$.

(7.7.1) A connection ∇ is called *regular* at the point $0 \in S$ (or $0 \in S$ is called a regular singularity of a connection ∇), if the corresponding system (7.2.2) has a regular singularity at 0. Taking into account the relation between regular and simple singular points and the fact that $\Gamma(t)$ is the matrix of operator $\partial_t = \nabla_{d/dt}$ in basis e, we obtain the following reformulation:

(7.7.2) *Definition* A lattice $L \subset M$ is called *saturated*, if it is stable under the operator $t\partial_t$, i.e. $t\partial_t(L) \subset L$. A meromorphic connection ∇ on M is called *regular*, if there exists a saturated lattice in M.

Indeed, if a lattice L is stable under $t\partial_t$, then the matrix of the operator ∂_t, i.e. $\Gamma(t)$, has at most a pole of the first order. The converse follows from a commutator relation

(7.7.3) $$[t\partial_t, g] = t \cdot g' \text{ for } g \in \mathcal{O}$$

Indeed, if in a basis $e_1, \ldots, e_\mu$ of a lattice L the matrix $\Gamma(t)$ has at most a pole of the first order, then for base elements $t\partial_t(e_j) \in L$. Then for $\sum g_j e_j \in L$ we have

$$t\partial_t\left(\sum g_j e_j\right) = \sum_j t\partial_t g_j e_j = \sum_j g_j t\partial_t e_j + \sum_j t g'_j e_j \in L,$$

i.e. L is stable under $t\partial_t$. Again from a commutator relation $[t\partial_t, t] = t$ it follows that for a saturated lattice L its $\mathcal{O}$-submodule tL is stable under $t\partial_t$.

(7.7.4) *Definition* The *residue* of a meromorphic connection ∇ on M wrt a saturated lattice L is the endomorphism

$$\mathrm{Res}_0 \nabla = t\partial_t \quad L/tL \to L/tL$$

of the vector space $\overline{L} = L/tL$, $\dim_{\mathbb{C}} \overline{L} = \mu$.

If a connection ∇ has a matrix of connection coefficients $\Gamma(t) = \Gamma_0/t + \Gamma_1(t)$ in a basis of L, where $\Gamma_1(t)$ is a holomorphic matrix, then $\mathrm{Res}_0\nabla$ has the matrix Γ_0 in the corresponding basis.

We can change the eigenvalues of the transformation Γ_0 by adding integers by means of a change of L. This is based on the commutator rule

$[\partial_t, t] = 1$, i.e. $\partial_t t = t\partial_t + 1$. It is straightforward to verify that the following commutator rule is satisfied

(7.7.5) $$[t\partial_t - (\alpha + 1)]^r \cdot t = t\cdot(t\partial_t - \alpha)^r.$$

From this we obtain:

(7.7.6) *Lemma about shift* Let L be a saturated lattice and let α be an eigenvalue of the linear transformation $\mathrm{Res}_0 \nabla = \overline{t\partial_t}$ on the space $\overline{L} = L/tL$. Consider a root subspace $\overline{E}$ in $\overline{L}$ corresponding to α. Namely, if r is the multiplicity of α (the multiplicity of α as a root of characteristic equation), then there is a partial Jordan decomposition

$$\overline{L} = \mathrm{Ker}\,(\overline{t\partial_t} - \alpha)^r \oplus \mathrm{Im}\,(\overline{t\partial_t} - \alpha)^r = \overline{E} \oplus \overline{F}.$$

Let $L = E \oplus F$ be a decomposition of L, which gives this decomposition mod (tL). Then a lattice

$$L' = E' \oplus F, \text{ where } E' = tE$$

is also saturated. The eigenvalues of $\mathrm{Res}_0 \nabla$ wrt L' i.e. the eigenvalues of $\mathrm{Res}_0 \nabla \colon L'/tL' \to L'/tL'$ are the same as wrt L except the eigenvalue α. And α for L becomes $\alpha + 1$ for L'.

Analogously, if we change E to $E' = (1/t)E$, then α is changed to $\alpha - 1$.

Proof Let $g \in E \subset L$. This means that $(t\partial_t - \alpha)^r g \in tL$, i.e.

$$(t\partial_t - \alpha)^r g = t\cdot h,\ h \in L.$$

Consider the element $tg \in E' = tE$. Then from (7.7.5) it follows that

$$(t\partial_t - (\alpha + 1))^r (tg) = t\cdot(t\partial_t - \alpha)^r g = t^2 h.$$

This means that $tg \in \mathrm{Ker}\,(t\partial_t - (\alpha + 1))^r \bmod (t^2 L)$, i.e. $\bmod (tE')$. Thus subspace $tE = E'$ gives a subspace in L'/tL' corresponding to an eigenvalue $\alpha + 1$ of the residue $\mathrm{Res}_0 \nabla$ on the lattice L'. ■

(7.7.7) To summarise: if ∇ is a meromorphic connection on a vector space M over $\mathscr{K}$, then there exists a basis e in M such that $M = \oplus M^{\alpha,q}$, where $M^{\alpha,q}$ denotes a subspace, generated by q basis vectors, with respect to which the matrix $-\Gamma$ is Jordan with the eigenvalue α. If we wish, we can change α by adding integers. For example, we can assume that all α belong to the interval $-1 < \alpha \leqslant 0$.

7.8 Residue and limit monodromy

Let $\mathscr{E} = \oplus_{j=1}^{\mu} \mathscr{O}_{S'} e_j$ be a free sheaf on S and ∇ be a connection on $\mathscr{E}_{|S'}$. The monodromy operator T is a linear transformation on the fibre

$\mathscr{E}(t) = \mathscr{E}_t \otimes \mathscr{O}_{S,t}/m$. Properly speaking, it depends on the point t, $T = T(t)$.

(7.8.1) *Proposition* If ∇ is a connection meromorphic at the point 0, i.e. it is a connection on the sheaf $\mathscr{E}[0] = \mathscr{E} \otimes_{\mathscr{O}} \mathscr{O}[1/t]$, and if $t = 0$ is a simple singular point, then the monodromy transformation T can be extended to the point 0, $\exists \lim_{t\to 0} T(t) = T_0$, and T_0 is connected with $\operatorname{Res}_0 \nabla$ by the relation

$$T_0 = e^{-2\pi i \operatorname{Res}_0 \nabla}.$$

Proof In a basis $e_1, \ldots, e_\mu$ the matrix of connection coefficients $\Gamma(t)$ has the form $\Gamma(t) = \Gamma_0/t + \Gamma_1(t)$, where Γ_0 is constant, and $\Gamma_1(t)$ is a holomorphic matrix, and $\Gamma_0 = \operatorname{Res}_0 \nabla$. Horizontal sections $s = ey$ satisfy a system of differential equations $y' = -\Gamma(t)y$, or $dy = -\Gamma(t)y\,dt$. To calculate the monodromy we restrict this equation to a circle. Let (ρ, φ) be polar coordinates on the complex plane $t = \rho\, e^{i\varphi}$. Then $dt = i\rho\, e^{i\varphi}\, d\varphi + e^{i\varphi}\, d\rho$, and on the circle $\rho = \rho_0 = \text{const}$ we have $dt = i\rho\, e^{i\varphi}\, d\varphi$. Substituting for t and dt to the equation for dy we obtain $dy = -(\Gamma_0/\rho\, e^{i\varphi} + \Gamma_1)y \cdot i\rho\, e^{i\varphi}\, d\varphi$, i.e. on the circle the differential equation has the form

$$\frac{dy}{d\varphi} = -(i\Gamma_0 + i\rho\, e^{i\varphi}\Gamma_1)y.$$

The monodromy transformation T at a point (ρ_0, φ_0) is obtained if we take the value of a solution Y of this equation at the point $(\rho_0, \varphi_0 + 2\pi)$, satisfying the initial condition $Y_{|(\rho_0,\varphi_0)} = E$, where E is a unit matrix. We see that the differential equation depends continuously on ρ and φ, and there exists $\lim_{\rho\to 0} T = T_0$, where T_0 is obtained from the equation with $\rho = 0$, i.e. $dy/d\varphi = -i\Gamma_0 y$. It follows that $Y = e^{-i\Gamma_0\varphi}$ and $T_0 = Y_{|\varphi=2\pi} = e^{-2\pi i\Gamma_0}$. ■

(7.8.2) *Remark* Properly speaking, it is wrong to say that T_0 is conjugated with $T(t)$ for t near to 0. However, if a characteristic polynomial of the monodromy T is independent of t, then it can be calculated by means of $\Gamma_0 = \operatorname{Res}_0 \nabla$.

8 Regularity of the Gauss–Manin connection

Let $f\colon X \to S$ be the Milnor fibration of an isolated singularity $f\colon (\mathbb{C}^{n+1}, \mathbb{C}) \to (\mathbb{C}, 0)$, and let $\nabla_{d/dt} = \partial_t\colon \mathscr{M} \to \mathscr{M}$ be the meromorphic

Gauss-Manin connection on the sheaf $\mathscr{M} = \mathscr{H}^{(0)} \otimes_{\mathscr{O}_S} \mathscr{O}_S[t^{-1}]$, a restriction of which to S' coincides with $\mathscr{H} = R^n f_* \mathbb{C}_X \otimes_{\mathbb{C}_{S'}} \mathscr{O}_{S'}$. A fibre at the point $0 \in S$ is a vector space $\mathscr{M}_0 = \mathscr{H}_0^{(0)} \otimes_{\mathscr{O}_{S,0}} \mathscr{K}$ of dimension μ over the field $\mathscr{K} = \mathbb{C}\{t\}[t^{-1}]$, see (4.6.6). A fundamental result about the Gauss–Manin connection $(\mathscr{M}, \partial_t)$ is its regularity. There is a filtration on $\mathscr{M}$ consisting of lattices $\mathscr{H}^{(-i)}$ (5.3.7), where

$$\mathscr{H}^{(-2)} = \mathscr{H}^n_{\mathrm{DR}}(X/S), \mathscr{H}^{(-1)} = f_*\Omega^n_{X/S}/\mathrm{d}(f_*\Omega^{n-1}_{X/S}),$$
$$\mathscr{H}^{(0)} = f_*\Omega^{n+1}_X/\mathrm{d}f \wedge \mathrm{d}(f_*\Omega^{n-1}_X)$$

To prove the regularity of the connection one can either construct a saturated lattice L in $\mathscr{M}$ (7.7.2) or show that in some basis any horizontal section has a moderate growth (7.4.3). We'll use both methods.

8.1 The period matrix and the Picard–Fuchs equation

Let $\mathscr{H}^* = \underline{H}^* \otimes_{\mathbb{C}_{S'}} \mathscr{O}_{S'}$, $\underline{H}^* = \bigcup_{t \in S'} H_n(X_t, \mathbb{C})$, be a sheaf dual to $\mathscr{H}$, and let ∇^* be a connection dual to the Gauss–Manin connection ∇ on $\mathscr{H}$. Horizontal sections of the sheaf $\mathscr{H}^*$ are families of cycles $\gamma(t) \in H_n(X_t, \mathbb{C})$, obtained by spreading a cycle in one fibre to nearby fibres.

(8.1.1) *Definition* Let ω be an n-form on X or a section of the sheaf $\mathscr{H}^{(-1)}$ represented by this form, and let $\gamma(t)$ be a horizontal (multivalued) section of the sheaf $\mathscr{H}^*$. A (multivalued) function on S'

$$I(t) = I_{\gamma,\omega}(t) = \int_{\gamma(t)} \omega$$

is called a *period* of the form ω.

(8.1.2) This name originates from the following classical situation. Let ω be a closed 1-form on a manifold X. Let us consider a multivalued function $F(t) = \int_{t_0}^t \omega$, the integral of ω over a path connecting a fixed point t_0 with a point t. Then the values of $F(t)$ at the point t differ by periods, i.e. integrals of ω over closed paths, i.e. over 1-cycles γ.

(8.1.3) Let $\omega_1, \ldots, \omega_\mu$ be a basis of (local) sections of the sheaf $\mathscr{H}$ represented by n-forms on X. If classes of these forms generate the germ at 0 of the sheaf $\mathscr{H}^{(-1)}$ (or $\mathscr{H}^{(-i)}$), then $\omega_1, \ldots, \omega_\mu$ is a basis of sections in a neighborhood of the point $0 \in S$. Let $\Gamma(t) = (\Gamma_{ij}(t))$ be a matrix of the Gauss–Manin connection coefficients wrt this basis (2.3.5). Then the horizontality condition of a section ω in coordinates $y_1, \ldots, y_\mu$ wrt a basis

$\omega_1, \ldots, \omega_\mu$ is written in the form of a system of linear differential equations (2.3.10)

$$y' = -\Gamma(t)y.$$

In a dual to the $\omega_1, \ldots, \omega_\mu$ basis of sections $\omega_1^*, \ldots, \omega_\mu^*$ of the sheaf $\mathscr{H}^*$ the matrix of the connection ∇^* is equal to $\Gamma^* = -\Gamma^t$ by (2.6.5). By (2.6.7) we obtain:

(8.1.4) *Proposition* If $\gamma(t)$ is a horizontal section of the sheaf $\mathscr{H}^*$, then the period integrals

$$I_1(t) = \int_{\gamma(t)} \omega_1, \ldots, I_\mu(t) = \int_{\gamma(t)} \omega_\mu$$

are solutions of the system

(8.1.5) $$y' = \Gamma^t(t)y.$$

This system is called *the Picard–Fuchs equation* of the Gauss–Manin connection in the basis $\omega_1, \ldots, \omega_\mu$.

Indeed, $I_1(t), \ldots, I_\mu(t)$ are coordinates of $\gamma(t)$ in the basis $\omega_1^*, \ldots, \omega_\mu^*$ and system (8.1.5) expresses a condition of horizontality of a section $\gamma(t)$ of the sheaf $\mathscr{H}^*$.

(8.1.6) If $\gamma_1, \ldots, \gamma_\mu$ is a basis of horizontal (multivalued) sections of the sheaf $\mathscr{H}^*$, then the *period matrix*

$$\Omega(t) = (\Omega_{ij}(t)), \text{ where } \Omega_{ij} = \int_{\gamma_j} \omega_i, \ 1 \leqslant i, j \leqslant \mu,$$

is a fundamental matrix of solutions of the Picard–Fuchs equation $y' = \Gamma^t y$.

(8.1.7) It is easy to see that the matrix $(\Omega^t)^{-1}(t)$ is a fundamental matrix of solutions of the system $y' = -\Gamma(t)y$ expressing a condition of horizontality of a section of the sheaf $\mathscr{H}$ in basis $\omega_1, \ldots, \omega_\mu$.

8.2 The regularity theorem follows from Malgrange's theorem

(8.2.1) *Regularity theorem* If $f\colon (\mathbb{C}^{n+1}, \mathbb{C}) \to (\mathbb{C}, 0)$ is an isolated singularity, then the singular Gauss–Manin connection ∇ on $\mathscr{H}^n_{\mathrm{DR}}(X/S)$ has a regular singular point at $0 \in S$.

(8.2.2) Originally the regularity theorem was proved by P. A. Griffiths (1968) in the global case, when $f\colon Y \to S$ is a projective morphism. Recall that the regularity theorem means that periods of integrals of the family $f\colon Y \to S$ have a polynomial growth $1/|t|^N$ at the point 0, and Griffiths obtained an estimate of the integrals' growth. Brieskorn [Br] obtained the desired estimate of integrals' growth in a local situation from the corresponding Griffiths's result about meromorphic differential forms on families of algebraic varieties.

(8.2.3) From (8.1.7) and from Malgrange's theorem (6.2.1) asserting that for any period $I(t) = \int_{\gamma(t)}\omega$, $\lim_{t\to 0} I(t) = 0$, it follows that horizontal sections of the Gauss–Manin connection have a moderate growth. This proves the regularity theorem (8.2.1).

(8.2.4) Afterwards other proofs of the regularity theorem were obtained, some using a resolution of the singularity and others not. There are two proofs in [M]. The second of these is based on Malgrange's criterion for the regularity of a singularity, using the index of the operator ∂_t. We'll sketch one more proof of the regularity theorem (P. Deligne, see also [H]), using a resolution of the singularity and logarithmic differentials.

8.3 The regularity theorem and connections with logarithmic poles

(8.3.1) The Gauss–Manin connection is a meromorphic connection ∇ on the sheaf

$$\mathscr{H}[0] = \mathscr{H}_{\mathrm{DR}}(X/S)[0] = \mathbb{R}^n f_*\Omega^{\cdot}_{X/S} \underset{\mathscr{O}_S}{\otimes} \mathscr{O}_S[0],$$

where $\mathscr{O}_S[0] = \mathscr{O}_S[t^{-1}]$, and $\nabla_{\mathrm{d}/\mathrm{d}t} = \partial_t$ is a connecting homomorphism in the hypercohomology exact sequence (4.7.4). Now let $f\colon X \to S$ be a morphism but not necessarily the Milnor fibration. When f is Stein, we have

$$\mathscr{H}[0] = \mathscr{H}^n(f_*\Omega^{\cdot}_{X/S}) \underset{\mathscr{O}_S}{\otimes} \mathscr{O}_S[0].$$

To prove the regularity of the connection ∇ we have to find a saturated lattice in $\mathscr{H}[0]$ (7.7.2).

(8.3.2) Denote by $\Omega^{\cdot}_{X/S}[X_0] = \Omega^{\cdot}_{X/S} \otimes_{f^{-1}\mathscr{O}_S} f^{-1}\mathscr{O}_S[0]$ a complex of relative meromorphic differential forms on X with poles at $X_0 = f^{-1}(0)$. We have an isomorphism

$$\mathbb{R}^n f_*\Omega^{\cdot}_{X/S}[X_0] \simeq (\mathbb{R}^n f_*\Omega^{\cdot}_{X/S}) \underset{\mathscr{O}_S}{\otimes} \mathscr{O}_S[0] = \mathscr{H}[0].$$

The sequences of complexes

$$0 \to \Omega^{\cdot -1}_{X/S} \overset{df\wedge}{\to} \Omega^{\cdot}_X \to \Omega^{\cdot}_{X/S} \to 0$$

is exact on X', because the restriction f to $X' = X \setminus X_0$, $f'\colon X' \to S'$, is a smooth morphism. Hence the kernel $\mathrm{Ker}\,(df\wedge)$ is concentrated on X_0 and is killed by $\otimes_{f^{-1}\mathcal{O}_S} f^{-1}\mathcal{O}_S[0]$. We obtain an exact sequence of complexes

$$0 \to \Omega^{\cdot -1}_{X/S}[X_0] \to \Omega^{\cdot}_X[X_0] \to \Omega^{\cdot}_{X/S}[X_0] \to 0$$

and the Gauss–Manin connection $\partial_t = \nabla_{d/dt}$ is a connecting homomorphism

$$\nabla\colon \mathbb{R}^n f_* \Omega^{\cdot}_{X/S}[X_0] \to \mathbb{R}^n f_* \Omega^{\cdot}_{X/S}[X_0]$$

in the exact hypercohomology sequence corresponding to this exact sequence.

(8.3.3) Consider a resolution of the singularity of the morphism f

$$\begin{array}{ccc} \tilde{X}_0 \subset \tilde{X} & \overset{\pi}{\to} & X \supset X_0 \\ & \tilde{f} \searrow \downarrow f & \\ & S & \end{array}$$

i.e. a proper morphism $\pi\colon \tilde{X} \to X$ such that $\pi\colon \tilde{X} \setminus \tilde{X}_0 \rightrightarrows X \setminus X_0$, where $\tilde{X}_0 = \pi^{-1}(X_0)$ and $(\tilde{X}_0)_{\mathrm{red}}$ is a divisor with normal crossings.

Let $\Omega^p_{\tilde{X}}\langle \tilde{X}_0 \rangle = \Omega^p_{\tilde{X}}(\log \tilde{X}_0) \subset \Omega^p_{\tilde{X}}[\tilde{X}_0]$ be sheaves of differential forms with logarithmic poles, and let

$$\Omega^p_{\tilde{X}/S}\langle \tilde{X}_0 \rangle = \Omega^p_{\tilde{X}}\langle X_0 \rangle / \frac{d\tilde{f}}{\tilde{f}} \wedge \Omega^{p-1}_{\tilde{X}}\langle \tilde{X}_0 \rangle$$

be the corresponding locally free sheaves of relative differential forms. Locally we have $\tilde{f} = z_1^{\nu_1}, \ldots, z_k^{\nu_k}$ and $d\tilde{f}/\tilde{f} = \sum_{j=1}^k \nu_j dz_j/z_j \in \Omega^1_{\tilde{X}}\langle X_0 \rangle$ can be completed to be a basis of $\Omega^1_{\tilde{X}}\langle X_0 \rangle$. Hence the complex $(\Omega^{\cdot}_{\tilde{X}}\langle X_0 \rangle, (d\tilde{f}/\tilde{f})\wedge$ is exact. We have an exact sequence of complexes

$$0 \to \Omega^{\cdot -1}_{\tilde{X}/S}\langle \tilde{X}_0 \rangle \overset{d\tilde{f}/f}{\to} \Omega^{\cdot}_{\tilde{X}}\langle \tilde{X}_0 \rangle \to \Omega^{\cdot}_{\tilde{X}/S}\langle \tilde{X}_0 \rangle \to 0$$

(8.3.4) Furthermore, the sheaves $\mathbb{R}^i \tilde{f}_* \Omega^{\cdot}_{\tilde{X}/S}\langle \tilde{X}_0 \rangle$ are coherent $\mathcal{O}_S$-modules and there are isomorphisms

$$\mathbb{R}^i \tilde{f}_* \Omega^{\cdot}_{\tilde{X}/S}\langle \tilde{X}_0 \rangle \underset{\mathcal{O}_S}{\otimes} \mathcal{O}_S[0] \simeq \mathbb{R}^i \tilde{f}_* \Omega^{\cdot}_{\tilde{X}/S}[\tilde{X}_0] \simeq \mathbb{R}^i f_* \Omega^{\cdot}_{X/S}[X_0].$$

Thus we can change the sheaf $\mathcal{H}[0] = \mathbb{R}^n f_* \Omega^{\cdot}_{X/S}[X_0]$ to the sheaf $\mathbb{R}^n \tilde{f}_* \Omega^{\cdot}_{\tilde{X}/S}[\tilde{X}_0]$.

(8.3.5) We assert that a sheaf $L = \mathbb{R}^n\tilde{f}_*\Omega^{\cdot}_{\tilde{X}/S}\langle\tilde{X}_0\rangle/\mathrm{Tors}$ is the desired saturated lattice in $\mathbb{R}^n\tilde{f}_*\Omega^{\cdot}_{\tilde{X}/S}[\tilde{X}_0]$, i.e. it is stable wrt the operator $t\partial_t$.

(8.3.6) Indeed, we have a commutative diagram

$$\begin{array}{ccccccc} 0 \to \Omega^{\cdot -1}_{\tilde{X}/S}\langle\tilde{X}_0\rangle & \stackrel{(\mathrm{d}\tilde{f}/f)\wedge}{\longrightarrow} & \Omega^{\cdot}_{\tilde{X}}\langle\tilde{X}_0\rangle & \to & \Omega^{\cdot}_{\tilde{X}/S}\langle\tilde{X}_0\rangle \to 0 \\ \downarrow \cdot\frac{1}{f} & & \downarrow & & \downarrow \\ 0 \to \Omega^{\cdot -1}_{\tilde{X}/S}[\tilde{X}_0] & \stackrel{\mathrm{d}f\wedge}{\longrightarrow} & \Omega^{\cdot}_{\tilde{X}}[\tilde{X}_0] & \to & \Omega^{\cdot}_{\tilde{X}/S}[\tilde{X}_0] \to 0 \end{array}$$

where the columns are exact sequences of complexes. We obtain a homomorphism of the exact hypercohomology sequence of the upper line to the analogous exact sequence of the low line. A fragment containing connecting homomorphism is

$$\begin{array}{ccc} \mathbb{R}^n\tilde{f}_*\Omega^{\cdot}_{\tilde{X}/S}\langle\tilde{X}_0\rangle & \stackrel{\partial_t}{\to} & \mathbb{R}^n\tilde{f}_*\Omega^{\cdot}_{\tilde{X}/S}\langle\tilde{X}_0\rangle \\ \downarrow\varphi & & \downarrow\frac{1}{t} \\ \mathbb{R}^n\tilde{f}_*\Omega^{\cdot}_{\tilde{X}/S}[\tilde{X}_0] & \stackrel{\partial_t}{\to} & \mathbb{R}^n\tilde{f}_*\Omega_{\tilde{X}/S}[\tilde{X}_0]. \end{array}$$

We have that $\operatorname{Im}\varphi$ is the desired lattice L because this diagram shows that $\partial_t(L) \subset (1/t)L$. ■

(8.3.7) *Remark* Arguing as above Hamm [H] proved the regularity of the Gauss–Manin connection on sheaves $R^i f'_*\mathbb{C}_{X'} \otimes_{\mathbb{C}_{S'}} \mathscr{O}_{S'} \simeq R^i f'_*(f^{-1}\mathscr{O}_{S'}) = \mathbb{R}^i f'_*\Omega^{\cdot}_{X'/S'}$ in the case of a non-isolated singularity f.

9 The monodromy theorem

9.1 Two parts of the monodromy theorem

(9.1.1) Let f: $X \to S$ be the Milnor fibration of an isolated singularity, and T: $H^n(X_{t_0}, \mathbb{C}) \to H^n(X_{t_0}, \mathbb{C})$, $t_0 \in S'$ be the monodromy transformation. This is an andomorphism of a vector space of dimension μ. As is well known, T can be reduced to a Jordan normal form, i.e. one can choose a basis in which T consists of Jordan blocks

$$\begin{pmatrix} \lambda & 1 & & 0 \\ & \ddots & \ddots & \\ 0 & & & 1 \\ & & & \lambda \end{pmatrix}$$

where λ is an eigenvalue of the monodromy on cohomology.

The monodromy theorem consists of two parts:

(i) the first concerns eigenvalues of T and asserts that all λ are roots of unity;

(ii) the second concerns the sizes of Jordan blocks and asserts that they do not exceed $n+1$ (we'll indicate a more exact estimate later).

In other words (i) means that the monodromy transformation T is quasiunipotent, i.e. $\exists$ integer N, such that T^N is unipotent, and (ii) means that the index of quasiunipotency of T, i.e. the least k such that $(T^N - E)^k = 0$, does not exceed $n+1$, $(T^N - E)^{n+1} = 0$.

9.2 Eigenvalues of monodromy

Theorem I All eigenvalues of T are roots of unity.

Proof We give a beautiful proof due to Brieskorn [Br], based on the regularity of the Gauss–Manin connection and on the positive solution of the seventh Hilbert problem.

(9.2.1) Eigenvalues λ_j of the monodromy T are algebraic numbers because they are roots of a characteristic polynomial $\Delta(\lambda)$ with integer coefficients (because T originates from an automorphism of a lattice $H^n(X_{t_0}, \mathbb{Z})$).

(9.2.2) The eigenvalues λ_j are of the form $\lambda_j = \mathrm{e}^{2\pi \mathrm{i}\alpha_j}$, where α_j are eigenvalues of the matrix $A_0 = -\mathrm{Res}_0 \nabla$. Indeed, the characteristic polynomial $\Delta(t)$ of the monodromy $T = T(t_0)$ has integer coefficients, thus it is constant under varying t_0, and hence it is the same as for $T_0 = \mathrm{e}^{-2\pi \mathrm{i} \mathrm{Res}_0 \nabla}$.

(9.2.3) The positive solution of the seventh Hilbert problem (0. A. Gelfond and Th. Schneider, 1934) is given by the following theorem: If α and $\lambda = \mathrm{e}^{2\pi \mathrm{i}\alpha}$ are both algebraic numbers, then $\alpha \in \mathbb{Q}$. Therefore it remains to prove that α_j are algebraic numbers. If they are, $\alpha_j \in \mathbb{Q}$ and $\lambda = \mathrm{e}^{2\pi \mathrm{i}\alpha_j}$ is a root of unity. The algebraicity of α_j follows from the algebraicity theorem.

(9.2.4) Algebraicity theorem of the Gauss–Manin connection. Let $x_0, \ldots, x_n$ be coordinates on X at the point 0 and let t be a coordinate at the point $0 \in S$. Then functions of $\mathcal{O}_{S,0} \simeq \mathbb{C}\{t\}$ and of $\mathcal{O}_{X,0} \simeq \mathbb{C}\{x_0, \ldots, x_n\}$ are identified with converging power series. Now let $\varphi\colon \mathbb{C} \to \mathbb{C}$ be any automorphism of the field $\mathbb{C}$. Applying φ to coefficients of series, we obtain an extension of φ to an automorphism of functions $\mathcal{O}_{X,0} \xrightarrow{\sim} \mathcal{O}_{X,0}$. Let φf be a function obtained from f in a such way. Denote by $H_f = H^n(\Omega^{\cdot}_{X/S,0})$ a germ of the De Rham cohomology, and similarly for the function φf. The algebraicity theorem asserts that the diagram

$$\begin{array}{ccc} H_f & \xrightarrow{\nabla} & H_f \otimes \Omega^1_S[0] \\ \downarrow\varphi & & \downarrow\varphi \\ H_{\varphi f} & \xrightarrow{\nabla} & H_{\varphi f} \otimes \Omega^1_S[0] \end{array},$$

where ∇ is the Gauss–Manin connection, is commutative.

(9.2.5) It follows from this theorem that if a matrix $\Gamma_0 = \mathrm{Res}_0 \nabla$ is the residue of the Gauss–Manin connection for f, then the corresponding matrix $\varphi\Gamma_0$ for φf is obtained by applying φ to all elements of Γ_0. Hence, if $e^{2\pi i \alpha_j}$ is an eigenvalue of the monodromy for f, then $e^{2\pi i \varphi(\alpha_j)}$ is the eigenvalue of the monodromy for φf, and consequently, $e^{2\pi i \varphi(\alpha_j)}$ is also an algebraic number for any automorphism φ. Now if α_j were transcendental, then we could transform α_j to any transcendental number $\varphi(\alpha_j)$ by some automorphism $\varphi\colon \mathbb{C} \to \mathbb{C}$, and (because the set of algebraic numbers is countable), we could choose φ such that $e^{2\pi i \varphi(\alpha_j)}$ is transcendental. However, because $e^{2\pi i \varphi(\alpha_j)}$ is an algebraic number for any φ, it follows that α_j is an algebraic number. This concludes the proof of the monodromy theorem I. ■

(9.2.6) *Remark* The theorem about eigenvalues of a monodromy was conjectured by Milnor (1968) based on calculation of $\Delta(\lambda)$ for Pham singularities. This theorem can be obtained from the corresponding theorem in the global case, when $f\colon X \to S$ is a proper morphism. In the global case the monodromy theorem was proved by A. Grothendieck, A. Landman (Thesis, 1967), C. H. Clemens (1969) and others. As noted by P. Deligne (see [Br]) the above proof (integrality of $\Delta(\lambda)$, regularity, algebraicity, seventh Hilbert problem) also gives a proof of the theorem about eigenvalues of a monodromy in the global algebraic case.

9.3 The size of Jordan blocks

(9.3.1) *Theorem* II (the second theorem about monodromy). If $f^{r+1} \in J_f = (f'_{x_0}, \ldots, f'_{x_n})$, then the size of the Jordan blocks of the monodromy T is at most $r+1$.

(9.3.2) *Remark* Initially it was proved that the size of the Jordan blocks is at most $n+1$ (N. M. Katz, 1970). Brieskorn conjectured that the estimate $n+1$ is connected with the following result of J. Briançon and H. Skoda

(1974): For an isolated singularity, $f^{n+1} \in J_f$. J Scherk conjectured theorem (9.3.1) in [Sc1] and proved it in [Sc2]. His proof is based on an embedding of the Milnor fibration $f\colon X \to S$ in a projective family $Y \to S$ and on an application of the limit MHS on $H^n(Y_t, \mathbb{C})$. We'll return to this question in chapter 2, (II.3.5.9). Now we only note that the condition $f^{r+1} \in J_f$ means that $f^{r+1}\Omega^{n+1}_{X,0} \subset df \wedge \Omega^n_{X,0}$ and is equivalent to the condition $t^{r+1}\cdot\mathscr{H}_0^{(0)} \subset \mathscr{H}_0^{(-1)}$.

When $r = 0$ the condition $f \in J_f$ is equivalent to quasihomogeneity of a singularity f, and the monodromy theorem asserts that all Jordan blocks are of size 1, i.e. the monodromy T is semisimple. We have already obtained this in (5.5.5). B. Malgrange [M2] gave an example of singularities who monodromy T contains a Jordan block of maximal possible size $n+1$. Such an example gives the singularity $x^8 + y^8 + z^8 + x^2y^2z^2$ for which $N^2 \neq 0$ (see also [S3] and (II.8.5.15)).

9.4 Consequences of the monodromy theorem. Decomposition of integrals into series

Let $\Lambda = \{\lambda_1, \ldots, \lambda_r\}$ be a set of eigenvalues of monodromy T on the cohomological fibration $\underline{H} = R^n f'_* \mathbb{C}_{X'} = \bigcup_{t\in S'} H^n(X_t, \mathbb{C})$. Then $\lambda_1^{-1}, \ldots, \lambda_r^{-1}$ are eigenvalues of the monodromy M on the homological fibration $\underline{H}^* = \bigcup_{t\in S'} H_n(X_t, \mathbb{C})$ which is dual to $\underline{H}$. Denote by

(9.4.1) $$\alpha_j = -\frac{1}{2\pi i} \ln \lambda_j$$

the eigenvalues of a matrix R, $M = e^{2\pi i R}$. By monodromy theorem I (§9.2) the eigenvalues $\lambda_j = e^{-2\pi i \alpha_j}$ are roots of unity, and so $\alpha_j \in \mathbb{Q}$ are rational number defined mod $\mathbb{Z}$. Denote by

(9.4.2) $$L(\lambda_j) = \{\alpha_j^0, \alpha_j^0 + 1, \alpha_j^0 + 2, \ldots\}$$

an arithmetical progression with one (suitable) value of α_j.

Let $\omega_1, \ldots, \omega_\mu$ be a basis of (local) sections of the sheaf $\mathscr{H} = \underline{H} \otimes \mathscr{O}_{S'}$, and let $\gamma(t)$ be either the local or multivalued section of the fibration $\underline{H}^*$, i.e. a horizontal section of the sheaf $\mathscr{H}^* = \underline{H}^* \otimes_{\mathbb{C}_{S'}} \mathscr{O}_{S'}$. Then the column

$$y = \Omega_\gamma(t) = \left(\int_{\gamma(t)} \omega_1, \ldots, \int_{\gamma(t)} \omega_\mu\right)^t$$

is a solution of the system $y' = \Gamma^t(t)y$ (8.1.5), where $\Gamma(t)$ is the matrix of coefficients of the Gauss–Manin connection ∇ on $\mathscr{H}$ in the basis $\omega_1, \ldots, \omega_\mu$. By (7.3.4) we have $y = Z(t)\cdot t^R C$, where $e^{2\pi i R} = M$ and C is a column of constants. By the regularity of the Gauss–Manin connection, the matrix $Z(t)$ is a matrix of functions meromorphic at the point $0 \in S$. From the structure of the matrix t^R (7.3.6), we obtain:

(9.4.3) *Corollary* If ω is a section of the sheaf $\mathscr{H}$ and $\gamma(t)$ is a horizontal section of the homological fibration $\underline{H}^*$, then the integral

$$I(t) = I_{\gamma,\omega} = \int_{\gamma(t)} \omega$$

is decomposed into the series

$$I(t) = \sum_{\lambda\in\Lambda} \sum_{\alpha\in L(\lambda)} \sum_{k=0}^{n} a_{k,\alpha}\cdot\frac{1}{k!}t^{\alpha}(\log t)^k, \tag{9.4.4}$$

and, moreover, if for an eigenvalue λ the monodromy T (or M) has Jordan blocks of dimension $\leqslant r+1$, then the inner summation goes only up to r, $\sum_{k=0}^{r}$.

(9.4.5) *Corollary* If in (9.4.3) ω is a section of the sheaf $\mathscr{H}^{(-1)}$ in a neighborhood of the point $0 \in S$, or ω is a holpmorphic n-form in a neighborhood of the point $0 \in X$ representing this section, then in the decomposition (9.4.4) all $\alpha \geqslant 0$ (all progressions $L(\lambda_j)$ consist of non-negative numbers), and, moreover, if $\alpha = 0$, then $a_{k,\alpha} = 0$ for $k \geqslant 1$.

Proof This follows from Malgrange's theorem (6.2.1), where it is proven that $\lim_{t\to 0} I(t) = 0$. ■

(9.4.6) *Corollary* If in (9.4.3) ω is a section of the sheaf $\mathscr{H}^{(0)}$ in a neighborhood of the point $0 \in S$, or ω is a holpmorphic $(n+1)$-form in a neighborhood of the point $0 \in X$ representing this section, and $w = s[\omega]$ is a geometrical section of the sheaf $\mathscr{H}$ corresponding to the Gelfand–Leray form $\omega/\mathrm{d}f = \mathrm{Res}\,[\omega/(f-t)]$, then in the decomposition (9.4.4) for integrals

$$I(t) = \int_{\gamma(t)} w = \int_{\gamma(t)} \omega/\mathrm{d}f$$

all $\alpha > -1$.

Proof By proposition (5.3.4) the operator ∂_t: $\mathscr{H}^{(-1)} \rightrightarrows \mathscr{H}^{(0)}$ is an isomorphism. Let $[\omega] = \partial_t([\eta])$, where η is an n-form. We have calculated the derivative of the function $F(t) = \int_{\gamma(t)}\eta$,

$$F'(t) = \frac{\mathrm{d}}{\mathrm{d}t}\int_{\gamma(t)} \eta = \langle \eta, \gamma \rangle' = \langle \partial_t[\eta], \gamma \rangle = \int_{\gamma(t)} \frac{\omega}{\mathrm{d}f} = I(t).$$

In the decomposition of $F(t)$ all $\alpha \geqslant 0$, and for $\alpha = 0$ we have only constants by (9.4.5), hence in the decomposition of $I(t) = F'(t)$ all $\alpha > -1$. ■

10 Gauss–Manin connection of a non-isolated hypersurface singularity

Let $f\colon (\mathbb{C}^{n+1}, 0) \to (\mathbb{C}, 0)$ be any, not necessarily isolated, singularity of a holomorphic function f. In this section we'll consider how the notion of the Gauss–Manin connection of the cohomology of the Milnor fibration, the sheaves $\mathscr{H}^{(-i)}$ and other notions and results obtained above for isolated singularities are transferred to a general case [Sr].

10.1 De Rham cohomology sheaves

(10.1.1) The Milnor fibration $f\colon X \to S$ of a singularity f is defined word for word as in the case of an isolated singularity [Di3]. X is an intersection of a ball $B = \{|x| < \varepsilon\} \subset \mathbb{C}^{n+1}$ of radius ε with the preimage of a disk $S = \{|t| < \delta\} \subset \mathbb{C}$ of radius $\delta \ll \varepsilon$. Here $X_0 = f^{-1}(0)$ is the single singular fibre. The morphism f is a smooth locally trivial fibration over $S' = S\backslash\{0\}$, $f'\colon X' \to S'$. It defines cohomology fibrations $\underline{H}^p \to S'$, where $\underline{H}^{p\prime} = \bigcup_{t \in S'} H^p(X_t, \mathbb{C}) = R^p f_* \mathbb{C}_{X'}$, defining topological Gauss–Manin connections ∇ on locally free sheaves of sections

$$\mathscr{H}^p = \underline{H}^p \underset{\mathbb{C}_{S'}}{\otimes} \mathscr{O}_{S'} = R^p f_* \mathbb{C}_{X'} \underset{\mathbb{C}_{X'}}{\otimes} \mathscr{O}_{S'}.$$

The monodromy transformation T corresponding to a standard generator of $\pi_1(S')$ acts on a fibration $\underline{H}^p$. The Betti numbers $b_p = \dim H^p(X_t, \mathbb{C})$ are called *Milnor numbers* μ^p of the singularity f.

(10.1.2) By the relative Poincaré lemma the sheaf $f^{-1}\mathscr{O}_{S'}$ has a resolution $\Omega^{\cdot}_{X'/S'}$, a complex of sheaves of relative differential forms. Thus sheaves $\mathscr{H}^p$ are identified with sheaves of relative De Pham cohomology

$$\mathscr{H}^p_{\mathrm{DR}}(X'/S') = \mathbb{R}^p f_*(\Omega^{\cdot}_{X'/S'}) = R^p f_*(f^{-1}\mathscr{O}_S)$$
$$\simeq R^p f_* \mathbb{C}_{X'} \underset{\mathbb{C}_{S'}}{\otimes} \mathscr{O}_{S'} = \mathscr{H}^p.$$

These sheaves are naturally extended to sheaves $\mathscr{H}^p_{\mathrm{DR}}(X/S) =$

$R^p f_*(\Omega^{\cdot}_{X/S})$ on S, which we'll denote by $\mathscr{H}^p_{(-2)}$ in analogy with the case of isolated singularities.

(10.1.3) The morphism f (as well as f') being Stein, these sheaves

$$\mathscr{H}^p_{\mathrm{DR}}(X/S) \simeq \mathscr{H}^p(f_*\Omega^{\cdot}_{X/S})$$

are isomorphic to the cohomology sheaves of complex $f_*\Omega^{\cdot}_{X/S}$. Thus we obtain (De Rham theory) a connection between the sheaves $\mathscr{H}^p$ and sheaves of differential forms (the connection of topology with algebra and analysis).

10.2 Coherence

(10.2.1) There is a question about the coherence of sheaves $\mathscr{H}^p_{\mathrm{DR}}(X/S)$. This is one of the fundamental technical results. Since it is technical, we'll not consider this question here and we'll assume that the singularities under consideration are such that these sheaves are coherent. We mention only the papers of Hamm [H], R. O. Buchweitz and G. M. Greuel (1980), and also that of van Straten [Sr], in which Brieskorn's proof of the coherence of sheaves $\mathscr{H}^p_{\mathrm{DR}}(X/S)$ for isolated singularities is transferred to the case of non-isolated concentrated singularities. We restrict ourselves by a definition of this notion.

(10.2.2) Let $X \xrightarrow{f} S$ be the Milnor fibration, where $X = X_{\varepsilon,\delta}$, with ε and δ radii of a ball and a disk, respectively. Denote by $\partial X = \partial B_\varepsilon \cap f^{-1}(S_\delta)$ and $\overline{X} = X \cup \partial X$ the relative boundary and closure of X. A sheaf L of $\mathbb{C}$-vector spaces on X is called transversally constant (relative to the boundary ∂X), if it is constant along transverses to ∂X. This means that there exists an open neighborhood U of the closure $\partial \overline{X}$ in $\mathbb{C}^N$ and a C^∞-vector field θ on U such that: (1) θ is transverse to ∂B; (2) the local θ-flow in U leaves X and the fibres X_t invariant; (3) the restriction of L to the integral curves of θ is a constant sheaf.

Consider a finite complex $(K^{\cdot}, \mathrm{d})$ of $\mathscr{O}_X$-coherent modules with $f^{-1}(\mathscr{O}_S)$-linear differential d. Van Straten proved that if the cohomology sheaves $\mathscr{H}^i(K^{\cdot})$ are transversally constant, then the sheaves $\mathbb{R}^i f_*(K^{\cdot})$ are $\mathscr{O}_S$-coherent modules.

A complex of sheaves $(K^{\cdot}, \mathrm{d})$ is called concentrated if for all $\varepsilon' \in (0, \varepsilon]$ there exists $\delta' \in (0, \delta]$ such that the cohomology sheaves $\mathscr{H}^i(K^{\cdot})$ are transversally constant on $X_{\varepsilon',\delta'}$. A singularity f is called concentrated if the complex $\Omega^{\cdot}_{X/S}$ is concentrated. Examples of concentrated singularities are:

(1) isolated singularities; (2) quasihomogeneous singularities (with a good $\mathbb{C}^*$-action); (3) singularities such that for some $X \to S$ there are only a finite number of isomorphism classes of germs among germs $(X, x) \to (S, f(x))$. Note that the sheaves $\mathcal{H}^i_{\mathrm{DR}}(X/S)$ may be non-coherent, as is the case, for example, for $f = y^4 + xy^2z^2 + z^4$.

10.3 Relation between $\mathcal{H}^p(f_*\Omega^\cdot_f)$ and $f_*\mathcal{H}^p(\Omega^\cdot_f)$

(10.3.1) Proposition (3.6.1) can be generalized to the case of non-isolated singularities. As in the case of isolated singularities, from the second hypercohomology spectral sequence (3.2.4) we obtain an exact sequence containing the De Rham cohomology sheaf $\mathcal{H}^i_{\mathrm{DR}}(X/S) = \mathcal{H}^i(f_*\Omega^\cdot_f)$,

(10.3.2) $$0 \to (R^i f_* \mathbb{C}_X) \underset{\mathbb{C}_X}{\otimes} \mathcal{O}_S \to \mathcal{H}^i(f_*\Omega^\cdot_f) \to f_*\mathcal{H}^i(\Omega^\cdot_f) \to 0$$

where the first sheaf has a zero germ at $0 \in S$, and the last one is concentrated at $0 \in S$ because $\mathcal{H}^i(\Omega^\cdot_f)_{|S'} = 0$ by virtue of the smoothness of f over S'. It follows that $\mathcal{H}^i(\Omega^\cdot_f)_0 \simeq f_*\mathcal{H}^i(\Omega^\cdot_f)_0$.

If f has a concentrated singularity at the point $x_0 = 0 \in X$, then

(10.3.3) $$f_*(\mathcal{H}^i(\Omega^\cdot_f))_0 \simeq \mathcal{H}^i(\Omega^\cdot_f)_{x_0} \simeq H^i(\Omega^\cdot_{f,x_0})$$

Van Straten [Sr] adapted Brieskorn's proof of this fact for isolated singularities to the case of concentrated singularities.

(10.3.4) If the De Rham cohomology sheaf $\mathcal{H}^i(f_*\Omega^\cdot_f)$ is torsion free, then it is locally free of rank μ_i and we get a formula for the calculation of Milnor numbers

$$\mu_i = \dim_{\mathbb{C}} H^i(\Omega^\cdot_{f,x_0})/tH^i(\Omega^\cdot_{f,x_0}).$$

Below we'll obtain a criterion for the sheaves $\mathcal{H}^i(f_*\Omega^\cdot_f)$ to be torsion free.

10.4 A general method of extension of a singular connection over the whole disk

(10.4.1) The topological connection on the sheaves $\mathcal{H}^p = R^p f_*\mathbb{C}_{X'} \otimes_{\mathbb{C}_{X'}} \mathcal{O}_{S'}$ is carried to the sheaves $\mathcal{H}^p_{\mathrm{DR}}$ under the isomorphism $\mathcal{H}^p \simeq \mathcal{H}^p_{\mathrm{DR}}(X'/S')$. The calculation of this connection ∇ on $\mathcal{H}^p_{\mathrm{DR}}(X'/S')$ is based on the Leray residue theorem and this was carried out in §4. We obtained the following result. If a local section $[\omega]$ of the sheaf $\mathcal{H}^p_{\mathrm{DR}}(X'/S') = \mathcal{H}^p(f_*\Omega^\cdot_{X'/S'})$ is represented by a p-form ω of X' then $\partial_t([\omega])$, $\partial_t = \nabla_{\mathrm{d}/\mathrm{d}t}$: $\mathcal{H}^p_{\mathrm{DR}}(X'/S') \to \mathcal{H}^p_{\mathrm{DR}}(X'/S')$, is defined in such a way. We take the form $\mathrm{d}\omega$; ω represents a cycle in $\Omega^\cdot_{X'/S'}$, and therefore

$\mathrm{d}\omega = \mathrm{d}f \wedge \eta$, where $\eta \in \Omega_X^p$. Then $\mathrm{d}f \wedge \mathrm{d}\eta = 0$ and hence $\mathrm{d}\eta = \mathrm{d}f \wedge \xi$ by the division lemma (morphism f' is smooth), i.e. η is a cycle in $\Omega^{\cdot}_{X'/S'}$. The class $[\eta] \in \mathscr{H}^p_{\mathrm{DR}}(X'/S')$ is correctly defined and just this class is $\partial_t([\omega]) = [\eta]$. Note that $[\eta] = \mathrm{d}\omega/\mathrm{d}t \in \mathscr{H}^p_{\mathrm{DR}}(X'/S')$ is a derivative of ω in t in the sense of Leray.

(10.4.2) We can interpret the operator ∂_t as a connecting homomorphism in an exact cohomology sequence. Namely, there is an exact sequence of complexes

$$0 \to \Omega^{\cdot -1}_{X'/S'} \xrightarrow{\mathrm{d}f\wedge} \Omega^{\cdot}_{X'} \to \Omega^{\cdot}_{X'/S'} \to 0$$

by virtue of the smoothness of $f'\colon X' \to S'$. The sequence

$$0 \to f_*\Omega^{\cdot -1}_{X'/S'} \xrightarrow{\mathrm{d}f\wedge} f_*\Omega^{\cdot}_{X'} \to f_*\Omega^{\cdot -1}_{X'/S'} \to 0 \qquad (10.4.3)$$

is exact since f' is Stein. The sheaf $\mathscr{H}^p_{\mathrm{DR}}(X'/S')$ is the cohomology sheaf $\mathscr{H}^p(f_*\Omega^{\cdot}_{X'/S'})$. A choice of a p-form $\omega \in f_*\Omega^p_{X'}$ representing the class $[\omega] \in \mathscr{H}^p_{\mathrm{DR}}(X'/S')$ is a lift of this class from $f_*\Omega^p_{X'/S'})$ to $f_*\Omega^p_{X'}$. We see that taking $\mathrm{d}\omega$, etc. is exactly the description of the construction of the connecting homomorphism $\delta\colon \mathscr{H}^p(f_*\Omega^{\cdot}_{X'/S'}) \to \mathscr{H}^p(f_*\Omega^{\cdot}_{X'/S'})$.

(10.4.4) Now it is clear how to extend the Gauss–Manin connection over the whole disk. We have to extend the exact sequence of complexes (10.4.3) to an exact sequence over S and then the connecting homomorphism in the exact cohomology sequence is an extension of the Gauss–Manin connection ∂_t.

(10.4.5) A meromorphic Gauss–Manin connection ∇ on the sheaf $\mathscr{H}^p[0] = \mathscr{H}^p(f_*\Omega^{\cdot}_f) \otimes_{\mathscr{O}_S} \mathscr{O}_S[0]$ is obtained if we take a connecting homomorphism ∂_t coming from the exact sequence

$$0 \to \Omega^{\cdot -1}_{X/S}[X_0] \to \Omega^{\cdot}_X[X_0] \to \Omega^{\cdot -1}_{X/S}[X_0] \to 0.$$

The regularity of the connection singularity at the point $0 \in S$ can be proved as in (8.3.1)–(8.3.6).

We proceed (see [Sr]) to the construction of sheaves $\mathscr{H}^p_{(-i)}$ on S – the analogs of the sheaves $\mathscr{H}^{(-i)}$ in the case of an isolated singularity f.

10.5 The sheaves $\mathscr{H}^p_{(-1)}$ and the Gauss–Manin connection ∂_t: $\mathscr{H}^p_{(-2)} \rightrightarrows \mathscr{H}^p_{(-1)}$

A complex $\Omega^{\cdot}_f = \Omega^{\cdot}_{X/S}$ of sheaves of relative differential forms is defined at the Coker of the homomorphism $\Omega^{\cdot} \xrightarrow{\mathrm{d}f\wedge} \Omega^{\cdot}$, where $(\Omega^{\cdot} = \Omega^{\cdot}_X, \mathrm{d})$ is the

De Rham complex of differential forms on X. Let $(\Omega^\cdot, \mathrm{d}f)$ be the Koszul complex of a function f

$$(10.5.1) \qquad 0 \to \mathscr{O}_X \xrightarrow{\mathrm{d}f\wedge} \Omega^1_X \xrightarrow{\mathrm{d}f\wedge} \ldots \to \Omega^{n+1}_X \to 0$$

This is really the Koszul complex $(K^\cdot, f)$ constructed from the sheaf $\Omega^1_X \simeq \mathscr{O}^{n+1}_X$ and the sequence of partial derivatives $f_i = f'_{x_i}$, $i = 0, \ldots, n$. Denote by $B^\cdot = \operatorname{Im}(\Omega^\cdot \xrightarrow{\mathrm{d}f\wedge} \Omega^\cdot) = \mathrm{d}f \wedge \Omega^{\cdot-1}$ the complex of sheaves of Koszul coboundaries. We have an exact sequence of complexes

$$0 \to B^\cdot \to \Omega^\cdot \to \Omega^\cdot_f \to 0$$

and we get an exact sequence on S

$$(10.5.2) \qquad 0 \to f_* B^\cdot \to f_*\Omega^\cdot \to f_*\Omega^\cdot_f \to 0$$

because f is Stein. This sequence extends the exact sequence (10.4.3) over S.

We set

$$(10.5.3) \qquad \mathscr{H}^p_{(-2)} = \mathscr{H}^p(f_*\Omega^\cdot_f) = \mathscr{H}^p_{\mathrm{DR}}(X/S).$$

In the case of an isolated singularity the extension of the Gauss–Manin connection $\partial_t\colon \mathscr{H}^{(-2)} \to \mathscr{H}^{(-1)} = f_*\Omega^n_f/\mathrm{d}(f_*\Omega^{n-1}_f)$ is found to be a connecting homomorphism in the exact cohomology sequence for (10.5.2) and by identification (using the division lemma) of $B^\cdot$ with $\overline{\Omega}^\cdot_{X/S}$ and the inclusion $B^\cdot \subset \Omega^\cdot$ with $\overline{\Omega}^\cdot_{X/S} \xrightarrow{\mathrm{d}f\wedge} \Omega^\cdot_X$. Therefore by definition we set

$$(10.5.4) \qquad \mathscr{H}^p_{(-1)} = \mathscr{H}^{p+1}(f_* B^\cdot),$$

and the Gauss–Manin connection $\partial_t\colon \mathscr{H}^p_{(-2)} \to \mathscr{H}^p_{(-1)}$ to be a connecting homomorphism δ in the exact cohomology sequence corresponding to the sequence (10.5.2). The next local description of ∂_t shows that ∂_t really is a connection, i.e. the Leibnitz identity is fulfilled.

(10.5.5) If $[\omega] \in \mathscr{H}^p(f_*\Omega^\cdot_f) = \mathscr{H}^p_{(-2)}$, where $\omega \in \Omega^p$, then

$$\partial_t([\omega]) = [\mathrm{d}\omega] \in \mathscr{H}^{p+1}(f_* B^\cdot) = \mathscr{H}^p_{(-1)},$$

where $\mathrm{d}\omega \in f_* B^\cdot$ and can be presented in the form $\mathrm{d}\omega = \mathrm{d}f \wedge \eta$, $\eta \in \Omega^p$.

(10.5.6) *Lemma* $\partial_t\colon \mathscr{H}^p_{(-2)} \overset{\sim}{\to} \mathscr{H}^p_{(-1)}$ is an isomorphism.

Proof Using the exact cohomology sequence corresponding to (10.5.2), we obtain

$$\ldots \to \mathscr{H}^p(f_*\Omega^\cdot) \to \mathscr{H}^p(f_*\Omega^\cdot_f) \xrightarrow{\partial_t} \mathscr{H}^{p+1}(f_* B^\cdot) \to \mathscr{H}^{p+1}(f_*\Omega^\cdot) \to \ldots$$

Since X is smooth, the complex $\Omega^\cdot$ is a resolution of the sheaf $\mathbb{C}_X$, i.e. it is acyclic for $p \geqslant 1$. Since f is Stein, the complex $f_*\Omega^\cdot$ is also acyclic for $p \geqslant 1$, $\mathscr{H}^p(f_*\Omega^\cdot) = 0$. ■

10.6 The sheaves $\mathscr{H}^p_{(0)}$ and the Gauss–Manin connection ∂_t: $\mathscr{H}^p_{(-1)} \rightrightarrows \mathscr{H}^p_{(0)}$

For an isolated singularity the sheaf $\mathscr{H}^{(0)} = \Omega^{n+1}/\mathrm{d}f \wedge \mathrm{d}(\Omega^{n-1})$ coincides with $\partial_t(\mathscr{H}^{(-1)})$. Therefore in order to generalize sheaves $\mathscr{H}^{(0)}$ to the case of non-isolated singularities following our recipe in (10.4.4), we have to extend the exact sequence (10.4.3) over S to have the complex $B^\cdot = \mathrm{d}f \wedge \Omega^{\cdot -1}$ on the right. Consider the exact sequence of complexes

(10.6.1) $$0 \to Z^\cdot \to \Omega^\cdot \overset{\mathrm{d}f\wedge}{\to} B^{\cdot +1} \to 0,$$

where

(10.6.2) $$Z^\cdot = \mathrm{Ker}\,(\Omega^\cdot \overset{\mathrm{d}f\wedge}{\to} \Omega^\cdot)$$

is the complex of sheaves of Koszul cocycles. As the morphism f is Stein, we have an exact sequence of complexes on S

(10.6.3) $$0 \to f_*Z^\cdot \to f_*\Omega^\cdot \overset{\mathrm{d}f\wedge}{\to} f_*B^{\cdot +1} \to 0.$$

Consider the exact cohomology sequence corresponding to (10.6.3)
(10.6.4)
$$\ldots \to \mathscr{H}^p(f_*\Omega^\cdot) \to \mathscr{H}^{p+1}(f_*B^\cdot) \overset{\partial_t}{\to} \mathscr{H}^{p+1}(f_*Z^\cdot) \to \mathscr{H}^{p+1}(f_*\Omega^\cdot) \to \ldots$$

Therefore, by definition we set

(10.6.5) $$\mathscr{H}^p_{(0)} \underset{\mathrm{d}f}{=} \mathscr{H}^{p+1}(f_*Z^\cdot)$$

and the Gauss–Manin connection ∂_t: $\mathscr{H}^p_{(-1)} \to \mathscr{H}^p_{(0)}$ to be the connecting homomorphism in the exact sequence (10.6.4). We can also see that ∂_t is really a connection, i.e. the Leibnitz rule is fulfilled, from the following local description of ∂_t.

(10.6.6) If $[\omega] \in \mathscr{H}^{p+1}(f_*B^\cdot) = \mathscr{H}^p_{(-1)}$ and $\omega = \mathrm{d}f \wedge \eta$ (η is a 'lifting' of ω to $f_*\Omega^\cdot$ in (10.6.3)), then

$$\partial_t([\omega]) = \partial_t([\mathrm{d}f \wedge \eta]) = [\mathrm{d}\eta] \in \mathscr{H}^{p+1}(f_*Z^\cdot) = \mathscr{H}^p_{(0)}.$$

Since $\mathscr{H}^p(f_*\Omega^\cdot) = 0$ for $p \geqslant 1$, we obtain from (10.6.4):

(10.6.7) *Lemma* ∂_t: $\mathscr{H}^p_{(-1)} \rightrightarrows \mathscr{H}^p_{(0)}$ is an isomorphism.

10.7 A generalization of diagram (5.3.4)

We can join the exact sequences of complexes (10.5.2) and (10.6.3) in one big diagram.

(10.7.1) *Definition* The sheaves $\mathscr{K}^p = \mathscr{H}^p(\Omega^\cdot, \mathrm{d}f\wedge)$ are called the sheaves of *Koszul cohomology* of the singularity f. We have a complex $(\mathscr{K}^\cdot, d)$ of sheaves of Koszul cohomology

$$\mathscr{K}^\cdot = Z^\cdot/B^\cdot = \mathrm{Ker}\,(\Omega^\cdot \xrightarrow{cf} \Omega^\cdot)/\mathrm{Im}\,(\Omega^\cdot \xrightarrow{\mathrm{d}f} \Omega^\cdot).$$

Obviously we have a commutative diagram of complexes with exact lines and columns

(10.7.2)
$$\begin{array}{ccccccccc}
 & & & & 0 & & 0 & & \\
 & & & & \uparrow & & \uparrow & & \\
 & & & & B^\cdot[1] & \simeq & B^\cdot[1] & & \\
 & & & & \uparrow \mathrm{d}f\wedge & & \uparrow \mathrm{d}f\wedge & & \\
0 & \to & B^\cdot & \subset & \Omega^\cdot & \to & \Omega^\cdot_f & \to & 0 \\
 & & \| & & \cup & & \cup & & \\
0 & \to & B^\cdot & \subset & Z^\cdot & \to & \mathscr{K}^\cdot & \to & 0 \\
 & & & & \uparrow & & \uparrow & & \\
 & & & & 0 & & 0 & &
\end{array}$$

The morphism $f: X \to S$ being Stein, we can apply the functor f_* to (10.7.2) and obtain the diagram of f_* (10.7.2) (in which all the sheaves are exchanged with their direct images), in which the lines and the columns remain exact. The exact sequences (10.5.2) and (10.6.3) are the exact sequences in diagram f_* (10.7.2) which go through the 'centre' $f_*\Omega^\cdot$. Now we consider two exact sequences which 'go' through $f_*\mathscr{K}^\cdot$. These give us the connection between the sheaves $\mathscr{H}^p_{(-i)}$, $i = 0, 1, 2$.

Write the exact cohomology sequence corresponding to the sequence $0 \to f_*B^\cdot \to f_*Z^\cdot \to f_*\mathscr{K}^\cdot \to 0$ in the upper line and the one corresponding to $0 \to f_*\mathscr{K}^\cdot \to f_*\Omega^\cdot_f \xrightarrow{\mathrm{d}f\wedge} f_*B^{\cdot+1} \to 0$ in the lower line of the following commutative diagram

(10.7.3)

$$\begin{array}{ccccccccccc}
 & & & \overset{\alpha}{\nearrow} & \mathscr{H}^{p+1}(f_*Z^\cdot) & \overset{\partial_t^{-1}}{\searrow} & & & & & \\
 & & & & \wr\uparrow \partial_t & & & & & & \\
\ldots \to \mathscr{H}^p(f_*Z^\cdot) & \to & \mathscr{H}^p(f_*\mathscr{K}^\cdot) & \xrightarrow{\partial} & \mathscr{H}^{p+1}(f_*B^\cdot) & \xrightarrow{j} & \mathscr{H}^{p+1}(f_*Z^\cdot) & \to & \mathscr{H}^{p+1}(f_*\mathscr{K}^\cdot) & \xrightarrow{\partial} & \ldots \\
\wr\uparrow & & \| & & \wr\uparrow \partial_t & \searrow \partial_t^{-1} & \wr\uparrow \partial_t & & \| & & \\
\ldots \to \mathscr{H}^p(f_*B^\cdot) & \xrightarrow{\partial} & \mathscr{H}^p(f_*\mathscr{K}^\cdot) & \to & \mathscr{H}^p(f_*\Omega^\cdot_f) & \xrightarrow{i} & \mathscr{H}^{p+1}(f_*B^\cdot) & \xrightarrow{\partial} & \mathscr{H}^{p+1}(f_*\mathscr{K}^\cdot) & \to & \ldots \\
 & & & & & \searrow \partial_t^{-1} & \wr\uparrow \partial_t & & & & \\
 & & & & & & \mathscr{H}^p(f_*\Omega^\cdot_f). & & & &
\end{array}$$

In this diagram the homomorphism

(10.7.4) $$\mathscr{H}^p_{(-1)} = \mathscr{H}^{p+1}(f_*B^\cdot) \xrightarrow{j} \mathscr{H}^{p+1}(f_*Z^\cdot) = \mathscr{H}^p_{(0)}$$

is induced by the inclusion $f_*B^\cdot \to f_*Z^\cdot$,

(10.7.5) $$\mathscr{H}^p_{(-2)} = \mathscr{H}^p(f_*\Omega^\cdot_f) \xrightarrow{i} \mathscr{H}^{p+1}(f_*B^\cdot) = \mathscr{H}^p_{(-1)}$$

is induced by the epimorphism $f_*\Omega^\cdot_f \xrightarrow{df\wedge} f_*B^{\cdot+1}$, and ∂ denotes the connecting homomorphisms. Moreover, in diagram (10.7.3) we included the operators

(10.7.6) $$\partial_t^{-1}\colon \mathscr{H}^p_{(-i)} \to \mathscr{H}^p_{(-i)},\ i = 0, 1, 2,$$

which make it commutative, and $\mathbb{C}$-linear homomorphisms

(10.7.7) $$\alpha\colon \mathscr{H}^p(f_*\mathscr{K}^\cdot) \to \mathscr{H}^{p+1}(f_*Z^\cdot) = \mathscr{H}^p_{(0)}.$$

(10.7.8) *Proposition* The following conditions are equivalent:

(i) the homomorphism $\alpha = 0$;
(ii) one of the homomorphisms ∂_t^{-1} (and thus all of them) is a monomorphism;
(iii) i or j are monomorphisms.

Proof This is obvious by virtue of the exactness of lines in diagram (10.7.3) and the fact that ∂_t are isomorphisms.

(10.7.9) *Corollary* If $\alpha = 0$, then diagram (10.7.3) falls to the following pieces with exact lines

(10.7.10)

$$\begin{array}{ccccccc} 0 \to & \mathscr{H}^{p+1}(f_*B^\cdot) & \xrightarrow{j} & \mathscr{H}^{p+1}(f_*Z^\cdot) & \to & \mathscr{H}^{p+1}(f_*\mathscr{K}^\cdot) \to 0 \\ & \wr\uparrow \partial_t & \searrow^{\partial_t^{-1}} & \wr\uparrow \partial_t & & \| \\ 0 \to & \mathscr{H}^p(f_*\Omega^\cdot_f) & \xrightarrow{i} & \mathscr{H}^{p+1}(f_*B^\cdot) & \to & \mathscr{H}^{p+1}(f_*\mathscr{K}^\cdot) \to 0 \end{array}$$

Diagram (10.7.10) generalizes diagram (5.3.4) to the case of non-isolated singularities.

(10.7.11) In [Sr] Van Straten proved that $\mathscr{H}^p_{(-i)}$ are $\mathscr{O}_S$-coherent modules if f is a concentrated singularity. He also obtained the following criterion.

Theorem The condition $\alpha = 0$ is equivalent to the absence of torsion in the sheaves $\mathscr{H}^p_{(-i)}$.

The proof of this consists of carrying Malgrange's proof of §6 to the case of non-isolated singularities.

Van Straten [Sr] verified the condition $\alpha = 0$ for the non-isolated singularities with a one-dimensional critical set, which is a reduced complete intersection. For these singularities

$$\mu^{n-1} = 1 \text{ or } 0$$

and

$$\mu^n = \dim H^{n+1}(\mathscr{K}^\cdot) = \dim(\Omega^{n+1}/\mathrm{d}f \wedge \Omega^n + \mathrm{d}\mathscr{K}^n).$$

This formula generalizes the celebrated formula for the Milnor number of an isolated singularity [Mi]

$$\mu = \dim \Omega^{n+1}/\mathrm{d}f \wedge \Omega^n = \dim \mathscr{O}_{\mathbb{C}^{n+1},0}/J_f.$$

II

Limit mixed Hodge structure on the vanishing cohomology of an isolated hypersurface singularity

1 Mixed Hodge structures. Definitions. Deligne's theorem

We first recall very briefly the necessary basic definitions (for a more detailed survey see, for example, [K-Ku].

1.1 Pure Hodge structure

(1.1.1) *Definition* A pure Hodge structure, or simply a *Hodge structure* (abbreviation HS) of weight n is a pair $\{H_{\mathbb{Z}}, H^{p,q}\}$ consisting of a lattice $H_{\mathbb{Z}}$ of finite rank and a direct sum decomposition of the space $H = H_{\mathbb{Z}} \otimes_{\mathbb{Z}} \mathbb{C}$

$$H = \bigoplus_{p+q=n} H^{p,q}$$

such that

$$H^{p,q} = \overline{H^{q,p}}.$$

The motivation for introducing such a concept lies in the Hodge decomposition into (p, q)-types of the cohomology of a smooth projective variety or a compact Köhler manifold.

(1.1.2) Defining the HS in the form of a Hodge decomposition $\{H_{\mathbb{Z}}, H^{p,q}\}$ is equivalent to giving the *Hodge filtration* $\{H_{\mathbb{Z}}, F^{\cdot}\}$, where $F^{\cdot}$ is a finite decreasing filtration of H such that

$$H = F^p \oplus \overline{F^{n-p+1}} \text{ for all } p \in \mathbb{Z}.$$

The filtration $F^{\cdot}$ defines a Hodge decomposition by

(1.1.3) $$H^{p,q} = F^p \cap \overline{F^q}.$$

The Hodge decomposition defines the Hodge filtration by

(1.1.4) $$F^p = \bigoplus_{r \geqslant p} H^{r,n-r} = H^{p,n-p} \oplus H^{p+1,n-p-1} \oplus \ldots \oplus H^{n,0},$$

i.e. $F^n = H^{n,0}$, $F^{n-1} = H^{n,0} \oplus H^{n-1,1}, \ldots, F^0 = H$.

In what follows we'll write an HS by its filtration $F^\cdot$ (assuming that a lattice $H_{\mathbb{Z}}$ is fixed).

1.2 Polarised HSs

(1.2.1) *Definition* A *polarized* HS of weight n is is an HS $\{H_{\mathbb{Z}}, F^\cdot\}$ together with a non-degenerate integral bilinear form

$$Q: H_{\mathbb{Z}} \times H_{\mathbb{Z}} \to \mathbb{Z},$$

symmetric, if n is even, and antisymmetric, if n is odd, and satisfying the *Riemann bilinear relations*:

in terms of a decomposition

(1.2.2) $$Q(H^{p,q}, H^{p',q'}) = 0 \text{ for } (p', q') \neq (p, q),$$
$$i^{p-q} Q(\psi, \overline{\psi}) > 0 \text{ for } \forall \psi \in H^{p,q}, \psi \neq 0$$

or in terms of a filtration

(1.2.3) $$Q(F^p, F^{n-p+1}) = 0,$$
$$Q(C\psi, \overline{\psi}) > 0, \forall \psi \neq 0,$$

where $C: H \to H$ is the Weil operator $C_{|H^{p,q}} = i^{p-q}$.

The bilinear form Q is called a *polarization* of the HS $F^\cdot$. The origin of this notion is the bilinear form $Q(\varphi, \psi) = (-1)^{n(n-1)/2} \int_X L^{d-n} \wedge \varphi \wedge \psi$ on the space of primitive n-cohomology of a smooth projective variety of dimension d.

1.3 Mixed Hodge structure

(1.3.1) *Definition* A *mixed Hodge structure* (abbreviation MHS) is a triple $(H_{\mathbb{Z}}, W_., F^\cdot)$, where $H_{\mathbb{Z}}$ is a lattice, $W_.$ is a finite increasing filtration on $H_{\mathbb{Q}} = H_{\mathbb{Z}} \otimes \mathbb{Q}$, and $F^\cdot$ is a finite decreasing filtration on $H = H_{\mathbb{Z}} \otimes \mathbb{C}$, such that the induced filtration on the quotient $Gr_k^W H = W_k / W_{k-1}$ defines an HS of weight k. Here $F^p Gr_k^W H$ is the image of $F^p \cap W_k H$ in $Gr_k^W H$,

$$F^p Gr_k^W H = W_k H \cap F^p + W_{k-1} H / W_{k-1} H.$$

$F^\cdot$ is called the *Hodge filtration* and $W_.$ the *weight filtration*. The notation for an MHS may be abbreviated to: a pair of filtrations $(W_., F^\cdot)$.

(1.3.2) Let $H^{p,q} = Gr_F^p Gr_{p+q}^W H$. Then $Gr_k^W = \bigoplus_{p+q=k} H^{p,q}$. The numbers $h^{p,q} = \dim H^{p,q}$ are called the *Hodge numbers* of the MHS $(W_., F^\cdot)$.

(1.3.3) If there are two MHSs on spaces H and H', then the space of linear mappings $\mathrm{Hom}(H, H')$ is endowed with a natural MHS,

$$W_m \mathrm{Hom}(H, H')_{\mathbb{Q}} = \{\varphi: H_{\mathbb{Q}} \to H'_{\mathbb{Q}} | \varphi(W_k H) \subset W_{k+m} H', \forall k\},$$
$$F^p \mathrm{Hom}(H, H') = \{\varphi: H \to H' | \varphi(F^k H) \subset F^{k+p} H', \forall k\}.$$

It is easy to verify that $\mathrm{Hom}(H, H')^{p,q} = \oplus \mathrm{Hom}(H^{r,s}, H'^{r+p,s+q})$. In particular, a morphism $\varphi: H \to H'$ is called an *MHS morphism* of type (r, r), if $\varphi(H_{\mathbb{Z}}) \subset H'_{\mathbb{Z}}$, $\varphi(W_k H) \subset W_{k+2r} H'$ and $\varphi(F^p H) \subset F^{p+r} H'$; φ is called an MHS morphism if $r = 0$.

One of the most important properties of an MHS morphism is its strict compatibility with both the filtrations $F^{\cdot}$ and $W_{\cdot}$, i.e.

(1.3.4) $$\varphi(W_k H) = W_{k+2r} H' \cap \mathrm{Im}\,\varphi,\ \varphi(F^p H) = F^{p+r} H' \cap \mathrm{Im}\,\varphi.$$

(1.3.5) It implies that if $H' \to H \to H''$ is an exact sequence of MHSs, then it remains exact after application of Gr_k^W, Gr_F^p and $Gr_F^p Gr_k^W$.

1.4 Deligne's theorem

(1.4.1) *Theorem* The complex cohomology groups $H^{\cdot}(X)$ of projective varieties$/\mathbb{C}$ carry MHSs which are functorial. In the case of a non-singular variety this MHS reduces to the ordinary HS of pure weight.

When proving this theorem Deligne introduced the filtrations $W_{.}$ and $F^{\cdot}$ on the level of resolutions $0 \to \mathbb{C}_X \to K^{\cdot}$ of a constant sheaf $\mathbb{C}_X$ (or $\mathbb{C}_{\mathbb{Q}}$), which induce filtrations on the hypercohomology $\mathbb{H}^k(X, K^{\cdot}) \simeq H^k(X, \mathbb{C})$.

2 The limit MHS according to Schmid

2.1 Variation of HS: geometric case

Let $\pi: Y \to S$ be a family of smooth projective varieties $Y_t = \pi^{-1}(t) \subset \mathbb{P}^m$, parametrized by a complex manifold S ($Y \subset S \times \mathbb{P}^m$ and π is induced by a projection). Then cohomology groups of fibres $H^n(Y_t, \mathbb{C})$, $0 \leqslant n \leqslant 2d$ ($d = \dim Y_t$), are glued to a locally constant fibration (local system) $\underline{H} = R^n \pi_* \mathbb{C}_Y = \bigcup_{t \in S} H^n(Y_t, \mathbb{C})$ on S. Let $\mathscr{H} = \mathscr{O}_S(\underline{H}) = \underline{H} \otimes_{\mathbb{C}_S} \mathscr{O}_S$ be a sheaf of holomorphic sections of this fibration. The local system $\underline{H} \subset \mathscr{H}$ defines a connection ∇ on $\mathscr{H}$. This connection is called the Gauss–Manin connection of the family π. The local system $\underline{H}$ contains a real subsystem $\underline{H}_{\mathbb{R}} = R^n \pi_* \mathbb{R}_Y$ and even a sublattice $\underline{H}_{\mathbb{Z}}$, fibres of which are images of $H^n(Y_t, \mathbb{Z})$ in $H^n(Y_t, \mathbb{R})$. There is a Hodge filtration (a flag) $F_t^{\cdot}$ on each fibre $\underline{H}_t = H^n(Y_t, \mathbb{C})$. Griffiths showed that this family of flags depends on t holomorphically. We obtain a decreasing filtration of holomorphic subsheaves in a locally free sheaf $\mathcal{H}$

$$\ldots \subset \mathcal{F}^{p+1} \subset \mathcal{F}^{p} \subset \ldots \subset \mathcal{H}.$$

The Griffiths transversality theorem asserts that $\nabla(\mathcal{F}^{p}) \subset \mathcal{F}^{p-1} \otimes \Omega^1_S$.

This geometric situation motivates the following definition.

2.2 Variation of HS: definition

(2.2.1) *Definition* A *variation of Hodge structure* (VHS) of weight n on a complex manifold is a pair $(\underline{H}, \mathcal{F}^{\cdot})$, where $\underline{H}$ is a locally constant fibration (or a locally free sheaf $\mathcal{H}$ with an integrable connection ∇, $\operatorname{Ker}\nabla = \underline{H}$) defined over $\mathbb{R}$ (or $\mathbb{Q}$), $\underline{H} = \underline{H}_{\mathbb{R}} \otimes \mathbb{C}$, and a decreasing filtration $\mathcal{F}^{\cdot} = \{\mathcal{F}^{p}\}$ of locally free subsheaves $\mathcal{F}^{p} \subset \mathcal{H} = \underline{H} \otimes_{\mathbb{C}_S} \mathcal{O}_S$. These objects have to satisfy the following two conditions:

(i) for each point $t \in S$ the filtration $\mathcal{F}^{\cdot}$ induces a filtration $\mathcal{F}^{\cdot}_t$ on the fibre $\underline{H}_t$ at t constituting an HS of weight n;
(ii) $\nabla(\mathcal{F}^{p}) \subset \Omega^1_S \otimes_{\mathcal{O}_S} \mathcal{F}^{p-1}$ for each p.

(2.2.2) Moreover, if there is a flat, non-degenerate bilinear form Q: $\underline{H} \times \underline{H} \to \mathbb{Q}_S$ defining a polarized HS, then the VHS $(\underline{H}, \mathcal{F}^{\cdot})$ said to be *polarized*.

If in the geometric case we consider $\underline{P} = \bigcup_{t \in S} P^n(Y_t, \mathbb{C})$ instead of $\underline{H} = \bigcup_{t \in S} H^n(Y_t, \mathbb{C})$, where $P^n(Y_t, \mathbb{C})$ is a primitive cohomology, then we obtain a polarized VHS. It is convenient to consider the polarized VHS for technical reasons. However, an HS on $H^n(Y_t, \mathbb{C})$ is completely defined by HSs on $P^k(Y_t, \mathbb{C})$ and so no information is lost.

2.3 Classifying spaces and period mappings

(2.3.1) Let $\mathcal{F}\ell$ be a set of all filtrations (flags) $F^{\cdot}$ in a vector space H of fixed dimensions $f^p = \dim F^p = \sum_{r \geqslant p} h^{r,s}$. It is a closed subvariety of a product of Grassman manifolds. Let Q be a bilinear form on H defining a polarization. Then the set of polarized HSs of given type is parametrized by the set of $F^{\cdot} \in \mathcal{F}\ell$ satisfying condition (1.2.3). The first bilinear relation defines a closed subvariety $\check{D} \subset \mathcal{F}\ell$, and second bilinear relation defines an open subset $D \subset \check{D}$; D is called a *classifying space* of the polarized HS of given type.

(2.3.2) Let $G = \{g \in GL(H) | Q(gu, gv) = Q(u, v), \ \forall u, v \in H\}$ be an orthogonal group of the bilinear form Q, and $G_{\mathbb{R}} \subset GL(H_{\mathbb{R}})$ the corresponding real subgroup, $G_{\mathbb{Z}} = \{g \in G_{\mathbb{R}} | gH_{\mathbb{Z}} = H_{\mathbb{Z}}\}$ an arithmetic subgroup of $G_{\mathbb{R}}$. The group G transitively acts on $\check{D}$, $\check{D} \simeq G/B$, and the group $G_{\mathbb{R}}$ transitively acts on D, $D = G_{\mathbb{R}}/V$, $V = G_{\mathbb{R}} \cap B$. For clarity it is useful to have in mind the simplest example of an HS of weight $n = 1$, $H = H^{0,1} \oplus H^{1,0}$, $h^{0,1} = h^{1,0} = 1$. In this case $G_{\mathbb{C}} = SL(2, \mathbb{C})$, $G_{\mathbb{R}} = SL(2, \mathbb{R})$, $\check{D} = \mathbb{P}^1(\mathbb{C})$, and $D = U \subset \mathbb{P}^1(\mathbb{C})$ is the upper half-plane.

(2.3.3) The group $\Gamma = G_{\mathbb{Z}}$ acts on D properly discontinuously. Hence there is a structure of a normal analytic space on D/Γ; D/Γ is called a *period space* of a polarized HS of the given type.

(2.3.4) Now assume that there is a variation of the polarized HS $(\underline{H}, \mathscr{F}^{\cdot})$ on a manifold S, for example a VHS corresponding to primitive n-dimensional cohomology of a family of projective varieties π: $Y \to S$. Let e: $\tilde{S} \to S$ be a universal covering. We can lift the VHS to $\tilde{S}$: $e^{-1}\underline{H} = \underline{H} \times_S \tilde{S} = \tilde{\underline{H}}$ is a locally constant fibration on $\tilde{S}$ defining a connection $\tilde{\nabla}$ on the sheaf $\tilde{\mathscr{H}} = e^*\mathscr{H}$ and $\tilde{\mathscr{F}}^{\cdot} = e^*\mathscr{F}^{\cdot}$ is a decreasing filtration on $\tilde{\mathscr{H}}$. Since $\tilde{S}$ is simply connected, $\tilde{\nabla}$ defines a canonical trivialization of the flat locally constant fibration $\tilde{\underline{H}}$, $\tilde{\underline{H}} \simeq H \times \tilde{S}$, where H is a vector space (a canonical fibre). This trivialization defines a family of filtrations $F^{\cdot}_s \subset \tilde{\underline{H}}_s \simeq H$ on the space H ($\tilde{\underline{H}}_s \simeq \underline{H}_t$, $t = e(s)$, $F^{\cdot}_s = F^{\cdot}_t$). We obtain a mapping $\tilde{\Phi}$: $\tilde{S} \to D$, $s \mapsto F^{\cdot}_s$ to a classifying space D. If s and $s' \in \tilde{S}$, $e(s) = e(s') = t$, then it is easy to see that $F^{\cdot}_s$ and $F^{\cdot}_{s'}$ differ by an element $\gamma \in \Gamma$, $F^{\cdot}_{s'} = \gamma F^{\cdot}_s$. Hence, the mapping $\tilde{\Phi}$ is reduced to a mapping Φ: $S \to D/\Gamma$; Φ is called a *period mapping* of a given variation of Hodge structures.

We'll be mainly interested in one-dimensional variations of the HS. Let us consider more carefully a period mapping in this case.

2.4 The canonical Milnor fibre

(2.4.1) Let $(\underline{H}, \mathscr{F}^{\cdot})$ be a variation of HS over a punctured disk $S' = S\backslash 0$, $S = \{|t| < 1\}$, where $\mathscr{F}^{\cdot}$ is a decreasing filtration on the the sheaf $\mathscr{H} = \underline{H} \otimes_{\mathbb{C}_{S'}} \mathscr{O}_{S'}$. Let ∇ be a connection on $\mathscr{H}$ defined by the locally constant fibration $\underline{H} = \operatorname{Ker} \nabla$.

In a geometric situation we have a projective morphism π: $Y \to S$, which is smooth over S', and $\underline{H} = R^n\pi_*\mathbb{C}_{Y'} = \bigcup_{t \in S'} H^n(Y_t, \mathbb{C})$ (to have a polarized VHS we need to consider primitive cohomology $P^n(Y_t, \mathbb{C}) \subset H^n(Y_t, \mathbb{C})$ instead of $H^n(Y_t, \mathbb{C})$). The connection ∇ on the sheaf

$\mathscr{H} = \underline{H} \otimes_{\mathbb{C}_{S'}} \mathscr{O}_{S'} = \mathbb{R}^n\pi_*(\Omega^{\cdot}_{Y'/S'}) = \mathscr{H}^n_{\mathrm{DR}}(Y'/S')$ is the Gauss–Manin connection.

(2.4.2) Denote by h_t: $Y_t \to Y_t$ the Picard–Lefschitz monodromy corresponding to a canonical generator of $\pi_1(S', t)$. This is a diffeomorphism of a fibre $Y_t = \pi^{-1}(t)$, $t \neq 0$. Let

$$M = (h_t)_*\colon H_n(Y_t) \to H_n(Y_t)$$

and

$$T = (h_t^*)^{-1}\colon H^n(Y_t) \to H^n(Y_t)$$

be, respectively, monodromy transformations for homology and cohomology, $T = (M^*)^{-1}$.

(2.4.3) To define a period mapping we have to identify fibres at different points $t \in S'$. This can be done by fixing a fibre Y_{t_0}, $t_0 \in S'$, and transferring filtrations to this fibre. To avoid dependence on the choice of a point t_0 and to make the construction more invariant, we introduce the notion of a canonical fibre Y_∞.

Let $U = \{u \in \mathbb{C} | \operatorname{Im} u > 0\}$ be the upper half-plane

(2.4.4) $$e\colon U \to S',\ u \mapsto t = e(u) \underset{\mathrm{df}}{=} \mathrm{e}^{2\pi i u}$$

be the universal covering of S'. Denote by Y_∞ an inverse image of $Y' = \pi^{-1}(S')$ via e

(2.4.5) $$\begin{array}{cccccc} Y \supset Y' & \leftarrow & Y_\infty = Y' \underset{S'}{\times} U & \overset{j_u}{\leftarrow} & Y_t = Y_{e(u)} & \\ \pi\downarrow\ \pi'\downarrow & & \downarrow & & & \\ S \supset S' & \leftarrow & U \ni u & & & ,\ j_u(x) = (x, u). \end{array}$$

Denote by j_u: $Y_t \subset Y_\infty$ an embedding of the fibre Y_t onto a fibre over a point $u \in U$, $j_u(x) = (x, u) \in Y_\infty$, $t = e(u)$. Here $Y_\infty \to U$ is a locally trivial fibration with fibres diffeomorphic to Y_t. Since U is contractible, the embeddings j_u are homotopy equivalences and, consequently,

(2.4.6) $$H_n(Y_t) \overset{(j_u)_*}{\rightleftarrows} H_n(Y_\infty),\ H = H^n(Y_\infty) \overset{j_u^*}{\rightleftarrows} H^n(Y_t)$$

are isomorphisms. Via these isomorphisms we'll identify $H^n(Y_t)$ with a fixed space $H = H^n(Y_\infty)$. For this reason Y_∞ is called a *canonical fibre* of a family π.

For an abstract variation of the HS we have

(2.4.7) $$\begin{array}{ccc} \underline{H} \leftarrow \underline{H}_U = e^{-1}\underline{H} = \underline{H} \underset{S'}{\times} U \underset{j_u}{\supset} \underline{H}_t & & \\ \downarrow\qquad \downarrow & & \\ S' \overset{e}{\leftarrow} U & & ,\ j_u(\omega) = (\omega, u). \end{array}$$

(2.4.8) We obtain an identification $H = H^0(U, \underline{H}_U) \overset{j_u^*}{\rightrightarrows} \underline{H}_t$ of the space H with a fibre $\underline{H}_t$. We can consider the space H as a space of (single-valued) horizontal sections of the fibration $\underline{H}_U$ or as a space of multivalued horizontal sections of the fibration $\underline{H}$. The identification j_u^*: $H \rightrightarrows \underline{H}_t$ consists of a restriction of a section $A \in H = H^0(U, \underline{H}_U)$ to a fibre $\underline{H}_u = \underline{H}_t$, $t = e(u)$. The inverse isomorphism consists of an extension of $A_t \in \underline{H}_t$ to a multivalued section A over U with the initial condition $A(u) = A_t$; H also is called a *canonical fibre* of the fibration $\underline{H}$.

We obtain a canonical trivialization

(2.4.9)
$$\underline{H}_U = \underline{H} \underset{S'}{\times} U \rightrightarrows H \times U,$$
$$(A_t, u) \mapsto ((j_u^*)^{-1} A_t, u).$$

(2.4.10) Let us carry the monodromy T: $\underline{H}_t \to \underline{H}_t$, which exists on every fibre $\underline{H}_t$ and is obtained by parallel transference along the canonical generator of $\pi_1(S', t)$, to the canonical fibre H by means of the identification

$$H \overset{j_u^*}{\rightrightarrows} \underline{H}_t.$$

Consider a diffeomorphism

(2.4.11)
$$h\colon \underline{H}_U \to \underline{H}_U,\ (\omega, u) \mapsto (\omega, u+1),$$

and the corresponding linear transformation

$$h^*\colon H \to H,$$

transferring a multivalued section A such that $A(u+1) = A_0$ to a multivalued section $h^* A$ such that $h^* A(u) = A_0$. Obviously, this implies:

(2.4.12) *Lemma* Under the identification $H \overset{j_u^*}{\rightrightarrows} \underline{H}_t$ for any u the monodromy T on $\underline{H}_t$ corresponds to the transformation h^*, i.e. there is a commutative diagram

$$\begin{array}{ccc} h^* A \in H & \overset{j_u^*}{\rightrightarrows} & \underline{H}_t \\ h^* \uparrow & & \uparrow T \\ A \in H & \underset{j_u^*}{\rightrightarrows} & \underline{H}_t \ni A_0. \end{array}$$

Proof. Let $A_0 \in \underline{H}_t$. Then $(j_u^*)^{-1}(A_0) = A \in H$ is a multivalued section such that $A(u) = A_0$ or $A(u+1) = TA_0$. By the definition of h^* we have $h^* A(u) = A(u+1) = TA_0$, and this means that $h^* A = (j_u^*)^{-1}(TA_0)$. ■

Thus, on the canonical fibre H the monodromy $T = h^*$. Moreover, two identifications j_u^* and j_{u+1}^* of the canonical fibre H with $\underline{H}_t$ are connected with the monodromy on H in the following way: there is a commutative diagram

(2.4.13)
$$\begin{array}{ccc} & H & \searrow j_u^* & \\ h^* = T & \uparrow & & \underline{H}_t \\ & H & \nearrow j_{u+1}^* . & \end{array}$$

The proof is obvious: if $A_0 \in \underline{H}_t$, then $(j_{u+1}^*)^{-1} A_0 = A$ is a multivalued section such that $A(u+1) = A_0$, and this means that $h^* A(u) = A_0$, i.e. $h^* A = (j_u^*)^{-1} A_0$.

2.5 The Schmid limit Hodge filtration $F^{\cdot}_S$

A variation of the HS $(\underline{H}, \mathscr{F}^{\cdot})$ defines a family of filtrations $F^{\cdot}_t$ on fibres $\underline{H}_t$, $t \in S'$. We can lift this family onto U and by means of a trivialization (2.4.9) $\underline{H}_U \simeq H \times U$ we get a family of filtrations $F^{\cdot}_u$ on the space H

(2.5.1)
$$u \mapsto F^{\cdot}_u = (j_u^*)^{-1} F^{\cdot}_t, \; t = e(u).$$

We obtain a period mapping

(2.5.2)
$$U \xrightarrow{\tilde{\varphi}} D, \; u \mapsto F^{\cdot}_u.$$

(2.5.3) *Lemma* The mapping $\tilde{\varphi}$ satisfies a condition
$$\tilde{\varphi}(u+1) = T^{-1} \tilde{\varphi}(u).$$

Proof. We have $j_{u+1}^* = j_u^* h^*$ (2.4.13), hence
$$F^{\cdot}_{u+1} = (j_{u+1}^*)^{-1} F_t = (h^*)^{-1} (j_u^*)^{-1} F_t = T^{-1} F^{\cdot}_u. \qquad \blacksquare$$

If we take a quotient of D by a cyclic group $\{T^k\}$, we obtain a period mapping φ

(2.5.4)
$$\begin{array}{ccc} U & \xrightarrow{\tilde{\varphi}} & D \\ e \downarrow & & \downarrow \\ S' & \xrightarrow[\varphi]{} & D/T \end{array}, \quad \varphi(t) = \tilde{\varphi}(u) = \tilde{\varphi}\left(\frac{1}{2\pi \mathrm{i}} \log t\right).$$

By the monodromy theorem ([Sm] and Theorem I of §I.9.2 all eigenvalues of T are roots of unity. Hence, if $T = T_s T_u$ is a Jordan decomposition, where T_{s} and T_{u} are semisimple and unipotent parts of the monodromy, then there exists m such that $T_{\mathrm{s}}^m = 1$.

Let $T = e^{2\pi i R}$. By means of a family of automorphisms $g(u) = e^{2\pi i Ru}$ of the space H having monodromy T, $g(u+1) = Tg(u)$, we can 'untwist' the mapping $\tilde{\varphi}$ and make it periodic: set $\tilde{\psi}(u) = g(u)\tilde{\varphi}(u)$, then $\tilde{\psi}(u+1) = g(u+1)\cdot\tilde{\varphi}(u+1) = g(u)TT^{-1}\tilde{\varphi}(u) = \tilde{\psi}(u)$ and hence, $\tilde{\psi}$ is 'descended' to S',

(2.5.5)
$$\begin{array}{ccc} U & \xrightarrow{\tilde{\psi}} & \check{D} \\ e\downarrow & \nearrow \psi & \\ S' & & \end{array}, \quad \tilde{\psi}(u) = e^{2\pi i Ru}\cdot\tilde{\varphi}(u).$$

A matrix $R(= (1/2\pi i)\log T)$ such that $T = e^{2\pi i R}$ is not defined uniquely, more exactly its semisimple part is not defined uniquely. Let $R = -R_s - N$, where $-R_s$ is a semisimple part, and $-N$ is a nilpotent part of R. Then $T = e^{2\pi i(-R_s - N)} = T_s\cdot T_u$, i.e. $T_s = e^{-2\pi i R_s}$, $T_u = e^{-2\pi i N}$. The nilpotent operator

(2.5.6)
$$N = -\frac{1}{2\pi i}\log T_u,$$

the *logarithm of the unipotent part of the monodromy* (divided by $-2\pi i$), is defined by T uniquely. Here for an unipotent operator T_u we set by definition

(2.5.7)
$$\log T_u = \log[E + (T_u - E)] = \sum_{k\geqslant 1}(-1)^{k+1}\frac{(T_u - E)^k}{k}.$$

(2.5.8) *Theorem–definition* Let $(\underline{H}, \mathscr{F}^{\cdot})$ be a variation of an HS over a punctured disk S', let $T = T_s T_u$ be the monodromy and let $T^m = 1$. Consider a covering $\tilde{S}' \to S'$, $t = \tilde{t}^m$, killing the semisimple part of the monodromy. Consider a mapping ψ: $\tilde{S}' \to \check{D}$, defined for a variation of HS on $\tilde{S}'$ in (2.5.2) (to simplify notation we can assume that the monodromy $T = T_u$ is unipotent, $R = -N$). then the mapping ψ: $S' \to \check{D}$ can be continued holomorphically over the point $0 \in S$. The point $\psi(0) \in \check{D}$ is a fixed point of T_s. The filtration $F_S = F^{\cdot}_{\infty}$ corresponding to this point is called a *limit Hodge filtration (according to Schmid)*. By construction $F^{\cdot}_S$ is invariant wrt T_s.

A generalization of this theorem for the case $\dim S > 1$ is part of Schmid's nilpotent orbit theorem [Sm]. In our one-dimensional case this theorem follows from the regularity of the connection ∇ (I.8.2.1). Indeed, the regularity of a connection means in essence that ψ is meromorphic, i.e. it has at most poles (see below). Then the map ψ is regular at the point $0 \in S$ because $\check{D}$ is a projective variety (we can 'get rid' of denominators).

2.6 An interpretation of $F^{\cdot}_S$ in terms of a canonical extension of $\mathscr{H}$

(2.6.1) It is known that any locally free $\mathscr{O}_X$-module $\mathscr{H}$ on a punctured disk S' is free. Trivializations of the sheaf $\mathscr{H}$ are in one to one correspondence with extensions $\overline{\mathscr{H}}$ of the sheaf $\mathscr{H}$ over the whole disk S: if a trivialization of $\mathscr{H}$ is chosen, i.e. a basis of sections $\omega_1, \ldots, \omega_\mu$ of $\mathscr{H}$ over S', then the fibre $\overline{\mathscr{H}}_o$ consists of those $\omega = \sum g_j(t)\omega_j$, for which $g_j(t) \in \mathscr{O}_{S,0}$. A sheaf $\mathscr{H} = \underline{H} \otimes_{\mathbb{C}_{S'}} \mathscr{O}_{S'}$ with a connection ∇ has a *canonical* (privileged) *extension* over S. We'll denote this by $\mathscr{L}$. We'll consider the canonical extension in more detail in §6. Now we assume that the monodromy $T = T_u$ is unipotent, and let $T = e^{-2\pi i N}$, $N = -(1/2\pi i) \log T_u$.

Let $A \in H$, i.e. $A = A(t)$ is a multivalued horizontal section of $\underline{H}$ (or $A = A(e^{2\pi i u})$ is a horizontal section of $\underline{H}_U$). The generator $\gamma \in \pi_1(S, t)$ acts on A by means of monodromy $\gamma A(t) = A(e^{2\pi i}t) = TA(t)$. Consider a family of automorphisms 'untwisting' multivalued sections

(2.6.2) $$g(t) = t^{N \log t} \text{ or } g(u) = e^{2\pi i N u},$$

where $u = (1/2\pi i) \log t$. Then γ transforms $g(t)$ to $e^{2\pi i N} g(t) = T^{-1} g(t)$. Therefore, sections

(2.6.3) $$s = s[A] = t^N A(t)$$

are invariant wrt γ and, consequently, define single-valued sections of the sheaf $\mathscr{H}$.

(2.6.4) *Definition* The *canonical extensions* $\mathscr{L}$ of the sheaf $\mathscr{H}$ (with ∇) onto S is an extension corresponding to a trivialization of $\mathscr{H}$ by means of the basis $s_j = s[A_j]$, $j = 1, \ldots, \mu$, where $A_1, \ldots, A_\mu \in H$ is a basis of horizontal sections.

Thus, we have a trivialization

(2.6.5) $$\mathscr{O}_S(H) = H \otimes_{\mathbb{C}} \mathscr{O}_S \xrightarrow{\sim} \mathscr{L}, \ A \mapsto s[A].$$

Now we can interpret the limit Hodge fibration $F^{\cdot}_S$ (2.5.8) in the following way.

(6.6) *Theorem* Let $(\underline{H}, \mathscr{F}^{\cdot})$ be a variation of HS and let $\mathscr{F}^{\cdot} = \{F^{\cdot}_t\}$ also denote a filtration on $\mathscr{H} = \mathscr{L}_{|S'} \simeq H \otimes_{\mathbb{C}} \mathscr{O}_{S'}$. Then the filtration $\mathscr{F}^{\cdot}$ is extended to a filtration of $\mathscr{L}$, and $F^{\cdot}_S$ is the filtration on the zero fibre $\mathscr{L}_0 \simeq H$.

$$F^{\cdot}_S = \lim_{t \to 0} F^{\cdot}_t.$$

Proof. Defining $F^{\cdot}_S$ we first lifted the family of filtrations $F^{\cdot}_t \subset \underline{H}_t$ on U

and, using a trivialization $\underline{H}_U = \underline{H} \times_{S'} U \simeq H \times U$, we obtained a family of filtrations on the space H, $\tilde{\varphi}(u) = F^{\cdot}_u = (j^*_u)^{-1} F^{\cdot}_t$, $t = e(u)$, satisfying the condition $\tilde{\varphi}(u+1) = T^{-1}\tilde{\varphi}(u)$.

In the same way, this trivialization permits us to consider any section $s \in \mathscr{H}$ of the fibration $\underline{H}$, after its lift $s = s(\mathrm{e}^{2\pi \mathrm{i} u})$ to a section of $\underline{H}_U$, as a section $\tilde{s} \in \mathscr{O}_U(H)$ of a constant fibration $H \times U$: $u \mapsto \tilde{s}_u = (j^*_u)^{-1} s(t) \in H$, where $\tilde{s}_u = \tilde{s}_u(w)$ (considered as a horizontal section of $\underline{H}_U$, $w \in U$) is such a horizontal section, which satisfies an initial condition $\tilde{s}_u(w)_{|w=u} = s(t) = s(\mathrm{e}^{2\pi \mathrm{i} u})$. As for filtrations we have $\tilde{s}_{u+1} = T^{-1}\tilde{s}_u$.

In particular, if $s = s[A] = t^N A(t) = \mathrm{e}^{2\pi \mathrm{i} N u} A(\mathrm{e}^{2\pi \mathrm{i} u}) \in \mathscr{H}$ is a section constructed by a section $A \in H$, then

$$\tilde{s}_u(w) = \mathrm{e}^{2\pi \mathrm{i} N u} A(e^{2\pi \mathrm{i} w}).$$

Indeed, it is a horizontal section of $\underline{H}_U$ as a function in w, and it satisfies the required initial condition. (This shows why $\tilde{s}_{u+1}$ is transformed to $\tilde{s}_{u+1} = T^{-1}\tilde{s}_u$ (and not to $T\tilde{s}_u$) – it is a dependence upon the initial condition.)

The same formulae show that horizontal sections $s = A(\mathrm{e}^{2\pi \mathrm{i} u})$ of $\underline{H}_U$ correspond to constant sections $\tilde{s} = A = \mathrm{const}$ of the fibration $H \times U$.

Now we return to a family of filtrations $F^{\cdot}_t$, i.e. to a family of subspaces $F^p_t \subset \underline{H}_t$. Here F^p_t is generated by base elements. Let ω be a base section of $\mathscr{F}^p$, $\omega(t) \in F^p_t$. Let us express it by a basis $s_j = s[A_j]$, $j = 1, \ldots, \mu$, of the canonical trivialization of $\mathscr{H} = \mathscr{L}_{|S'}$. Let $\omega = \sum_{j=1}^{\mu} c_j(t) s_j(t)$, where $c_j(t) \in \mathscr{O}_{S'}$. Then the family of filtrations $\tilde{\varphi}(u) = F^{\cdot}_u$ on H consists of subspaces $F^{\cdot}_u$, generated by base elements $\tilde{\omega} = \sum_{j=1}^{\mu} c_j(\mathrm{e}^{2\pi \mathrm{i} u})(\tilde{s}_j)_u$. The 'untwisted' family of filtrations $\tilde{F}^{\cdot}_u = \tilde{\psi}(u) = \mathrm{e}^{-2\pi \mathrm{i} N u}\tilde{\varphi}(u)$ is generated by base sections $\mathrm{e}^{-2\pi \mathrm{i} N u} . \tilde{\omega} = \sum_{j=1}^{\mu} c_j(\mathrm{e}^{2\pi \mathrm{i} u}) A_j$, where A_j is a constant section of $H \times U$. Now by definition F^p_S is generated by sections $\lim_{t\to 0} \sum_{j=1}^{\mu} c_j(t) A_j$, and this means that F^p_S corresponds to a filtration $\lim_{t\to 0} F^p_t$ on the zero fibre $\mathscr{L}_0$ of the canonical extension, which is generated by base elements $\lim_{t\to 0}\omega = \lim_{t\to 0}\sum_{j=1}^{\mu} c_j(t) s_j$, where s_j are base vectors of a canonical trivialization of $\mathscr{L}$. ■

2.7 The weight filtration of a nilpotent operator

(2.7.1) *Lemma–definition* Let N: $H \to H$ be a nilpotent operator on a vector space H, $N^{k+1} = 0$. Then there exists a unique increasing filtration $W_{\cdot} = W(N)$ on H,

$$\ldots \subset W_{-1} \subset W_0 \subset W_1 \subset \ldots,$$

characterized by properties

(i) $N(W_i) \subset W_{i-2}$,
(ii) $N^r\colon Gr^W_r \rightrightarrows Gr^W_{-r}$ is an isomorphism for $\forall\, r$.

This filtration $W_. = W(N)$ is called the *weight filtration of a nilpotent operator N* (with centre 0 or with central index 0).

The filtration $W(N)$ can be constructed by different means, for example, by means of an iteration process. However, the most down-to-earth method of construction of the filtration $W(N)$ is the following.

Choose a Jordan basis $u_1, \ldots, u_\mu$ of the operator N on H, i.e. such a basis that wrt it the matrix of N has a Jordan normal form, i.e. consists of blocks of the form

$$N = \begin{pmatrix} 0 & & \cdots & 0 \\ 1 & \ddots & & \vdots \\ \vdots & \ddots & & \\ 0 & \cdots & 1 & 0 \end{pmatrix}, \; Nu_1 = u_2, \ldots, Nu_{q-1} = u_q, \; Nu_q = 0.$$

Number basis vectors correspond to each block in such a way that their numbers are symmetric relative to 0 and that N decreases the index by 2 (i.e. $N(u_i) = u_{i-2}$):

if $q = 2k + 1$ is odd,
  then $u_{2k}, \ldots, u_2, u_0, u_{-2}, \ldots, u_{-2k}$;
if $q = 2k$ is even,
  then $u_{2k-1}, \ldots, u_{-1}, u_1, \ldots, u_{-(2k-1)}$.

In such way we attach a weight to each base vector – this is its number in the numeration, and we have $N(u_i) = u_{i-2}$. Let W_k = a subspace generated by vectors of weight $\leqslant k$. Obviously, properties (i) and (ii) are then fulfilled. ■

We can visualize the filtration $W_.$ by means of the following schematic diagram. In this a base vector is represented by a square with a weight written in it. Base vectors of any block are set in a line at a distance from a fixed axis which is equal to their weight:

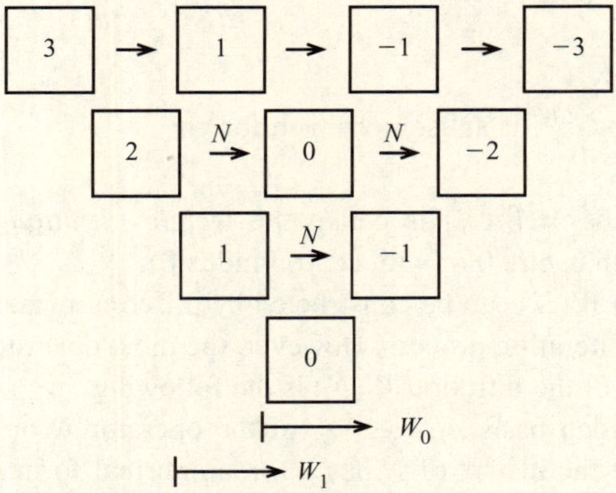

(2.7.2) *Definition* A *weight filtration of a nilpotent operator N with centre n* (or with central index *n*) is a filtration ${}^nW_.$ obtained from $W_. = W(N)$ by a shift of numeration

$$ {}^nW_. = W_.[-n], \text{ i.e. } {}^nW_k = W_{-n+k}. $$

For the filtration ${}^nW_.$ property (ii) can be written as:

(2.7.3) $\qquad N^r\colon Gr^{{}^nW}_{n+r} \rightrightarrows Gr^{{}^nW}_{n-r}$ is an isomorphism for all r.

Let us make a simple, but very useful, note needed in the next section.

(2.7.4) Let $\overline{H} = H/\operatorname{Ker} N$, and let $\overline{N}$ be an operator induced by N on $\overline{H}$. Denote by $\overline{W}$ a weight filtration (with centre 0) of the operator $\overline{N}$ on $\overline{H}$, and by $\overline{{}^nW}$ a filtration induced by the filtration ${}^nW$ on $\overline{H}$.

(2.7.5) *Proposition*

$$ \overline{{}^nW} = {}^{n+1}\overline{W}, $$

i.e. a filtration induced on $\overline{H}$ by the weight filtration on H with centre n coincides with the weight filtration of the operator $\overline{N}$ on $\overline{H}$ with centre $n+1$.

Proof This is shown in the following diagram

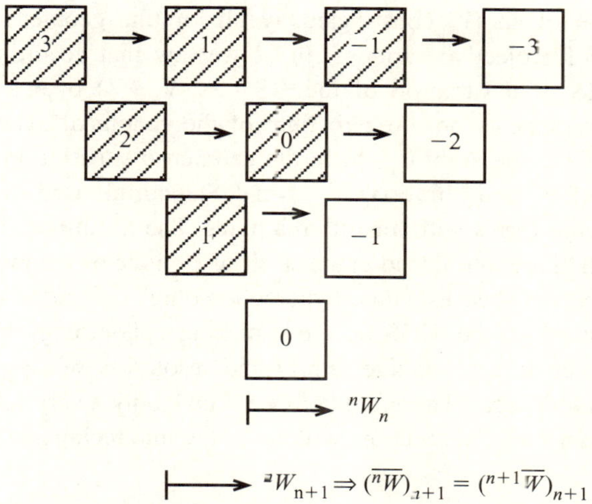

and so on. Here $\operatorname{Ker} N$ is generated by base vectors represented by the unshaded squares, and $\overline{H}$ by shaded squares. ■

2.8 Schmid's theorem

(2.8.1) *Theorem* Let $(\underline{H}, \mathscr{F}^{\cdot})$ be a polarized variation of an HS of weight n over a punctured disk, H be a space of horizontal (multivalued) sections of $\underline{H}$, and T be the monodromy operator. Let $F_S^{\cdot}$ be a limit Hodge filtration introduced in (2.5.8), $W_{.}$ be a weight filtration of the nilpotent operator $N = -(1/2\pi \mathrm{i}) \log T_{\mathrm{u}}$ (the logarithm of the unipotent part of the monodromy) with centre n. Then the filtrations $(W_{.}, F_S^{\cdot})$ determine an MHS on H, which is called the *limit MHS* of the given variation of *HS*. With respect to this MHS the operator N: $H \to H$ is a morphism of the MHS of type $(-1, -1)$, i.e. $NF_S^p \subset F_S^{p-1}$, $NW_k \subset W_{k-2}$.

The proof of this theorem follows from the Schmid's SL_2-orbit theorem [Sm].

3 The limit MHS according to Steenbrink

In this chapter we are mainly interested in the problem of introducing the MHS on the cohomology of a canonical fibre $H^n(X_\infty) = H$ of the Milnor fibration f: $X \to S$ associated with an isolated singularity f: $(\mathbb{C}^{n+1}, 0) \to (\mathbb{C}, 0)$. This was first done by Steenbrink [S3] by using an embedding of f into a family of projective hypersurface π: $Y \to S$ and then resolving the singularities of this family. He used the limit MHS of a variation of an HS constructed by him in the geometric situation of a family of projective varieties. In §2 we saw that according to Schmid a limit HS of a variation of the HS $(\mathcal{H}, \nabla, \mathcal{F}^{\cdot})$ over a punctured disk naturally appears on the zero fibre of the canonical extension $\mathcal{L}$ of the sheaf $\mathcal{H}$ to the point $0 \in S$. In the geometric situation in order to extend the sheaf $\mathcal{H}$ and filtrations $\mathcal{F}^{\cdot}$ over S Steenbrink used sheaves of relative differential forms with logarithmic poles. The filtrations $F^{\cdot}$ and $W_{.}$ of the limit MHS are introduced in the spirit of Deligne by means of filtrations on complexes of sheaves whose hypercohomology coincides with H. Our goal is to introduce the MHS on the vanishing cohomology without using an embedding in a projective family and resolution of singularities (this is Varchenko's idea). Therefore here we'll give only a very schematic account of Steenbrink's construction, without going into technical details.

3.1 The limit MHS for projective families: the case of unipotent monodromy

Let $\pi\colon Y \to S$ be a family of projective varieties (induced by a projection $Y \subset \mathbb{P}^N \times S \to S$), which is smooth over a punctured disk $S' = S\backslash\{0\}$, and let $Y_0 = \pi^{-1}(0)$ be a degenerate fibre. Resolving singularities of Y_0, we can assume Y_0 to be a divisor with normal crossings. First we'll consider the case in which Y_0 is a reduced divisor.

(3.1.1) The family π defines a variation of HS $(\mathcal{H}, \nabla, \mathcal{F}^{\cdot})$ on the sheaf $\mathcal{H} = R^n\pi_*\mathbb{C}_{Y'} \otimes_{\mathbb{C}} \mathcal{O}_{S'} \simeq \mathbb{R}^n\pi_*\Omega^{\cdot}_{Y'/S'}$, where $\pi\colon Y' \to S'$ is a restriction of π onto $Y' = Y\backslash Y_0$. The connection ∇ is defined by a local system $\underline{H} = R^n\pi_*\mathbb{C}_{Y'}$, and the Hodge filtration $\mathcal{F}^{\cdot}$, as is well known, corresponds to a stupid filtration on the De Rham complex $\Omega^{\cdot}_{Y'/S'}$, which is a resolution of the sheaf $\pi^{-1}\mathcal{O}_{S'}$ (the relative Poincaré lemma), $\mathcal{F}^p = \mathbb{R}^n\pi_*F^p\Omega^{\cdot}_{Y'/S'}$, where $F^p = \sigma_{\geqslant p}\colon \ldots \to 0 \to \Omega^p_{Y'/S'} \to \Omega^{p+1}_{Y'/S'} \to \ldots$.

(3.1.2) Steenbrink proves that a sheaf $\overline{\mathcal{H}} = \mathbb{R}^n\pi_*\Omega^{\cdot}_{Y/S}(\log Y_0)$ is locally free on S and, consequently, is a natural extension of $\mathcal{H}$ over S. The connection ∇ is extended to a connection on $\overline{\mathcal{H}}$ with a logarithmic pole at $0 \in S$, $\nabla\colon \overline{\mathcal{H}} \to \Omega^1_S(\log 0) \otimes \overline{\mathcal{H}}$, (i.e. $\overline{\mathcal{H}}$ is a saturated lattice in the sense of (I.7.7.2)). The eigenvalues α of the connection residue $\operatorname{Res}_0 \nabla$ on $\overline{\mathcal{H}}$ satisfy the condition $0 \leqslant \alpha < 1$ (N. Katz). (In our case, when Y_0 is reduced, the monodromy $T = T_{\mathrm{u}}$ is unipotent and hence $\alpha \in \mathbb{Z}$ since $\lambda = \mathrm{e}^{2\pi i\alpha} = 1$, and, consequently, $\overline{\mathcal{H}}$ coincides with the canonical extension $\mathcal{L}$ (6.3.5) for which $-1 < \alpha \leqslant 0$.) Sheaves $\overline{\mathscr{F}^p} = \overline{\mathscr{H}} \cap j_* \mathscr{F}^p$, where j: $S' \to S$, extending the filtration $\mathscr{F}^\cdot$ over S according to Schmid, coincide in our geometric case with sheaves $R^n f_* F^p \Omega^\cdot_{Y/S}(\log Y_0)$.

(3.1.3) The limit MHS in a natural way 'lives' in the vector space $H = \mathbb{H}^n(Y_0, \Omega^\cdot_{Y/S}(\log Y_0) \otimes \mathscr{O}_{Y_0})$ isomorphic to the zero fibre of the sheaf $\overline{\mathscr{H}}$. The Hodge filtration $F^\cdot$ on H is induced by the filtration $\overline{\mathscr{F}}^\cdot$ on $\overline{\mathscr{H}}$. It corresponds to a stupid filtration on $\Omega^\cdot_{Y/S}(\log Y_0) \otimes \mathscr{O}_{Y_0}$. The weight filtration $W_\cdot$ on H can be defined in the same way as in §2 (the weight filtration of the logarithm of unipotent part of the monodromy). Steenbrink proceeds in the following way. He defines a complex $A^\cdot$, quasiisomorphic to the complex $\Omega^\cdot_{Y/S}(\log Y_0) \otimes \mathscr{O}_{Y_0}$, and defines two filtrations $W_\cdot$ and $F^\cdot$ on it. Then he proves that $(A^\cdot, W, F)$ is a cohomological mixed Hodge complex in the sense of Deligne, which induces an MHS on $H = \mathbb{H}^n(Y_0, A^\cdot)$ coinciding with Schmid's limit MHS.

3.2 The limit MHS for projective families: the general case

Let π: $Y \to S$ be a family of projective varieties and $\pi^{-1}(0) = Y_0 = m_0 E_0 + \ldots + m_k E_k = E$ be a divisor with normal crossings. Let us perform a base change $\tilde{S} \xrightarrow{\rho} S$, $t = \tau^m$, where $m = \operatorname{lcm}(m_0, \ldots, m_k)$. Let $\tilde{Y}$ be a normalization of $Y \times_S \tilde{S}$, and $\tilde{\pi}$: $\tilde{Y} \to \tilde{S}$ and φ: $\tilde{Y} \to Y$ be natural mappings:

$$
(3.2.1)\quad \begin{array}{ccccccc}
\sum m_j E_j = Y_0 \subset Y & \xleftarrow{\varphi} & \tilde{Y} & \leftarrow & Y_\infty = Y' \underset{S'}{\times} U = \tilde{Y} \underset{\tilde{S}'}{\times} U \\
\pi\downarrow & & \downarrow\tilde{\pi} & & \downarrow \\
0 \in S_\rho & \leftarrow & \tilde{S} & \xleftarrow{\tilde{e}} & U.
\end{array}
$$

Let $D_j = \varphi^{-1}(E_j)$, $j = 0, \ldots, k$. Denote by $Y_\infty = Y' \times_{S'} U = \tilde{Y}' \times_{\tilde{S}'} U$ the canonical ('general') fibre of the family π. Here U is the upper half-plane, and $e = \rho \cdot \tilde{e}$: $U \to S'$, $t = e^{2\pi i u}$, is the universal covering.

(3.2.2) *Lemma* [S3] $\tilde{Y}$ is a V-manifold, i.e. it has at most finite quotient singularities, and $\pi^{-1}(0) = D = D_1 + \ldots + D_k$ is a reduced divisor with V-normal crossings on $\tilde{Y}$. ■

(3.2.3) Steenbrink extends the theory of the previous item [S1] to V-manifolds [S3]. The role of sheaves of differential forms $\tilde{\Omega}^p_{\tilde{Y}}$ on a V-manifold $\tilde{Y}$ is played by sheaves $(\alpha_x \Omega^p_z)^G$, where (locally) $\tilde{Y} = Z/G$, α: $Z \to \tilde{Y}$, i.e. sheaves of invariant differential forms on a smooth manifold Z covering $\tilde{Y}$. Analogously one defines sheaves $\tilde{\Omega}^p_{\tilde{Y}}(\log \tilde{Y}_0)$, $\tilde{\Omega}^p_{\tilde{Y}/\tilde{S}}$, $\tilde{\Omega}^p_{\tilde{Y}/\tilde{S}}(\log \tilde{Y}_0)$.

Projective V-manifolds behave in all respects as non-singular projective varieties. In spite of the fact that the $\tilde{Y}$ have weak singularities, the MHS on $H^n(\tilde{Y}_t, \mathbb{C})$ is pure of weight n, there is a Lefshetz decomposition and so on.

(3.2.4) The sheaf $\tilde{\bar{\mathscr{H}}} = \mathbb{R}^n\tilde{\pi}_*\tilde{\Omega}^{\cdot}_{\tilde{Y}/\tilde{S}}(\log \tilde{Y}_0)$ is locally free and gives a natural extension of the sheaf $\tilde{\mathscr{H}} = \rho^*\mathscr{H} = \mathbb{R}^n\tilde{\pi}_*\tilde{\Omega}^{\cdot}_{\tilde{Y}'/\tilde{S}'}$ to the point $0 \in \tilde{S}$. The limit MHS in a natural way 'lives' in the zero fibre of this sheaf $\tilde{\bar{\mathscr{H}}}_0 = \mathbb{H}^n(\tilde{Y}_0, \tilde{\Omega}^{\cdot}_{\tilde{Y}/\tilde{S}}(\log \tilde{Y}_0) \otimes_{\mathscr{O}_{\tilde{Y}}} \mathscr{O}_{\tilde{Y}_0})$. In the same way as in (3.1.3) one defines a complex $\tilde{A}^{\cdot}$ quasiisomorphic to the complex $\tilde{\Omega}^{\cdot}_{\tilde{Y}/\tilde{S}}(\log \tilde{Y}_0) \otimes_{\mathscr{O}_{\tilde{Y}}} \mathscr{O}_{\tilde{Y}_0}$ and also two filtrations $W_{\cdot}$ and $F^{\cdot}$, etc. We thus obtain an MHS on the space $\mathbb{H}^n(\tilde{Y}_0, \tilde{A}^{\cdot})$.

A choice of local coordinates t and τ, $\tau^m = t$, defines the isomophisms

(3.2.5) $$\mathbb{H}^n(\tilde{Y}_0, \Omega^{\cdot}_{\tilde{Y}/\tilde{S}}(\log \tilde{Y}_0) \otimes_{\mathscr{O}_{\tilde{Y}}} \mathscr{O}_{\tilde{Y}_0}) \rightleftarrows H^n(Y_\infty, \mathbb{C}) \leftrightarrows \mathbb{H}^n(Y_0, \Omega^{\cdot}_{Y/S}(\log Y_0) \otimes_{\mathscr{O}_Y} \mathscr{O}_{Y_0}).$$

This enables us to carry the MHS to the cohomology of the canonical fibre $H^n(Y_\infty, \mathbb{C})$ of the family π.

Denote by T and $\tilde{T} = T^m$ the monodromy operators on $H^n(Y_\infty, \mathbb{C})$ corresponding to generators of $\pi_1(S')$ and $\pi_1(\tilde{S}')$. Then the monodromy $\tilde{T} = \tilde{T}_u$ is unipotent. Denote the logarithm of the unipotent part of the monodromy by

(3.2.6) $$N = \log \tilde{T} = \frac{1}{m} \log T^m = \frac{1}{m} \sum_{k \geqslant 1} (-1)^{k+1} \frac{(T^m - 1)^k}{k}.$$

(3.2.7) *Theorem* [S3] There exists an MHS on $H^n(Y_\infty)$ such that

(i) N: $H^n(Y_\infty) \to H^n(Y_\infty)$ is an endomorphism of the MHS $(H^n(Y_\infty), W_{\cdot}, F^{\cdot})$ of type $(-1, -1)$, i.e. $N(W_k) \subset W_{k-2}$, $N(F^p) \subset F^{p-1}$.

(ii) For all $r \geqslant 0$,
$$N^r\colon Gr^W_{n+r} H^n(Y_\infty) \to Gr^W_{n-r} H^n(Y_\infty)(-r)$$
is a MHS isomorphism.

(iii) The semisimple part T_s of the monodromy is an MHS isomorphism.

■

Let us consider Hodge numbers of the MHS on $H^n(Y_\infty)$

(3.2.8) $$h^{p,q} = \dim Gr^p_F Gr^W_{p+q} H^n(Y_\infty).$$

By the monodromy theorem (A. Landman and others, cf §I.9) we have $N^{n+1} = 0$ and, consequently, from the definition of $W_{\cdot}$ we have

$$0 \subset W_0 \subset W_1 \subset \ldots \subset W_{2r-1} \subset W_{2n} = H^n(Y_\infty).$$

For the limit Hodge filtration $F^\cdot$ as well as for the Hodge filtration on the cohomology of a non-singular projective variety of dimension n we have

$$0 \subset F^n \subset F^{n-1} \subset \ldots \subset F^0 = H^n(Y_\infty).$$

From this and by the definition of Hodge numbers it follows that:

(3.2.9) $h^{p,q}$ not equal to zero are contained in a quadrate $[0, n] \times [0, n]$;

(3.2.10) $h^{p,q} = h^{q,p}$ (this is a symmetry relative to the diagonal – the bisector of the coordinate angle);

(3.2.11) $h^{p,q} = h^{n-q,n-p}$ (this is a symmetry relative to the second diagonal); it follows from the second part of theorem (3.2.7) because there is an isomorphism

$$N^{p+q-n}\colon Gr_F^p Gr_{p+q}^W \xrightarrow{\sim} Gr_F^{n-q} Gr_{2n-p-q}^W.$$

3.3 Brieskorn construction

Let $f\colon (\mathbb{C}^{n+1}, 0) \to (\mathbb{C}, 0)$ be an isolated singularity of a holomorphic function. Brieskorn [Br] proposed the following construction of an embedding of f to a family of projective varieties $\overline{\pi}\colon Y \to S$. The singularity f being isolated, we can assume after a change of coordinates that f is a polynomial. If we add a homogeneous polynomial $h(x_0, \ldots, x_n)$ of sufficiently large degree d to f, then, on one hand, we obtain a singularity analytically isomorphic to the initial singularity at $0 \in \mathbb{C}^{n+1}$ because f is isolated. On the other hand, if d is sufficiently large and h is chosen sufficiently general, then the hypersurface in $\mathbb{P}^{n+1}$ with the equation $f(x) = 0$ will have singularities only at the point $0 \in \mathbb{C}^{n+1}$ (the Bertini theorem). Let us consider a family of hypersurfaces $\overline{Y} \subset \mathbb{P}^{n+1} \times \mathbb{C}$ with affine equations $f(x_0, \ldots, x_n) = t$ or with projective equations

$$x_{n+1}^d f\left(\frac{x_0}{x}, \ldots, \frac{x_n}{x}\right) - tx_{n+1}^d = 0.$$

Let $\overline{\pi}\colon \overline{Y} \to \mathbb{C}$ be induced by projection. Then $\overline{\pi}$ has at most a finite number of degenerate fibres $\overline{Y}_t$, and, choosing a disk $S \ni 0$ of sufficiently small radius, we can assume that $\overline{\pi}$ has a single degenerate fibre Y_0 over S. We obtain the following picture

$$\text{(3.3.1)} \qquad \begin{array}{ccc} \mathbb{P}^{n+1} \times S \supset \overline{Y} & \supset & X \\ \overline{\pi}\downarrow & \swarrow f, & \\ S & & \end{array}$$

where $\overline{\pi}$ is a projective morphism, smooth over S', which has in the fibre $\overline{\pi}^{-1}(0) = \overline{Y}_0$ a single singular point $0 \in \overline{Y}_0$ analytically equivalent to the germ of hypersurface X_0: $f(x) = 0$, and $X = Y \cap B$ is the intersection with a ball B of small radius centred at $0 \in \mathbb{C}^{n+1}$, and f: $X \to S$ is the Milnor fibration of the singularity f.

3.4 Limit MHS on a vanishing cohomology

Let f: $(\mathbb{C}^{n+1}, 0) \to (\mathbb{C}, 0)$ be an isolated singularity. Embed it in a projective family by means of the Brieskorn construction (3.3.1). Let σ: $Y \to \overline{Y}$ be a resolution of the singularity of the map $\overline{\pi}$, i.e. $Y \backslash \sigma^{-1}(0) \overset{\sigma}{\rightrightarrows} \overline{Y} \backslash 0$ is an isomorphism, and the zero fibre of the map $\pi = \overline{\pi} \cdot \sigma$: $Y \to S$ is a divisor with normal crossings $\pi^{-1}(0) = Y_0 = E_0 + m_1 E_1 + \ldots + m_k E_k$. Then π: $E_o \to \overline{Y}_0$ is a good resolution of the singularity $(Y_0, 0)$ with exceptional divisor $C = \bigcup C_j$, where $C_j = E_0 \cap E_j$, $j = 1, \ldots, k$. We find ourselves with a projective family π: $Y \to S$, in the cohomology of the canonical fibre $Y_\infty = \overline{Y}_\infty$ of which the MHS was defined in (3.2.7),

$$\text{(3.4.1)} \qquad \begin{array}{c} \overline{Y}_0 \subset \overline{Y} \overset{\sigma}{\leftarrow} Y \supset Y_0 = E_0 + \sum_{j=1}^{k} m_j E_j \\ \overline{\pi} \searrow \downarrow \pi \\ S \end{array} .$$

Outside the ball $B \ni 0$ the map $\overline{\pi}$: $\overline{Y} \backslash B = \overline{Y} \backslash X \to S$ is a locally trivial C^∞-fibration. The fibre $X_0 = f^{-1}(0)$ is a cone over $K = X_0 \cap \partial B$ with a vertex at the point 0. Therefore the fibre $\overline{Y}_0$ has the homotopy type of a quotient space $\overline{Y}_t / X_t = Y_t / X_t$, $t \neq 0$. Thus $\tilde{H}^i(\overline{Y}_0) \simeq H^i(Y_t, X_t)$. The exact cohomology sequence of the pair (Y_t, X_t) gives an exact sequence:
(3.4.2)

$$0 \to \tilde{H}^n(\overline{Y}_0) \overset{r^*}{\to} H^n(Y_t) \overset{i^*}{\to} H^n(X_t) \to H^{n+1}(\overline{Y}_0) \to H^{n+1}(Y_t) \to 0$$

because X_t has the homotopy type of a bouquet of μ spheres S^n and $H^i(X_t) = 0$ for $i \neq 0$ and n. Here i^* is induced by the embedding i: $X \subset Y$, and r^* is induced by the retraction r: $\overline{Y} \to \overline{Y}_0$. Homomorphisms in this exact sequence are equivariant wrt the monodromy T because one may take the geometric monodromy h_t: $Y_t \to Y_t$ to be the identity outside X_t.

Let $Y_\infty = Y \times_{S'} U = \tilde{Y} \times_{\tilde{S}'} U$ and $X_\infty = X \times_{S'} U = \tilde{X} \times_{\tilde{S}'} U$ be the canonical fibres of the maps π (which is the same as of maps $\overline{\pi}$ or $\tilde{\pi}$) and f. The exact sequence (3.4.2) can be rewritten in the form

(3.4.3)

$$0 \to \tilde{H}^n(\overline{Y}_0) \xrightarrow{r^*} H^n(Y_\infty) \xrightarrow{i^*} H^n(X_\infty) \to H^{n+1}(\overline{Y}_0) \to H^{n+1}(Y_\infty) \to 0.$$

All the terms of this sequence except $H^n(X_\infty)$ already carry a MHS structure. $\overline{Y}_0$ is a projective variety with a single singular point $0 \in \overline{Y}_0$ and a complement $\overline{Y}_0 \backslash 0 \simeq E_0 \backslash C$, where C is a divisor with normal crossings. In the spirit of Deligne Steenbrink [S3] constructs an exact sequence $0 \to A^\cdot(\overline{Y}_0) \to A^\cdot \to A^\cdot(X_\infty) \to 0$ of cohomological mixed Hodge complexes, the hypercohomologies of which are $\tilde{H}^q(\overline{Y}_0)$, $H^q(Y_\infty)$ and $H^q(X_\infty)$. This implies the following theorem:

(3.4.4) *Theorem* The vanishing cohomology $H^n(X_\infty)$ carries an MHS such that the exact sequence (3.4.3) is an exact sequence of the MHS. The MHS on $H^n(X_\infty)$ is invariant wrt the semisimple part T_s of the monodromy. ■

(3.4.5) One can show [S3] that the MHS on $H^{n+1}(\overline{Y}_0)$ is pure of weight $n+1$.

3.5 The weight filtration on $H^n(X_\infty)$. Symmetry of Hodge numbers

Scherk and Steenbrink [Sc–S] showed that if in the Brieskorn construction we choose $\deg f = d$ sufficiently large, then i^* is an epimorphism, i.e. we have an exact MHS sequence

$$0 \to H^n(\overline{Y}_0) \xrightarrow{r^*} H^n(Y_\infty) \xrightarrow{i^*} H^n(X_\infty) \to 0, \tag{3.5.1}$$

and we can consider the MHS on $H^n(X_\infty)$ to be a quotient of the MHS on $H^n(Y_\infty)$.

(3.5.2) The weight filtration W on $H^n(X_\infty)$ as well as that on $H^n(Y_\infty)$ is the weight filtration of the nilpotent operator N logarithm of the unipotent part of monodromy T_u.

The local invariant cycle theorem [S1] asserts that the MHS sequence

$$H^n(\overline{Y}_0) \xrightarrow{r^*} H^n(Y_\infty) \xrightarrow{N} H^n(Y_\infty) \tag{3.5.3}$$

is exact (it is also a part of the Clemens–Schmid exact sequence). Therefore the image of r^* coincides with $\operatorname{Ker} N = \operatorname{Ker}(T_u - id) =$

$\operatorname{Ker} N \cap H^n(Y_\infty)_1$ (and $\operatorname{Ker} N \cap H^n(Y_\infty)_{\neq 1} = 0$). Here $H^n(Y_\infty) = \oplus_\lambda H^n(Y_\infty)_\lambda$ is a root decomposition corresponding to the eigenvalues λ of the monodromy T, $H^n(Y_\infty)_{\neq 1} = \oplus_{\lambda \neq 1} H^n(Y_\infty)_\lambda$. Since the MHS is invariant wrt the semisimple part of monodromy T_s, we have from (3.5.1):

(3.5.4) (1) $H^n(X_\infty)_{\neq 1} = H^n(Y_\infty)_{\neq 1}$ and the weight filtration $W_.$ on $H^n(X_\infty)_{\neq 1}$ as well as on $H^n(Y_\infty)$ is the weight filtration of the operator N with centre n;

(2) $H^n(X_\infty)_1 = H^n(Y_\infty)_1/\operatorname{Ker}(N)_1$ and the weight filtration $W_.$ on $H^n(X_\infty)_1$, as was noted in (2.7.5), is the weight filtration of the operator N with centre $n+1$ (!).

Now let $h^{p,q}$ be Hodge numbers of the MHS on $H^n(X_\infty)$

(3.5.5) $$h^{p,q} = \dim Gr_F^p Gr_{p+q}^W H^n(X_\infty).$$

We have a decomposition into a sum $h^{p,q} = \sum_\lambda h_\lambda^{p,q} = h_1^{p,q} + h_{\neq 1}^{p,q}$ over Hodge numbers corresponding to different eigenvalues of the monodromy T. As with $H^n(Y_\infty)$ the Hodge numbers $h^{p,q} \neq 0$ are contained in a quadrate $[0, n] \times [0, n]$.

There are two types of symmetry for Hodge numbers, and for the second type the symmetries for $\lambda = 1$ and $\lambda \neq 1$ are different:

(i) Since the Hodge filtration $F^\cdot$ induces on Gr_{p+q}^W an HS of weight $p+q$, we have

(3.5.6) $$h^{p,q} = h^{q,p},$$

or in more detail $h_\lambda^{p,q} = h_{\bar{\lambda}}^{q,p}$,

$$h_1^{p,q} = h_1^{q,p}, \; h_{\neq 1}^{p,q} = h_{\neq 1}^{q,p};$$

in the matrix $(h^{p,q})$ this is a symmetry relative to the accessory diagonal.

(ii) From the second part of theorem (3.7) we have

$$N^{p+q-n}\colon Gr_{p+q}^W H^n(X_\infty)_{\neq 1} \xrightarrow{\sim} Gr_{2n-p-q}^W H^n(X_\infty)_{\neq 1},$$

and (taking into account the shift of the central index)

$$N^{p+q-(n+1)}\colon Gr_{p+q}^W H^n(X_\infty)_1 \xrightarrow{\sim} Gr_{2(n+1)-p-q}^W H^n(X_\infty)_1.$$

We obtain isomorphisms

$$N^{p+q-n}\colon Gr_F^p Gr_{p+q}^W H^n(X_\infty)_{\neq 1} \xrightarrow{\sim} Gr_F^{n-q} Gr_{2n-p-q}^W H^n(X_\infty)_{\neq 1}$$

and

$$N^{p+q-(n+1)}\colon Gr_F^p Gr_{p+q}^W H^n(X_\infty)_1 \xrightarrow{\sim} Gr_F^{n+1-q} Gr_{2(n+1)-p-q}^W H^n(X_\infty)_1,$$

and, consequently,

(3.5.7) $$h^{p,q}_{\neq 1} = h^{n-q,n-p}_{\neq 1},$$

(3.5.8) $$h^{p,q}_{1} = h^{n+1-q,n+1-p}_{1}.$$

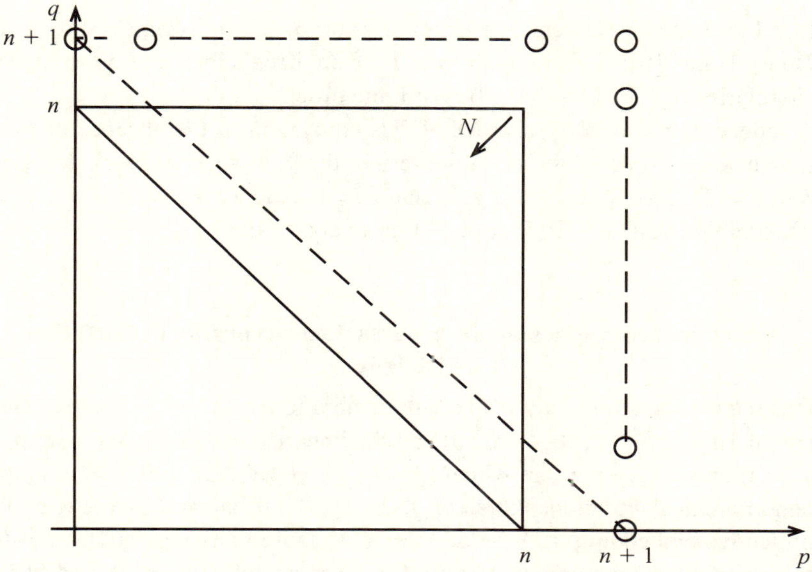

In the case of $h^{p,q}_{\neq 1}$ this is a symmetry relative to the main diagonal of the matrix $(h^{p,q})$, $0 \leqslant p,\ q \leqslant n$, and in the case of $h^{p,q}_{1}$ it is a symmetry relative to the main diagonal of a $(n+1) \times (n+1)$-matrix. It follows that $W_0 H^n(X_\infty)_1 = W_1 H^n(X_\infty)_1 = 0$ and we obtain the following more precise version of the monodromy theorem (I.9.1.1).

(3.5.9) *Theorem* The Jordan blocks of the monodromy T on $H^n(X_\infty)$ are of size at most $n+1$. The Jordan blocks for eigenvalue $\lambda = 1$ of T are of size at most n. ■

It is clear that the following conditions are equivalent: (i) there is a Jordan block of size $n+1$ (and therefore necessarily with $\lambda \neq 1$); (ii) $Gr^W_{2n} H^n(X_\infty)_{\neq 1} \neq 0$ (correspondingly: (i) there is a Jordan block of size n with $\lambda = 1$; (ii) $Gr^W_{2n} H^n(X_\infty)_1 \neq 0$).

Van Doorn and Steenbrink [DS] obtained the following:

(3.5.10) *Supplement to the monodromy theorem:*

$$Gr^W_{2n} H^n(X_\infty)_{\neq 1} \neq 0 \Rightarrow Gr^W_{2n} H^n(X_\infty)_1 \neq 0,$$

i.e. if the monodromy T on $H^n(X_\infty)$ has a Jordan block of size $n+1$

(necessary for an eigenvalue $\lambda \neq 1$), then T also has a Jordan block of size n for the eigenvalue $\lambda = 1$.

(3.5.11) This supplement is a generalization of the following result of Le Dung Trang [L]: The monodromy T of an irreducible one-dimensional singularity f: $(\mathbb{C}^2, 0) \to (\mathbb{C}, 0)$ is of finite order.

Indeed, the finiteness of order of T is equivalent to the absence of two-dimensional Jordan blocks. However, if there is such a block (of size $n+1=2$), then T has the eigenvalue $\lambda = 1$. And as is known, $f(x, y)$ is irreducible if and only if T has $\lambda = 1$ as an eigenvalue.

4 Hodge theory of a smooth hypersurface according to Griffiths–Deligne

Our main subject of study is the Milnor fibration f: $X \to S$ of an isolated singularity f: $(\mathbb{C}^{n+1}, 0) \to (\mathbb{C}, 0)$ and the linearization associated with it – a local system $\underline{H} \to S'$, in which $\underline{H} = \bigcup_{t \in S'} H^n(X_t, \mathbb{C}) = R^n f_* \mathbb{C}_{X'}$ is the cohomological fibration. Fibres of f, $X_t = f^{-1}(t)$, as well as fibres of its projective embedding π: $Y \to S$, $Y_t = \pi^{-1}(t)$, are non-singular (for $t \neq 0$) hypersurfaces in a complex manifold X (correspondingly, in Y), and fibres $\underline{H}_t = H^n(X_t, \mathbb{C})$ are a cohomology of such hypersurfaces. Therefore before we proceed, we consider how to calculate the cohomology of a hypersurface or its complement by means of differential forms on the ambient manifold with poles on this hypersurface and how the Hodge filtration is connected with the pole order filtration. The answer to this problem is supplied by Griffiths–Deligne theory ([Gr] or in the MHS context [D3]). Temporarily we change our notation.

4.1 The Gysin exact sequence

Let X be a complex manifold, $\dim X = n+1$, and $Y \subset X$ be a hypersurface defined by the (local) equation $s = 0$, $U = X \backslash Y$, and j: $U \subset X$ be an embedding. The cohomologies of X, Y and U are connected by an exact sequence

$$(4.1.1) \qquad \ldots \to H^{k-2}(Y, \mathbb{C}) \to H^k(X, \mathbb{C}) \to H^k(U, \mathbb{C}) \to H^{k-1}(Y) \to H^{k+1}(X) \to \ldots,$$

which is called the *Gysin sequence*.

From the topological point of view this sequence may be considered in one of the following three ways:

(1) As an exact sequence obtained from the degenerate Leray spectral sequence for the embedding $j\colon U \to X$

(4.1.2) $$E_2^{p,q} = H^p(X, R^q j_* \mathbb{C}_U) \Rightarrow H^{p+q}(U, \mathbb{C}_U).$$

In our case $j_*\mathbb{C}_U = \mathbb{C}_X$, $R^1 j_* \mathbb{C}_U = \mathbb{C}_Y$ (locally), and $R^q j_* \mathbb{C}_U = 0$ for $q > 1$ because for a ball neighborhood $V \subset X$ at a point $y \in Y$ the inverse image $j^{-1}(V) = V \backslash Y$ has the homotopy type of a circle S^1.

(2) As an exact cohomology sequence of a pair (X, U) or an exact local cohomology sequence

(4.1.3) $$\ldots \to H_Y^k(X) \to H^k(X) \to H^k(U) \to \ldots,$$

where $H_Y^k(X) \underset{\text{Df}}{=} H^k(X, U)$. The cohomology $H_Y^k(X, \mathbb{C})$ may be considered as derived functors of the functor $H_Y^0(X, \cdot)$ of sections with a support in Y of a sheaf on X. From a spectral sequence connecting sheaves of local cohomology $\mathscr{H}_Y^q(\mathscr{F})$ with local cohomology

(4.1.4) $$E_2^{p,q} = H^p(X, \mathscr{H}_Y^q(\mathscr{F})) \Rightarrow H_Y^{p+q}(X, \mathscr{F}).$$

In our case for $\mathscr{F} = \mathbb{C}_X$ we have $\mathscr{H}_Y^q(X, \mathbb{C}_X) = 0$ for $q \neq 2$ and $\mathscr{H}_Y^2(X, \mathbb{C}) \simeq \mathbb{C}_Y$ (locally) because for a neighborhood X of a point y we have $H_Y^q(X, \mathbb{C}_X) = H^q(X, X\backslash Y) = H^q(D^2, D^2\backslash\{0\})$, where D^2 is a 2-disk. The terms $E_2^{p,q}$ vanish except when $q = 2$ and this gives an identification

(4.1.5) $$H_Y^k(X, \mathbb{C}_X) = H^{k-2}(X, \mathbb{C}_Y) = H^{k-2}(Y, \mathbb{C}_Y).$$

(3) As an exact Gysin sequence associated with a spectral sequence of a fibration with fibre S^1. One needs to consider a tubular neighborhood of Y in X.

(4.1.6) *Remark* A non-singular hypersurface $Y \subset X$ is a particular case of a divisor with normal crossings in X considered by Deligne. The Gysin sequence (4.1.1) is a particular case of a Leray spectral sequence for an embedding j.

4.2 Hodge theory for a complement $U = X\backslash Y$. Hodge filtration and pole order filtration

Deligne considered a case in which Y is a divisor with normal crossings. Any non-singular quasiprojective variety U may be presented in the form $X\backslash Y$, where X is a non-singular projective variety and Y is a divisor with normal crossings. This enables us to introduce an MHS on $H^\cdot(U)$.

(4.2.1) By the Poincaré lemma we have a resolution $0 \to \mathbb{C}_U \to \Omega_U^\cdot$ and hence $H^{\cdot}(U, \mathbb{C}) = \mathbb{H}^{\cdot}(\Omega^{\cdot}_U)$. The embedding j is a Stein morphism (because Y is a hypersurface) and hence $\Omega^{\cdot}_U$ is a j_*-acyclic resolution, $R^q j_* \mathscr{F} = 0$ for $q > 0$ for coherent sheaves, and, consequently, $H^{\cdot}(U, \mathbb{C}) = \mathbb{H}^{\cdot}(j_* \Omega^{\cdot}_U)$.

Denote by $\Omega^{\cdot}_X(*Y) \subset j_* \Omega^{\cdot}_U$ a subcomplex of sheaves of meromorphic differential forms on X with poles along Y. Let $\Omega^{\cdot}_X(\log Y)$ be a complex of sheaves of differential forms on X with logarithmic poles along Y. If Y is a divisor with normal crossings, then these three complexes

$$\Omega^{\cdot}_X(\log Y) \subset \Omega^{\cdot}_X(*Y) \subset j_* \Omega^{\cdot}_U$$

are quasiisomorphic, i.e. their cohomology sheaves are isomorphic. Consequently,

$$H^{\cdot}(U, \mathbb{C}) \simeq \mathbb{H}^{\cdot}(\Omega^{\cdot}_X(\log Y)). \tag{4.2.2}$$

If $Y \subset X$ is a non-singular hypersurface, then there is an exact sequence of complexes of sheaves on X

$$0 \to \Omega^{\cdot}_X \to \Omega^{\cdot}_X(\log Y) \xrightarrow{R} j_* \Omega^{\cdot}_Y[-1] \to 0, \tag{4.2.3}$$

where R is the Poincaré residue which takes the form $\omega = \psi \wedge \mathrm{d}s/s \in \Omega^k_X(\log Y)$ to the form $\psi_{|Y} \in \Omega^{k-1}_Y$. The exact hypercohomology sequence associated with the sequence (4.2.3) is nothing but the exact Gysin sequence (4.1.1).

(4.2.4) *Remark* The inclusion $0 \subset \Omega^{\cdot}_X = W_0 \subset \Omega^{\cdot}_X(\log Y) = W_1$ in the case a non-singular hypersurface Y is a particular case of the weight filtration $W_{.}$ on $\Omega^{\cdot}_X(\log Y)$.

(4.2.5) Now we come to the Hodge filtration $F^{\cdot}$ on $H^{\cdot}(U)$. As is known, under the isomorphism $H^{\cdot}(U, \mathbb{C}) \simeq \mathbb{H}^{\cdot}(\Omega^{\cdot}_X(\log Y))$, filtration $F^{\cdot}$ on $H^{\cdot}(U, \mathbb{C})$ corresponds to a stupid filtration $\sigma_{\geqslant p}$ on the complex $\Omega^{\cdot}_X(\log Y)$, where for a complex $K^{\cdot}$ by definition $\sigma_{\geqslant p} K^{\cdot}$ is a complex $0 \to \ldots \to 0 \to K^p \to K^{p+1} \to \ldots$.

Now we define a *pole order filtration* $P^{\cdot}$ on the complex $\Omega^{\cdot}_X(*Y)$. First we define an increasing filtration $P_{.}$ on the zero term of this complex $\mathscr{O}_X(*Y)$:

$$\begin{array}{ccccccccc} 0 & \subset & \mathscr{O}_X(Y) & \subset & \mathscr{O}_X(2Y) & \subset \ldots \subset & \mathscr{O}_X((k+1)Y) & \subset \ldots \\ \| & & \| & & \| & & \| & \\ P_{-1} & & P_0 & & P_1 & & P_k\mathscr{O}_X(*Y). & \end{array} \tag{4.2.6}$$

Let $P^{\cdot}$ be the corresponding decreasing filtration, $P^k = P_{-k}$,

$$P^k \mathscr{O}_X(*Y) = \mathscr{O}_X((-k+1)Y),\ P^k = 0 \text{ for } k > 0. \tag{4.2.7}$$

A differentiation leads to an increase in the pole order of 1. Hence, to spread this filtration to $\Omega^{\cdot}_X(*Y)$ respecting d, it is natural to set

(4.2.8) $$P^k\Omega^p_X(*Y) = P^{k-p}\mathcal{O}_X(*Y)\otimes\Omega^p_X = \Omega^p_X((p-k-1)Y).$$

We obtain a filtration on $\Omega^{\cdot}_X(*Y)$ by subcomplexes P^k

(4.2.9)

$$\begin{array}{lllllll}
\vdots & \vdots & & \vdots & & & \\
\cup & \cup & & \cup & & & \\
P^{-1}: & \mathcal{O}_X(2Y) & \to & \Omega^1_X(3Y) & \to \dots & & \\
\cup & \cup & & \cup & & & \\
P^0: & \mathcal{O}_X(Y) & \to & \Omega^1_X(2Y) & \to \dots & & \to \Omega^{n+1}_X((n+2)Y) \to 0 \\
\cup & \cup & & \cup & & & \\
P^1: & 0 & \to & \Omega^1_X(Y) & \to \dots & & \\
\cup & & & & & & \\
P^2: & 0 & \to & 0 & \to \Omega^2_X(Y) \to \dots & & \\
\cup & & & & & & \\
\vdots & \dots & & & & & \\
\cup & & & & & & \\
P^k: & 0 & \to & 0 & \to \dots 0 \to \Omega^k_X(Y) \to \dots & & \to \Omega^{n+1}_X((n+2-k)Y) \to 0 \\
\cup & & & & & & \\
\vdots & & & & & &
\end{array}$$

It is obvious that on the subcomplex $\Omega^{\cdot}_X(\log Y) \subset \Omega^{\cdot}_X(*Y)$ the filtration $P^{\cdot}$ coincides with the stupid filtration $\sigma_{\geqslant p}$,

(4.2.10) $$P^p = \sigma_{\geqslant p}.$$

(4.2.11) *Theorem* The inclusion $(\Omega^{\cdot}_X(\log Y), \sigma_{\geqslant p}) \to (\Omega^{\cdot}_X(*Y), P^{\cdot})$ is a filtered quasiisomorphism. ■

4.3 De Rham complex of the sheaf $B_{[Y]X}$ and the cohomology of a hypersurface Y

We may use the inclusion $\Omega^{\cdot}_X(\log Y) \subset \Omega^{\cdot}_X(*Y)$ and embed the exact sequence (4.2.3) into a commutative diagram with exact lines

(4.3.1) $$\begin{array}{ccccccc}
0 \to \Omega^{\cdot}_X & \to & \Omega^{\cdot}_X(\log Y) & \xrightarrow{R} & j_*\Omega^{\cdot}_Y[-1] & \to 0 \\
\| & & b\cap & & \downarrow a & \\
0 \to \Omega^{\cdot}_X & \to & \Omega^{\cdot}_X(*Y) & \to & DR^{\cdot}(B_{[Y]X}) & \to 0,
\end{array}$$

where by definition

(4.3.2) $$DR^{\cdot}(B_{[Y]X}) \underset{\mathrm{Df}}{=} \Omega^{\cdot}_X(*Y)/\Omega^{\cdot}_X = \Omega^{\cdot}_X \underset{\mathcal{O}}{\otimes} (\mathcal{O}_X(*Y)/\mathcal{O}_X).$$

The quotient complex $DR^{\cdot}(B_{[Y]X})$ is the De Rham complex of the sheaf of the principle parts of meromorphic functions with poles along Y

(4.3.3) $$B_{[Y]X} = \mathcal{O}_X(*Y)/\mathcal{O}_X.$$

The mapping a is defined by commutativity: if $\omega \in \Omega^p_Y$ is a p-form on Y (locally), then first we lift it to $\Omega^p_X(\log Y)$ getting a $(p+1)$-form $\tilde{\omega} \wedge \mathrm{d}s/s$, where $\tilde{\omega}$ is an extension of ω to X, and then we consider a class of the form $\tilde{\omega} \wedge \mathrm{d}s/s \in \Omega^{p+1}(*Y) \operatorname{mod} \Omega^{p+1}_X$ as an element $a(\omega) \in DR^{p+1}(B_{[Y]X})$. The pole order filtration on $\Omega^{\cdot}_X(*Y)$ induces a pole order filtration $P^{\cdot}$ on $DR^{\cdot}(B_{[Y]X})$. Since the homomorphism b is a filtered quasiisomorphism (4.2.11), we obtain the following theorem:

(4.3.4) *Theorem* The homomorphism

$$a\colon (j_*\Omega^{\cdot}_Y[-1], \sigma_{\geqslant}) \to (DR^{\cdot} B_{[Y]X}, P^{\cdot})$$

is a filtered quasiisomorphism. ∎

(4.3.5) *Corollary*

$$H^k(Y, \mathbb{C}) \simeq \mathbb{H}^{k+1}(X, DR^{\cdot}(B_{[Y]X}),$$

and under this isomorphism the filtration F^p on $H^k(Y, \mathbb{C})$ is induced by the filtration P^{p+1} on the complex $DR^{\cdot}(B_{[Y]X})$.

4.4 The case of a smooth hypersurface Y in a projective space $X = \mathbb{P}^{n+1}$

On the cohomology of the complement $U = \mathbb{P}^{n+1}\backslash Y$

$$H^d(U, \mathbb{C}) = \mathbb{H}^d(X, \Omega^{\cdot}_X(\log Y)) = \mathbb{H}^d X, \Omega^{\cdot}_X(*Y))$$

the Hodge filtration $F^{\cdot}$ is induced by the stupid filtration $\sigma_{\geqslant}$ on $\Omega^{\cdot}_X(\log Y)$ or by the pole order filtration $P^{\cdot}$ on $\Omega^{\cdot}_X(*Y)$, i.e. subspaces $F^pH^d(U)$ are images of spaces $\mathbb{H}^d(X, P^p\Omega^{\cdot}_X(*Y)$, where

(4.4.1) $$P^p\Omega^{\cdot}_X(*Y)\colon 0 \to \ldots \to 0 \to \Omega^p_X(Y) \to \Omega^{p+1}_X(2Y) \to \ldots \to \Omega^d_X((d-p+1)Y) \to \ldots$$

($\Omega^p_X(Y)$ stands on the pth place).

By Bott's theorem

(4.4.2) $$H^i(X, \Omega^p_X(kY) = 0 \text{ for } k \geqslant 1,\ i \geqslant 1,\ p \geqslant 0.$$

This means that the complex $P^p\Omega^{\cdot}_X(*Y)$ is acyclic wrt the functor of sections, and hence to calculate the hypercohomology we may take the corresponding complex of global sections

(4.4.3) $$\mathbb{H}^d(X, P^p\Omega^{\cdot}_X(*Y) = H^d(\Gamma(X, P^p\Omega^{\cdot}_X(*Y)).$$

It follows that the cohomology classes $c \in F^pH^d(U)$ are represented by closed d-forms $\alpha \in H^0(X, \Omega^d_X((d-p+1)Y)$ with poles of order $\leqslant d-p+1$.

Let us consider the middle cohomology $H^n(Y, \mathbb{C})$ of a hypersurface Y, $\dim Y = n$. In the Gysin sequence (4.1.1)

$$\ldots \to H^{n-1}(Y) \to H^{n+1}(\mathbb{P}^{n+1}) \to H^{n+1}(U) \xrightarrow{R} H^n(Y) \xrightarrow{g} H^{n+2}(\mathbb{P}^{n+1}) \to \ldots$$

the homomorphisms g sends c to $h \wedge c$ where h is the class of the hyperplane section. Denote by $P^k(Y, \mathbb{C}) \subset H^k(Y, \mathbb{C})$ the primitive cohomology of Y. Then $P^n(Y, \mathbb{C}) = \operatorname{Ker} g$ and we get an isomorphism $H^{n+1}(U) \overset{R}{\rightleftarrows} P^n(Y)$. From the above it follows that the Hodge filtration on $P^n(Y)$ is of the following sort:

(4.4.4)

$$\begin{array}{ccccccccccc} H^{n+1}(U) & = & \mathscr{H}^{n+1}_{n+1}(Y) & \supset & \mathscr{H}^{n+1}_{n}(Y) & \supset \ldots \supset & \mathscr{H}^{n+1}_{n-p+1} & \supset \ldots \supset & \mathscr{H}^{n+1}_{1}(Y) & \supset & 0 \\ R \downarrow\wr & & \downarrow\wr & & \downarrow\wr & & \downarrow\wr & & \wr & & \\ P^n(Y) & = & F^0 P^n(Y) & \supset & F^1 P^n(Y) & \supset \ldots \supset & F^p P^n(Y) & \supset \ldots \supset & F^n P^n(Y) & \supset & 0, \end{array}$$

where

$$\begin{aligned} \mathscr{H}^{n+1}_{n-p+1} &= H^{n+1}(\Gamma(\mathbb{P}^{n+1}, P^{p-1}\Omega^{\cdot}_{\mathbb{P}^{n+1}}(*Y))) \\ &= H^0(\Omega^{n+1}_{\mathbb{P}^{n+1}}((n-p+1)Y))/\mathrm{d}H^0(\Omega^n_{\mathbb{P}^{n+1}}((n-p)Y)). \end{aligned} \qquad (4.4.5)$$

4.5 Generalization to the case of a hypersurface with singularities

(4.5.1) Let X be a complex manifold, $\dim X = n+1$, and let $Y \subset X$ be a hypersurface, possibly with singularities, $U = X \backslash Y$. In the same way as in (4.2.1) we have $H^{\cdot}(U, \mathbb{C}) \simeq \mathbb{H}^{\cdot}(\Omega^{\cdot}_X(*Y))$. There is a pole order filtration $P^{\cdot}$ (4.2.9) on the complex $\Omega^{\cdot}_X(*Y)$, which induces a *pole order filtration* $P^{\cdot}$ on the cohomology of the complement $H^{\cdot}(U, \mathbb{C})$. We want to compare this with the Hodge filtration $F^{\cdot}$ from MHS theory.

In the global situation of (4.2.11) and (4.4.1) we make the following generalization:

(4.5.2) *Theorem* [D–Di] If Y is an hypersurface in a projective space $X = \mathbb{P}^{n+1}$ or even in a weighted projective space $\mathbb{P} = \mathbb{P}(w_0, \ldots, w_{n+1})$ and $F^{\cdot}$ is the Hodge filtration on $H^{\cdot}(U, \mathbb{C})$ ([D3]) then for all k and m we have

$$P^k H^m(U) \supset F^k H^m(U). \qquad \blacksquare$$

If a hypersurface Y is (quasi)smooth, then $P^{\cdot} = F^{\cdot}$. In general, this is not the case (see, for example, (VI.1.33) in [Di]).

The position is not quite the same in the local case as it is in the global one:

(4.5.3) Let $f\colon X \to S$ be the Milnor fibration of a hypersurface singularity and $Y = X_0 = f^{-1}(0)$ be the singular fibre. In this case $U = X \backslash Y$ plays the role of a knot of the singularity. Usually the intersection $K = f^{-1}(0) \cap S^{2n+1}$ with a sphere $S^{2n+1} = \partial B^{2n}$ is called a knot of a singularity. The pair (X, Y) is homeomorphic to the cone on (S^{2n+1}, K) with the vertex $x_0 = 0 \in X$. Hence, $Y \backslash x_0$ and $X \backslash Y$ are homotopy equivalent to K and $S^{2n+1} \backslash K$. Applying the Alexander duality to $K \subset S^{2n+1}$, we obtain

$$H^{i+1}(U) \simeq H^i(Y \backslash x_0) = H^i(K).$$

Thus from the homological point of view we may consider $U = X \backslash Y$ as a knot of the singularity.

(4.5.4) Embedding Y in a projective variety as in (3.3.1) and applying the excision axiom, we may assume that Y is a projective variety. In this case there is an MHS on $H^{\cdot}(Y \backslash x_0)$ ([D2]). Let $F^{\cdot}$ be the Hodge filtration of this MHS. The filtrations $F^{\cdot}$ and $P^{\cdot}$ are connected by the following relation:

(4.5.5) *Theorem* ([Ka1, Di2]) If $f\colon X \to S$ is the Milnor fibration of an isolated singularity $f\colon (\mathbb{C}^{n+1}, 0) \to (\mathbb{C}, 0)$ and $F^{\cdot}$ and $P^{\cdot}$ are the Hodge and the pole order filtrations on $H^{\cdot}(X \backslash X_0, \mathbb{C})$, then

(i) for all k:

$$P^k H^{n+1}(X \backslash X_0) \supset F^k H^{n+1}(X \backslash X_0);$$

(ii) for all k

$$P^k H^n(X \backslash X_0) \subset F^k H^n(X \backslash X_0). \qquad \blacksquare$$

5 The Gauss–Manin system of an isolated singularity

We now study the Gauss–Manin connection on the sheaf $\mathcal{H} = \underline{H} \otimes_{\mathbb{C}} \mathcal{O}_{S'}$, $\underline{H} = R^n f_* \mathbb{C}_{X'} = \bigcup_{t \in S'} H^n(X_t, \mathbb{C})$, associated with the Milnor fibration $f\colon X \to S$ of an isolated singularity $f\colon (\mathbb{C}^{n+1}, 0) \to (\mathbb{C}, 0)$. A fibre of the sheaf $\mathcal{H}$ is the cohomology $H^n(X_t)$ of a non-singular hypersurface $X_t \subset X$. In the previous section we saw how one can study the cohomology of the hypersurface by means of differential forms on X with poles along X_t, $H^n(X_t, \mathbb{C}) \simeq \mathbb{H}^{n+1}(X, DR^{\cdot}(B_{[X_t]X}))$. Now we want to extend this to the relative case – to a family of hypersurfaces X_t, $t \in S'$, and then to apply this to get a natural extension of the sheaf $\mathcal{H}$ onto the whole disk S.

In this section we try to explain of the appearance of the differential system $\mathcal{H}_X$ in [Sc-S], without using the language of the theory of D-modules.

5.1 Hodge theory of a smooth hypersurface in the relative case

Let $f: X \to S$ be a morphism of a complex manifold X, $\dim X = n + 1$, to a curve S. For any $t \in S$ we have an embedding of a hypersurface $f^{-1}(t) = X_t \subset X$. Varying t we glue the fibres X_t together into a graph Γ in the manifold $Z = X \times S$. A graph $\Gamma \subset X \times S$ is a smooth hypersurface in Z defined by the equation $f(x) - t = 0$. The mapping f factors to a composition of an embedding $i: X \subset Z$ onto the graph and a projection p

(5.1.1)
$$\begin{array}{c} i: X \tilde{\to} \Gamma \subset Z = X \times S \\ f \searrow \ \ \overline{p} \searrow \ \ \downarrow p \\ S \end{array}$$

The intersection of the fibre $p^{-1}(t) = X \times \{t\}$ with the hypersurface Γ is the fibre $X_t = f^{-1}(t)$, and the pair $X_t \subset X$ is isomorphic to the pair $\Gamma_t \subset Z_t$.

The sheaf

(5.1.2) $$B_{[f]X\times S} = B_{[\Gamma]Z} = \mathcal{O}_Z(*\Gamma)/\mathcal{O}_Z$$

of principal parts of the sheaf of meromorphic functions $\mathcal{O}_Z(*\Gamma) = \mathcal{O}_Z[1/[f(x) - t]]$ with poles along Γ: $f(x) - t = 0$ on Z is a relative analog of the sheaf $B_{[X_t]X}$ in (4.3.3). Consider the relative De Rham complex of this sheaf

(5.1.3) $$DR^{\cdot}_{Z/S}(B_{[\Gamma]Z}) \underset{\text{def}}{=} \Omega_{Z/S} \otimes_{\mathcal{O}_Z} B_{[\Gamma]Z} = \Omega^{\cdot}_{Z/S}(*\Gamma)/\Omega^{\cdot}_{Z/S}.$$

(5.1.4) *Proposition* If $f: X \to S$ is smooth, then
$$R^k f_* \mathbb{C}_X \otimes \mathcal{O}_S \simeq \mathbb{R}^{k+1} p_*(DR^{\cdot}_{Z/S}(B_{[\Gamma]Z}))$$
(this is a relative variant of corollary (4.3.5)).

Indeed, if f is smooth, then the sheaves $\Omega^p_{\Gamma/S}$ are locally free and we have the commutative diagram

$$\begin{array}{ccccccc} 0 \to \Omega^{\cdot}_{Z/S} & \to & \Omega^{\cdot}_{Z/S}(\log\Gamma) & \xrightarrow{R} & i_*\Omega^{\cdot}_{\Gamma/S}[-1] & \to 0 \\ \| & & b\cap & & \downarrow a & \\ 0 \to \Omega^{\cdot}_{Z/S} & \to & \Omega^{\cdot}_{Z/S}(*\Gamma) & \to & DR_{Z/S}(B_{[\Gamma]Z}) & \to 0 \end{array}$$

generalizing diagram (4.3.1) to the relative case. In this diagram a is a quasiisomorphism. From this and from the relative Poincaré lemma (for smooth f there is a resolution $0 \to f^{-1}\mathcal{O}_S \to \Omega^{\cdot}_{X/S}$ of the sheaf $f^{-1}\mathcal{O}_S$) we obtain

$$R^k f_* \mathbb{C}_X \otimes_{\mathbb{C}_S} \mathcal{O}_S = R^k f_*(f^{-1}\mathcal{O}_S) = R^k \overline{p}_*(\overline{p}^{-1}\mathcal{O}_S)$$
$$= \mathbb{R}^k \overline{p}_* \Omega^{\cdot}_{\Gamma/S} = \mathbb{R}^{k+1} p_*(DR^{\cdot}_{Z/S}(B_{[\Gamma]Z})). \quad \blacksquare$$

5.2 The Gauss–Manin differential system

In the general case, when f is not necessarily smooth, we define a complex

(5.2.1) $$\int_f^{\cdot} \mathcal{O}_X = \mathbb{R}p_*(DR^{\cdot}_{Z/S}(B_{[\Gamma]Z}))[n+1].$$

Its cohomology sheaves

(5.2.2) $$\int_f^k \mathcal{O}_X = \mathbb{R}^k p_*(DR^{\cdot}_{Z/S}(B_{[\Gamma]Z}))[n+1]$$
$$= \mathbb{R}^{n+1+k} p_*(DR^{\cdot}_{Z/S}(B_{[\Gamma]Z}))$$

are called the (kth) *Gauss–Manin differential systems of a morphism f.*

In particular, in the case of isolated singularities we are interested in the sheaf

(5.2.3) $$\mathcal{H}_X = \int_f^0 \mathcal{O}_X = \mathbb{R}^{n+1} p_*(DR^{\cdot}_{Z/S}(B_{[\Gamma]Z})),$$

where $f\colon X \to S$ is the Milnor fibration. This sheaf is called the *Gauss–Manin differential system of the singularity f.*

Since the morphism $f_{|X'}\colon X' \to S'$ is smooth, it follows from proposition (5.1.4) that

(5.2.4) $$\mathcal{H}_{X|S'} = R^n f_* \mathbb{C}_{X'} \otimes_{\mathbb{C}} \mathcal{O}_{S'} = \mathcal{H},$$

i.e. the sheaf $\mathcal{H}_X$ is a natural extension of the sheaf $\mathcal{H}$ to the point $0 \in S$.

(5.2.5) *Proposition* If f is a Steinian morphism, for example, if f is the Milnor fibration of a singularity, then

$$\mathcal{H} \simeq \mathcal{H}^{n+1}(p_*(DR^{\cdot}_{Z/S}(B_{[\Gamma]Z})),$$

where $\mathcal{H}^k(.)$ denotes the kth cohomology of a complex of sheaves.

A standard deduction of formula (5.2.5) from definition (5.2.3) is the following. Since $f = \overline{p}$ is Stein, we obtain by Cartan's theorem B that the complex $DR^{\cdot}_{Z/S}(B_{[\Gamma]Z})$ is acyclic relative the direct image functor. This enables us to calculate hypercohomology by applying p_* directly to this complex. However, here the sheaves $\Omega^q_{Z/S}(*\Gamma)/\Omega^q_{Z/S}$ are only quasicoherent (not coherent) $\mathcal{O}_Z$-modules, and Cartan's theorem B does not hold in general for quasicoherent sheaves, but one can overcome this difficulty [Sc-S].

5.3 Interpretation of the complex $DR^{\cdot}_{Z/S}(B_{[\Gamma]Z})$ in terms of the morphism $f\colon X \to S$

The De Rham complex $DR^{\cdot}_{Z/S}(B_{[\Gamma]Z}) = \Omega^{\cdot}_{Z/S}(*\Gamma)/\Omega^{\cdot}_{Z/S}$ (5.1.3) is concentrated on the graph Γ because $\Omega^{\cdot}_{Z/S}(*\Gamma)$ and $\Omega^{\cdot}_{Z/S}$ coincide outside Γ. By means of the isomorphism $i\colon X \xrightarrow{\sim} \Gamma \subset Z = X \times S$ we carry to X the complex $DR^{\cdot}_{Z/S}(B_{[\Gamma]Z})$ and all its natural structures (i.e. differential d, the structure of the D_S-module, the pole order filtration inducing an HS on fibres of f in the case of a projective morphism).

Obviously, we have an isomorphism

$$\Omega^{\cdot}_X\left[\frac{1}{f(x)-t}\right] \rightrightarrows \mathrm{i}^{-1}\Omega^{\cdot}_{Z/S}(*\Gamma)/\Omega^{\cdot}_{Z/S}. \tag{5.3.1}$$

Indeed, if a q-form $\omega/(f(x)-t)^k$ represents an element of the right-hand side, where ω is a q-form with coefficients depending on x and t, then changing t by $f(x)-(f(x)-t)$ and redecomposing coefficients we can assume that $\omega \in \Omega^q_X$.

Complex (5.3.1) naturally is a complex of D_S-modules, where on a form $\omega/(f(x)-t)^k$, $\omega \in \Omega^q_X$, the operator ∂_t acts as a differentiation wrt the parameter t:

$$\partial_t\left(\frac{\omega}{(f(x)-t)^k}\right) = k\frac{\omega}{(f(x)-t)^{k+1}}. \tag{5.3.2}$$

This implies that

$$\partial_t^i\left(\left[\frac{\omega}{f(x)-t}\right]\right) = \left[\frac{i!\omega}{(f(x)-t)^{i+1}}\right] \tag{5.3.3}$$

(here the brackets [ ] mean the class of an element in the quotient complex in the right-hand side of formula (5.3.1)).

Formula (5.3.3) shows that we can consider the complex $\Omega^{\cdot}_X[1/(f(x)-t)]$, elements of which are polynomials in $1/(f(x)-t)$, as a complex $\Omega^{\cdot}_X[D]$, elements of which are polynomials in a formal indeterminate D interpreted as the differentiation operator ∂_t, i.e. there is an isomorphism

$$\Omega^{\cdot}_X[D] \simeq \Omega^{\cdot}_X\left[\frac{1}{f(x)-t}\right], \tag{5.3.4}$$

$$\omega D^i \mapsto \partial_t^i\left(\left[\frac{\omega}{f(x)-t}\right]\right) = \left[\frac{i!\omega}{(f(x)-t)^{i+1}}\right].$$

The structure of the D_S-module $\Omega^{\cdot}_X[1/(f(x)-t)]$ carried to $\Omega^{\cdot}_X[D]$ is of the following form. Since

$$\partial_t\left(\frac{i!\omega}{(f(x)-t)^{i+1}}\right)=\frac{(i+1)!\omega}{(f(x)-t)^{i+2}},$$

$$t\cdot\frac{i!\omega}{(f(x)-t)^{i+1}}=(f(x)-(f(x)-t))\frac{i!\omega}{(f(x)-t)^{i+1}}$$

$$=f(x)\frac{i!\omega}{(f(x)-t)^{i+1}}-i\frac{(i-1)!\omega}{(f(x)-t)^{i}},$$

the action of the ring D_S on $\Omega^{\cdot}_X[D]$ is given by the formulae

(5.3.5) $$\partial_t\cdot\omega D^i=\omega D^{i+1},$$
$$t\cdot\omega D^i=f(x)\omega D^i-i\omega D^{i-1}.$$

The differential d on the complex $\Omega^{\cdot}_X[1/(f(x)-t)]$

$$\mathrm{d}\left(\frac{i!\omega}{(f(x)-t)^{i+1}}\right)=\frac{i!\omega}{(f(x)-t)^{i+1}}-\frac{(i+1)!\mathrm{d}f\wedge\omega}{(f(x)-t)^{i+2}}$$

($\mathrm{d}(f(x)-t)=\mathrm{d}f-\mathrm{d}t$, but $\mathrm{d}t=0$ in $\Omega^{\cdot}_{Z/S}$) transforms to a differential d on the complex $\Omega^{\cdot}_X[D]$ given by the formula

(5.3.6) $$\underline{d}(\omega D^i)=\mathrm{d}\omega\cdot D^i-\mathrm{d}f\wedge\omega D^{i+1}.$$

The pole order filtration $P^{\cdot}$ on complex (5.3.1) (see §4) transforms to a filtration $F^{\cdot}$ on $\Omega^{\cdot}_X[D]$ defined below in proposition (5.3.11).

Taking into account the identification (5.3.1) of the complex $DR^{\cdot}_{Z/S}(B_{[\Gamma]Z})$, concentrated on the graph Γ, with the complex $\Omega^{\cdot}_X[D]$ on X, we obtain an expression of the Gauss–Manin differential system (5.2.2) in terms of the complex $\Omega^{\cdot}_X[D]$

(5.3.7) $$\int_f^k\mathscr{O}_X=\mathbb{R}^{n+1+k}p_*(DR^{\cdot}_{Z/S}(B_{[\Gamma]Z}))$$
$$=\mathbb{R}^{n+1+k}f_*(i^{-1}DR^{\cdot}_{Z/S}(B_{[\Gamma]Z}))$$
$$=\mathbb{R}^{n+1+k}f_*(\Omega^{\cdot}_X[D]).$$

Let $f\colon X\to S$ be the Milnor fibration of a singularity $f\colon(\mathbb{C}^{n+1},0)\to(\mathbb{C},0)$. Then by proposition (5.2.5) we have

(5.3.8) $$\int_f^k\mathscr{O}_X=\mathscr{H}^{n+1+k}(f_*\Omega^{\cdot}_X[D]).$$

Consider the complex $(f_*\Omega^{\cdot}_X[D],\underline{\mathrm{d}})=(K^{\cdot},\underline{\mathrm{d}})$. From formula (5.3.6) it follows that this complex can be considered as the single complex associated with the double complex $(K^{\cdot\cdot},d_1,d_2)$, where

(5.3.9) $$K^{pq}=f_*\Omega_X^{p+q}D^q,\ q\geqslant 0,\ K^m=f_*\Omega_X^m[D],$$

and the differentials are: $\mathrm{d}_1 = \mathrm{d}$ is exterior differentiation, $\mathrm{d}_2 = -\mathrm{d}f\wedge$ is a Koszul differential, i.e. exterior multiplication by the form $-\mathrm{d}f$

(5.3.10)

$$
\begin{array}{llllllll}
 & 0 & & & & & & \\
 & \uparrow & & & & & & \\
\ldots\to & f_*\Omega_X^{n+1} & \to 0 & & & & & \\
 & \uparrow & \uparrow & & & & & \\
\ldots\to & f_*\Omega_X^{n} & \to f_*\Omega_X^{n+1} \to 0 & & & & & \\
 & \uparrow & & \searrow & & & & \\
 & \vdots & & & & & & \\
 & \uparrow & & & & & & \\
\ldots\to & f_*\Omega_X^{p} & \to\ldots & & \ldots f_*\Omega_X^{n+1} \to 0 & & & \\
 & \uparrow & \searrow & & \uparrow & \searrow & & \\
 & \vdots & & & \vdots & \ldots \xrightarrow{\mathrm{d}} & f_*\Omega_X^{n+1} & \to 0 \\
 & \uparrow -\mathrm{d}f\wedge & & & \uparrow & & \uparrow -\mathrm{d}f\wedge & \nwarrow \partial_t \\
0\to f_*\Omega_X^0 \to & f_*\Omega_X^1 & \xrightarrow{\mathrm{d}}\ldots & \searrow & \to f_*\Omega_X^{p+1} \to \ldots & \xrightarrow{\mathrm{d}} & f_*\Omega_X^{n} \xrightarrow{\mathrm{d}} & f_*\Omega_X^{n+1} \to 0 \\
 & \uparrow -\mathrm{d}f\wedge & & \searrow & \uparrow -\mathrm{d}f\wedge & & \uparrow -\mathrm{d}f\wedge & \uparrow -\mathrm{d}f\wedge \\
0\to & f_*\Omega_X^0 \xrightarrow{\mathrm{d}} & f_*\Omega_X^1 \xrightarrow{\mathrm{d}} \ldots \xrightarrow{\mathrm{d}} & f_*\Omega_X^{p} & \xrightarrow{\mathrm{d}} f_*\Omega_X^{p+1} \ldots & & f_*\Omega_X^{n-1} \to & f_*\Omega_X^{n} \xrightarrow{\mathrm{d}} f_*\Omega_X^{n+1} \to 0 \\
 & & & \searrow & & & & \\
 & & & & f_*\Omega_X^p[D]. & & &
\end{array}
$$

We summarize:

(5.3.11) *Proposition* Let $f\colon X \to S$ be the Milnor fibration of a singularity f. Then the Gauss–Manin differential system $\int_f^k \mathcal{O}_X = \mathcal{H}^{n+1+k}(f_*\Omega^{\cdot}_X[D])$ is the $(n+1+k)$th cohomology of the single complex $(f_*\Omega^{\cdot}_X[D], \underline{\mathrm{d}})$ associated with the double complex (5.3.10), $(K^{pq} = f_*\Omega_X^{p+q}D^q, \mathrm{d}, -\mathrm{d}f\wedge)$, and, in particular

$$\text{(5.3.12)} \qquad \mathcal{H}_X = \int_f^0 \mathcal{O}_X = \mathcal{H}^{n+1}(f_*\Omega^{\cdot}_X[D]).$$

The complex $f_*\Omega^{\cdot}_X[D]$ is a complex of D_S-modules with operators of multiplication by ∂_t and t defined by formulae (5.3.5). Under the identification (5.3.4) the pole order filtration $P^{\cdot}$ corresponds to the Hodge filtration $F^{\cdot}$ on the complex $f_*\Omega^{\cdot}_X[D]$, the second filtration of the double complex (5.3.10), $F^pK^{\cdot} = \oplus_{r\geqslant p+1}K^{r,\cdot}$,

$$\text{(5.3.13)} \quad F^pf_*\Omega_X^{n+1}[D] = f_*\Omega_X^{n+1} + f_*\Omega_X^{n+1}D + \ldots + f_*\Omega_X^{n+1}D^{n-p}.$$

This filtration induces a filtration $F^{\cdot}$ on $\mathcal{H}_X$. ■

The operator $\partial_t = D$ corresponds to a shift along the diagonal of the double complex $K^{\cdot\cdot}$. It takes F^p to F^{p-1},

(5.3.14) $$\partial_t(F^p) \subset F^{p-1}.$$

5.4 Connection between the differential system $\mathscr{H}_X$ and the Brieskorn lattice $\mathscr{H}^{(0)}$

The Brieskorn lattice $\mathscr{H}^{(0)} = {}''\mathscr{H} = f_*\Omega_X^{n+1}/\mathrm{d}f \wedge \mathrm{d}(f_*\Omega_X^{n-1})$ of an isolated singularity f is defined in (I.5.1.3). It is a locally free $\mathscr{O}_S$-module of rank μ extending $\mathscr{H}$ onto S. Obviously, $F^nK^\cdot = K^{n+1,0}$ coincides with $f_*\Omega_X^{n+1}$ ($F^\cdot$ is the second filtration of the double complex (5.3.10)).

(5.4.1) *Lemma* The inclusion $f_*\Omega_X^{n+1} \subset K^\cdot$ onto $F^nK^\cdot$ gives an isomorphism

$$\mathscr{H}^{(0)} \xrightarrow{\sim} F^n\mathscr{H}_X = F^n\mathscr{H}^{n+1}(K^\cdot).$$

Proof To prove this we have to find those elements $\omega \in f_*\Omega_X^{n+1} = F^nK^\cdot$ which are cohomologous to 0 in $K^\cdot$, $\omega = \underline{\mathrm{d}}\chi$. An easy search of the diagram (5.3.10) shows that these are exactly those ω having the form $\omega = \mathrm{d}f \wedge \mathrm{d}\theta$, where $\theta \in f_*\Omega_X^{n+1}$. ■

(5.4.2) *Lemma* The operator ∂_t is invertible on $\mathscr{H}_X$, $\partial_t\colon \mathscr{H}_X \xrightarrow{\sim} \mathscr{H}_X$. The Hodge filtration on $\mathscr{H}_X$ coincides with the filtration by powers of the operator ∂_t applied to the lattice $\mathscr{H}^{(0)}$:
$F^n\mathscr{H}_X = \mathscr{H}^{(0)}$, $F^{n-1}\mathscr{H}_X = \partial_tF^n\mathscr{H}_X, \ldots, F^{n-k}\mathscr{H}_X = \partial_t^kF^n\mathscr{H}_X, \ldots$

Proof This again is reduced to a diagram search in diagram (3.10). ■

Let

(5.4.3) $$\mathscr{H}^{(k)} = F^{n-k}\mathscr{H}_X = \partial_t^kF^n\mathscr{H}_X.$$

Then the decreasing Hodge filtration on $\mathscr{H}_X$, $0 \subset F^n\mathscr{H}_X \subset F^{n-1}\mathscr{H}_X \subset \ldots$, becomes the increasing filtration

(5.4.4) $$\mathscr{H}^{(0)} \subset \mathscr{H}^{(1)} \subset \ldots \mathscr{H}^{(k)} \subset \mathscr{H}^{(k+1)} \subset \ldots \subset \mathscr{H}_X$$

(5.4.5) $$\partial_t\colon \mathscr{H}^{(k)} \rightrightarrows \mathscr{H}^{(k+1)}.$$

We have already met an extension of this filtration to the left in (I.5.3.7), $\mathscr{H}^{(-1)} = {}'\mathscr{H}, \mathscr{H}^{(-2)} = \mathscr{H}^n_{DR}(X/S), \ldots$.

(4.6) *Remark* Scherk and Steenbrink [Sc-S] introduced the limit Hodge filtration of the MHS on the vanishing cohomology based upon the theory of D_S-modules with regular singularity at the point $0 \in S$. The invertibility of the operator ∂_t on $\mathcal{H}_{X,0}$ then implies [Ph1] that $\mathcal{H}_X \simeq \oplus_{j=1}\mathcal{M}^{\alpha_j,q_j}$, $\alpha_j \neq -1, -2, \dots$, where $\mathcal{M}^{\alpha,q} = D_S/D_S\cdot(t\partial_t - \alpha)^q$.

6 Decomposition of a meromorphic connection into a direct sum of the root subspaces of the operator $t\partial_t$. The $V^{\cdot}$-filtration and the canonical lattice

6.1 'Block' decomposition

Let $(\mathcal{M}, \nabla)$ be a meromorphic connection on a disk $S \subset \mathbb{C}$, regular singular at the point $0 \in S$ (see (I.4.6.1), (I.7.7.2)). $\mathcal{M}$ is a sheaf on S such that $\mathcal{M}_{|S'} = \mathcal{H}$ is a locally free sheaf of $\mathcal{O}_{S'}$-modules of rank μ, and the zero germ $\mathcal{M}_0 \simeq \mathcal{K}^\mu$ is a vector space over the field $\mathcal{K} = \mathbb{C}\{t\}[t^{-1}]$, the field of meromorphic functions with poles at 0, i.e. the field of fractions of the local ring $\mathcal{O}_{S,0}$ (for simplicity we'll often write $\mathcal{M}$ instead of $\mathcal{M}_0$). In the following applications $\mathcal{M}$ is the meromorphic Gauss–Manin connection and $\mathcal{M} = \mathcal{H}^{(0)}_{(t)} = \mathcal{H}^{(0)} \otimes_{\mathcal{O}_{S,0}} \mathcal{K}$ is a localization of the Brieskorn lattice $\mathcal{H}^{(0)}$ wrt the maximal ideal $m \subset \mathcal{O}_{S,0}$. Let $\partial_t = \nabla_{d/dt}\colon \mathcal{M} \to \mathcal{M}$ denote the operator of covariant differentiation (the sheaf $\mathcal{M}$ may be considered as a sheaf of D_S-modules). Let $\underline{H} = \operatorname{Ker}\nabla \subset \mathcal{H}$ be the local system of horizontal sections and T be the monodromy transformation.

Recall (I.7.7.2) that a connection ∇ is regular, if there exists a lattice $\mathcal{L}$ in $\mathcal{M}$ invariant wrt the operator $t\partial_t$. A residue of ∇ wrt such a lattice is a linear operator $\operatorname{Res}_{\mathcal{L}} \nabla$ on the vector space $\mathcal{L}/t\mathcal{L}$ induced by the operator $t\nabla_t$. From the classical Sauvage theorem (I.7.6.2) it follows from (I.7.7.7) that $\mathcal{M}$ decomposes into a direct sum $\mathcal{M} = \oplus_{i=1}^r \mathcal{M}^{\alpha_i,q_i}$ of standard blocks corresponding to the decomposition of the monodromy T into Jordan blocks. Here r is the number of blocks and q_i their sizes, $\sum_{i=1}^r q_i = \mu = \dim_{\mathcal{K}} \mathcal{M}$. Or in other words, there exists a basis $u_1, \dots , u_\mu$ in $\mathcal{M}$ in which the operator $t\partial_t$ matrix consists of blocks each of the form $\alpha E + N$, where E is a unity matrix, and N corresponds to a nilpotent operator, $Nu_1 = u_2, \dots , Nu_{q-1} = u_q$, $Nu_q = 0$ (units are under the main diagonal),

(6.1.1) $$t\partial_t\cdot(u_1, \dots , u_q) = (u_1, \dots , u_q)\cdot(\alpha E + N).$$

This also means that wrt the basis $u_1, \dots , u_q$ the connection ∇ has a matrix of connection coefficients of the form $\Gamma(t) = (1/t)(\alpha E + N)$.

If $y = (y_1, \dots , y_q)^t$ is a column of coordinate functions wrt the basis u, then the horizontality condition for a section $u\cdot y$ reduces to a system

(6.1.2) $$y' = \frac{R}{t} y, \quad \text{where } R = -(\alpha E + N).$$

This system has a fundamental matrix of solutions

(6.1.3) $$Y(t) = t^R = e^{R\cdot\log t}.$$

This implies that the monodromy on $\underline{H}$ is of the form

(6.1.4) $$T = e^{2\pi i R} = e^{-2\pi i\alpha}\cdot e^{-2\pi i N} = T_s T_u,$$

(6.1.5) $\lambda = e^{-2\pi i\alpha}$ is an eigenvalue of the monodromy T, and

(6.1.6) $$\alpha = -\frac{1}{2\pi i}\log\lambda,\ N = -\frac{1}{2\pi i}\log T_u.$$

From the definition of a residue (I.7.7.4) it also follows that wrt the basis $u_1, \ldots, u_q$ or wrt the lattice $\oplus_{i=1}^q \mathcal{O}_S\cdot u_i$, the residue is equal to

(6.1.7) $$\operatorname{Res}\nabla = \alpha E + N$$

and α is an eigenvalue of $\operatorname{Res}\nabla$ wrt the lattice $\oplus_{i=1}^q \mathcal{O}_S u_i$.

6.2 Decomposition of a meromorphic connection $\mathcal{M}$ into a direct sum of the root subspaces

First let $\mathcal{M} = \mathcal{M}^{\alpha,q} = \mathcal{K}u_1 \oplus \ldots \oplus \mathcal{K}u_q$, i.e. let T have only one Jordan block with the eigenvalue $\lambda = e^{-2\pi i\alpha}$.

Let ω be an element of $\mathcal{M}$ with coordinates $\varphi(t)$ wrt a basis u, $\omega = u\cdot\varphi(t)$, where $\varphi(t) = (\varphi_1(t), \ldots, \varphi_q(t))^t$ is a column of coordinates. Decomposing $\varphi(t)$ into the Laurent series $\varphi(t) = \sum \varphi^{(i)}\cdot t^i$, where $\varphi^{(i)}$ are constant columns, we obtain a decomposition

(6.2.1) $$\omega = \sum u\cdot\varphi^{(i)}\cdot t^i = \sum (ut^i)\cdot\varphi^{(i)}.$$

It appears that this decomposition has an invariant sense.

Recall that a root subspace L_α of a linear operator A on a vector space L corresponding to an eigenvalue α is the subspace $L_\alpha = \{\omega \in L\colon \exists\, r, (A-\alpha)^r\omega = 0\}$. Now note that on the subspace $B = \oplus_{i=1}^q \mathbb{C}u_i \subset \mathcal{M}$ the operator $t\partial_t = \alpha E + N$, and hence, the operator $t\partial_t - \alpha = N$ is nilpotent and hence $B \subset C_\alpha$, where

(6.2.2) $$C_\beta \overset{\text{def}}{=} \{\omega \in \mathcal{M}\colon \exists\, r, (t\partial_t - \beta)^r\omega = 0\}$$

is the root subspace of the operator $t\partial_t$ on $\mathcal{M}$ corresponding to an eigenvalue β. Moreover, the relation $[\partial_t, t] = 1$ implies by (I.7.7.6) the following lemma:

(6.2.3) *Lemma* On the space $B^{(k)} = t^k B = \oplus_{i=1}^q \mathbb{C}\cdot t^k u_i$ we have

$$t\partial_t = (\alpha + k)E + N.$$

This means that $B^{(k)} \subset C_{\alpha+k}$ and, consequently, the decomposition $\mathscr{M} = \hat{\oplus}_k t^k B$ is the root decomposition of the operator $t\partial_t$,

(6.2.4) $$\mathscr{M} = \hat{\oplus}_{\alpha} C_\alpha.$$

Here α runs over an arithmetical progression, $\alpha = -(1/2\pi\mathrm{i} \log\lambda$ with step 1, which is infinite on both sides. Here the sum sign $\hat{\oplus}$ means that there are inclusions $\oplus_\alpha C_\alpha \subset \mathscr{M} \subset \prod_\alpha C$ and $\mathscr{M}$ is generated by $\oplus C_\alpha$ over $\mathscr{O} = \mathscr{O}_{\mathbb{C},0}$. Note that all spaces C_α have the same dimension $\dim C_\alpha = q$.

In general, when $\mathscr{M} = \oplus_{i=1}^r \mathscr{M}^{\alpha_i,q_i}$ is a sum of blocks, we obtain:

(6.2.5) *Theorem* If $\mathscr{M}$ is a meromorphic connection with a regular singularity, then

$$\mathscr{M} = \hat{\oplus} C_\alpha,$$

where C_α is a root subspace of the operator $t\partial_t$ on $\mathscr{M}$, and α (for those $C_\alpha \neq 0$) runs over k arithmetical progressions $\alpha = -(1/2\pi\mathrm{i})\lambda_j$, where $\lambda_1, \dots, \lambda_k$ are all different eigenvalues of the monodromy T. ■

(6.2.6) *Corollary* Any element $\omega \in \mathscr{M}$ is uniquely represented by the form

$$\omega = \sum \omega_\alpha,$$

where $\omega_\alpha \in C_\alpha$. ■

We call this the *root decomposition of the element* ω, and the components ω_α are called the *root* (or homogeneous) *components of order* α. If $\omega = \omega_\alpha$, the element ω is called homogeneous.

Obviously, the operator of multiplication by t

(6.2.7) $$t\colon C_\alpha \rightrightarrows C_{\alpha+1}$$

is an isomorphism for all α. Since on C_α, $t\partial_t = \alpha E + N$ and $\partial_t = (1/t)(\alpha E + N)$, we have that

(6.2.8) $$\partial_t\colon C_\alpha \to C_{\alpha-1}$$

is an isomorphism for all $\alpha \neq 0$. In particular, on the subspace

(6.2.9) $$\mathscr{L} = \bigoplus_{\alpha > -1} C_\alpha$$

there exists an operator which is inverse to the operator ∂_t,

(6.2.10) $$\partial_t^{-1}\colon \mathscr{L} \to \mathscr{L},\ \partial_t \cdot \partial_t^{-1} = \mathrm{id}.$$

6.3 The order function α and the $V^{\cdot}$-filtration

Now let ∇ be a quasiunipotent connection as is the case for the Gauss–Manin connection, i.e. the monodromy operator T is quasiunipotent: there exist integers m and n such that $(T^m - E)^{n+1} = 0$. Thus all eigenvalues $\lambda = e^{2\pi i\beta}$ of the operator T are roots of unity, $\beta \in \mathbb{Q}$. Then all members of the arithmetical progressions $\alpha = -(1/2\pi i)\log\lambda$ are rational numbers and hence are naturally ordered.

(6.3.1) *Definition* If $\omega = \sum \omega_\alpha \in \mathscr{M}$, then a rational number
$$\alpha(\omega) = \min\{\alpha\colon \omega_\alpha \neq 0\}$$
is called the *order of the element* ω. The first non-vanishing term

(6.3.2) $$\omega_1 = \omega_{\alpha(\omega)}$$

is called the *main* or *leading part* of an element ω. The graduation $\mathscr{M} = \hat{\oplus} C_\alpha$ is called the *root graduation*.

(6.3.3) *Definition* Natural decreasing filtrations, the $V^{\cdot}$*-filtration and* the $V^{>}$*-filtration*, are associated with the root graduation on $\mathscr{M}$:
$$V^\alpha\mathscr{M} \overset{\text{def}}{=} \{\omega \in \mathscr{M}\colon \alpha(\omega) \geqslant \alpha\} = \hat{\bigoplus_{\beta \geqslant \alpha}} C_\beta$$
and
$$V^{>\alpha}\mathscr{M} \overset{\text{def}}{=} \{\omega \in \mathscr{M}\colon \alpha(\omega) > \alpha\} = \hat{\bigoplus_{\beta > \alpha}} C_\beta.$$

The corresponding joined objects are

(6.3.4) $$Gr_V^\alpha = V^\alpha / V^{>\alpha} \simeq C_\alpha.$$

From the description of $\mathscr{M} = \hat{\oplus} C_\alpha$ in (6.2.5) it is obvious that all terms of the filtration $V^\alpha\mathscr{M}$ and of $V^{>\alpha}\mathscr{M}$ are free $\mathscr{O}_S$-modules of rank μ

(6.3.5) $$V^\alpha\mathscr{M} = \bigoplus_{\alpha \leqslant \beta < \alpha+1} \mathscr{O}_S C_\beta,\ V^{>\alpha}\mathscr{M} = \bigoplus_{\alpha < \beta \leqslant \alpha+1} \mathscr{O}_S C_\beta,$$

invariant wrt $t\partial_t$, i.e. they are saturated lattices.

The lattice

(6.3.6) $$\mathscr{L} \underset{\text{df}}{=} V^{>-1}\mathscr{M} = \bigoplus_{-1 < \alpha \leqslant 0} \mathscr{O}_S C_\alpha$$

is called *canonical*.

(6.3.7) In other words, $\mathscr{L} \subset \mathscr{M}$ is a locally free $\mathscr{O}_S$-module extending the sheaf $\mathscr{H}$ to the whole disk S, $\mathscr{L}_{|S'} = \mathscr{H}$. The sheaf $\mathscr{L}$ is uniquely characterized by two conditions:

(1) the connection ∇ has a logarithmic pole on $\mathscr{L}$; and

(2) the eigenvalues of $\operatorname{Res}_{\mathscr{L}} \nabla$, i.e. the eigenvalues of the operator $t\partial_t$ on $\mathscr{L}/t\mathscr{L} = \oplus_{-1<\alpha\leqslant 0} C_\alpha$, satisfy the condition $-1 < \alpha \leqslant 0$.

The sheaf $\mathscr{L}$ is called the *canonical extension* of the sheaf $\mathscr{H}$

6.4 Identification of the zero fibre of the canonical extension $\mathscr{L}$ and the canonical fibre of the fibration $\underline{H}$

By a canonical fibre of the fibration $\underline{H}$ we mean a vector space H of its multivalued sections over S', or a space of multivalued horizontal sections of the sheaf $\mathscr{H} = \underline{H} \otimes_{\mathbb{C}} \mathscr{O}_{S'} = \mathscr{M}_{|S'}$, or a space of single-valued horizontal sections of the lift of $\mathscr{H}$ to the universal covering $e\colon U \to S'$ (2.4.8).

There is a natural isomorphism of the space H and the space $\mathscr{L}/t\mathscr{L} = \oplus_{-1<\alpha\leqslant 0} C_\alpha$, the zero fibre of the canonical extension $\mathscr{L}$.

First let us consider the case of one block. Let the monodromy T on the space H have one Jordan block with eigenvalue λ. Then for any choice of one of the values of logarithm $\alpha = -(1/2\pi\mathrm{i})\log\lambda$ we have an isomorphism

$$H \simeq C_\alpha.$$

The mutually inverse isomorphisms $\varphi_\alpha\colon C_\alpha \to H$ and $\psi_\alpha\colon H \to C_\alpha$ are arranged in just such a way.

A choice of α corresponds to a choice of the basis $u_1, \ldots, u_q$ in $\mathscr{M} = \mathscr{M}^{\alpha,q}$ in which $t\partial_t = \alpha E + N = -R_\alpha$. In this basis the space of horizontal sections H is a space of solutions of the system of differential equations $y' = (R_\alpha/t)y$ which has a fundamental system of solutions (6.1.3) t^{R_α}, and an element $A \in H$ is represented by the form $A = t^{R_\alpha}\cdot y_0$, where $y_0 \in \mathbb{C}^q$ is a constant column. Then:

(6.4.1) $\quad \varphi_\alpha\colon C_\alpha \to H$ takes $\omega = u\cdot y_0 \mapsto A = t^{R_\alpha} y_0 \in H,\ A = A^\omega_\alpha.$

Conversely, if $A \in H$ is a multivalued horizontal section, then the multivalued family of matrices $t^{-R_\alpha} = t^\alpha\cdot \mathrm{e}^{N\log t}$ having the monodromy $\mathrm{e}^{-2\pi\mathrm{i}R_\alpha} = T^{-1}$ 'unravels' the multivalued section A:

(6.4.2) $$s[A, \alpha] = t^{-R_\alpha}\cdot A$$

is a single-valued section of the sheaf $\mathscr{H}$. An immediate calculation shows (because $\partial_t A = 0$) that

$$t\partial_t s[A, \alpha] = \alpha s[A, \alpha] + s[NA, \alpha]$$

and hence, $s[A, \alpha] \in C_\alpha$. Then:

(6.4.3) $\quad \psi_\alpha\colon H \to C_\alpha$ takes $A \mapsto s[A, \alpha]$.

The same formulae are valid when the monodromy T consists of some Jordan blocks with the same eigenvalue λ. We obtain:

(6.4.4) Let $H_\lambda \subset H$ be a root subspace of the space of multivalued horizontal sections corresponding to an eigenvalue λ of the monodromy T, $\dim H_\lambda = \mu_\lambda$. Then for any value of the logarithm $\alpha = -(1/2\pi \mathrm{i}) \log \lambda$ all root spaces C_α have the dimension $\dim C_\alpha = \mu_\lambda$ and there are isomorphisms

$$\psi_\alpha\colon H_\lambda \rightrightarrows C_\alpha,\ A \mapsto \psi_\alpha(A) = s[A, \alpha] = t^\alpha t^N A, \tag{6.4.5}$$

where $t^N = \mathrm{e}^{N \log t}$. A section $s[A, \alpha]$ of the sheaf $\mathscr{H}$ is called an *elementary section of order* α corresponding to a multivalued section $A = A(t) \in H_\lambda$.

Calculation (6.4.2) shows that under the isomorphism ψ_α the operator $N = -(1/2\pi\mathrm{i}) \log T_\mathrm{u}$ on H_λ transforms to the operator $t\partial_t - \alpha$ on C_α, i.e. there is a commutative diagram

$$\begin{array}{ccc} C_\alpha & \overset{t\partial_t - \alpha}{\longrightarrow} & C_\alpha \\ \psi_\alpha\uparrow & & \uparrow\psi_\alpha \\ H_\lambda & \overset{N}{\longrightarrow} & H_\lambda \end{array} \qquad (t\partial_t - \alpha)\cdot\psi_\alpha = \psi_\alpha \cdot N. \tag{6.4.6}$$

(6.4.7) In the general case let $H = \oplus_{j=1}^k H_{\lambda_j}$ be the root decomposition of the monodromy T. Choosing for every λ_j the value $\alpha_j = -(1/2\pi\mathrm{i}) \log \lambda_j$ in the interval $-1 < \alpha_j \leqslant 0$ and applying the above construction to each block, we obtain an isomorphism

$$\psi\colon H \rightrightarrows \mathscr{L}/t\mathscr{L} = \bigoplus_{-1<\alpha\leqslant 0} C_\alpha, \tag{6.4.8}$$

which takes $A = \sum_{j=1}^k A_j \in H = \oplus H_j$ to

$$\psi(A) = s[A] = \sum_{j=1}^k s[A_j, \alpha_j]. \tag{6.4.9}$$

The isomorphism ψ is extended to an isomorphism of sheaves

$$\psi\colon \mathscr{O}_S(H) = H \otimes \mathscr{O}_S \rightrightarrows \mathscr{L} \tag{6.4.10}$$

and we obtain a natural trivialization of the sheaf $\mathscr{L}$.

(6.4.11) *Remark* The canonical extension $\mathscr{L}$ of the sheaf $\mathscr{H}$ constructed in (2.6.4) is the canonical extension (6.3.5) when all $\lambda_j = 1$, and then $\alpha = 0$ (cf. formulae (2.6.3) and (6.4.5)). Recall that in §2 we made a covering of the base S such that the monodromy became unipotent (all $\lambda_j = 1$).

6.5 The decomposition of sections $\omega \in \mathscr{M}$ into a sum of elementary sections

Let $\omega = \sum \omega_\alpha \in \mathscr{M}$, $\omega_\alpha \in C_\alpha$, be a root decomposition of the element ω.

By means of isomorphisms (6.4.5) $\psi_\alpha\colon H_\lambda \rightrightarrows C_\alpha$ we can represent homogeneous components ω_α in the form of elementary sections

(6.5.1) $$\omega_\alpha = s[A_\alpha^\omega, \alpha] = t^{-R_\alpha}\cdot A_\alpha^\omega = t^\alpha\, \mathrm{e}^{N\log t}\, A_\alpha^\omega,$$

where $A_\alpha^\omega \in H_\lambda$, $\lambda = \mathrm{e}^{-2\pi i\alpha}$, are multivalued horizontal sections. In more detail

(6.5.2) $$\omega_\alpha = t^\alpha\left(E + N\frac{\log t}{1!} + N^2\frac{\log^2 t}{2!} + \ldots + N^n\frac{\log^n t}{n!}\right)A_\alpha^\omega,$$

or using Varchenko's notation ([V1, AGV])

(6.5.3) $$\omega_\alpha = t^\alpha\sum_{k=0}^{n} A_{k,\alpha}^\omega\frac{\log^k t}{k!},$$

where $A_{k,\alpha}^\omega = N^k A_\alpha^\omega$.

Now if $\gamma \in H^*$ is a multivalued horizontal section of homological fibration $\underline{H}^*$, the dual to $\underline{H}$, then

(6.5.4) $$\begin{aligned} I_{\gamma,\omega}(t) = \langle\omega, \gamma\rangle &= \sum_\alpha\langle\omega_\alpha, \gamma\rangle = \sum_\alpha t^\alpha\langle \mathrm{e}^{N\log t}A_\alpha^\omega, \gamma\rangle \\ &= \sum_\alpha t^\alpha\left(\sum_{k=0}^{n}\langle A_{k,\alpha}^\omega, \gamma\rangle\frac{\log^k t}{k!}\right) = \sum_{\alpha,k} a_{k,\alpha}^\omega t^\alpha\frac{\log^k t}{k!}, \end{aligned}$$

where $a_{k,\alpha}^\omega = \langle A_{k,\alpha}^\omega, \gamma\rangle$.

We thus obtain the decomposition into series of period integrals $I_{\gamma,\omega}(t) = \int_{\gamma(t)}\omega$ (see (I.9.4.4)), from which one usually begins the exposition. The advantage of the above approach (the inverse order of exposition) lies in its invariant character.

From corollaries of the Malgrange theorem ((I.9.4.5) and (I.9.4.6)) we obtain

(6.5.5) *Corollary* If $\omega \in \mathscr{H}^{(-1)}$, then $\alpha(\omega) \geqslant 0$. If $\omega \in \mathscr{H}^{(0)}$, then $\alpha(\omega) > -1$.

Thus,

(6.5.6) $$\mathscr{H}^{(0)} \subset V^{>-1}\mathscr{M} = \mathscr{L}.$$

6.6 Transfer of automorphisms from the Milnor lattice H to the meromorphic connection $\mathscr{M}$

Let $\underline{H}$ be the cohomological fibration of a singularity $f\colon (\mathbb{C}^{n+1}, 0) \to (\mathbb{C}, 0)$. Then $H = H^n(X_\infty, \mathbb{C})$ is the cohomology of the canonical Milnor fibre (2.4.6). In the study of singularities which follows, the meromorphic connection $\mathscr{M}$, a topological invariant of a singularity, will play greater role. $\mathscr{M}$ is a 'universum' in which all Brieskorn lattices $\mathscr{H}^{(0)}$ of singularities f of a given topological type live.

(6.6.1) The isomorphism (6.4.8)

$$\psi\colon H \rightrightarrows \bigoplus_{-1<\alpha\leqslant 0} C_\alpha \subset \mathscr{M} = \hat{\bigoplus_\alpha} C_\alpha$$

enables us to carry different structures from H to $\mathscr{M}$.

(6.6.2) We will now carry automorphism of H to automorphism of $\mathscr{M}$. Obviously, any automorphism $g \in GL(H)$ is uniquely extended to an automorphism of $\mathscr{M}$ as a ($\mathscr{K} = \mathbb{C}\{t\}[t^{-1}]$)-module in such a way that the diagram

$$\begin{array}{cccl} H & \overset{\psi}{\rightrightarrows} & \bigoplus_{-1<\alpha\leqslant 0} & C_\alpha \subset \mathscr{M} \\ g\downarrow & & \downarrow & \downarrow\psi(g) \\ H & \overset{\psi}{\rightrightarrows} & \bigoplus_{-1<\alpha\leqslant 0} & C_\alpha \subset \mathscr{M}. \end{array}$$

is commutative. We denote by $\psi(g) \in \operatorname{Aut}\mathscr{M}$ this automorphism of $\mathscr{M}$ which on $\oplus_{-1<\alpha\leqslant 0} C_\alpha$ is equal to $\psi(g) = \psi\cdot g\cdot\psi^{-1}$.

(6.6.3) *Lemma* The automorphism on $\mathscr{M}$ corresponding to the monodromy T on H is

$$\psi(T) = \mathrm{e}^{-2\pi\mathrm{i}t}\,\partial_t.$$

Proof It follows from (6.4.6) that the operator $t\partial_t - \alpha$ on C_α corresponds to the operator $N = -(1/2\pi\mathrm{i})\log T_\mathrm{u}$ on H. A power of $t\partial_t - \alpha$ corresponds to a power of the operator N. Hence, the operator on C_α corresponding to the monodromy operator T and coinciding with $\lambda T_\mathrm{u} = \lambda\,\mathrm{e}^{-2\pi\mathrm{i}N}$ on H_λ is given by

$$\lambda\cdot\mathrm{e}^{-2\pi\mathrm{i}(t\partial_t-\alpha)} = \lambda\cdot\mathrm{e}^{2\pi\mathrm{i}\alpha}\,\mathrm{e}^{-2\pi\mathrm{i}t\partial_t} = \mathrm{e}^{-2\pi\mathrm{i}t\partial_t}.$$

It follows also from (6.4.6) that the operator $\mathrm{e}^{-2\pi\mathrm{i}t\partial_t}$ defines $\psi(T)$ in the sense (6.6.2) on $\mathscr{M}$. ■

$\mathscr{M}$ is in a natural way also a $\mathbb{C}[\partial_t]$-module.

(6.6.4) *Lemma* An automorphism $g \in GL(H)$ commutes with T if and only if $\psi(g)$ commutes with ∂_t, i.e. $\psi(g)$ is an automorphism of $\mathscr{M}$ as a $\mathbb{C}[\partial_t]$-module.

Proof If $\psi(g)$ commutes with ∂_t, then $\psi(g)$ also commutes with $t\partial_t$

because $\psi(g)$ commutes with t. Then $\psi(g)$ commutes with $e^{-2\pi i t\partial_t} = \psi(T)$, and, consequently, g commutes with T. Conversely, if g commutes with T, then $\psi(g)$ commutes with $e^{-2\pi t\partial_t}$, and, consequently, with $t\partial_t$. However, since $\psi(g)$, as an automorphism of the $\mathscr{K}$-module, commutes with t, $\psi(g)$ commutes also with ∂_t. ■

7 The limit Hodge filtration according to Varchenko and to Scherk–Steenbrink

At first the MHS on the vanishing cohomology of an isolated singularity $f\colon (\mathbb{C}^{n+1}, 0) \to (\mathbb{C}, 0)$, i.e. on the cohomology $H^n(X_\infty, \mathbb{C})$ of the (canonical) Milnor fibre, was introduced by Steenbrink (3.4.4) by means of an embedding of the Milnor fibration $X \to S$ in a projective fibration $Y \to S$. By a result of Scherk (3.5.1), this embedding can be chosen such that the homomorphism $H^n(Y_\infty, \mathbb{C}) \to H^n(X_\infty, \mathbb{C})$ is an epimorphism and the limit MHS on $H^n(X_\infty, \mathbb{C})$ is a quotient of the limit MHS on $H^n(Y_\infty, \mathbb{C})$. The weight filtration $W_.$ on $H^n(X_\infty, \mathbb{C})$ is a quotient weight filtration on $H^n(Y_\infty, \mathbb{C})$ and is defined by means of the nilpotent operator $N = -(1/2\pi i)\log T_u$, the logarithm of the unipotent part of the monodromy on $H^n(X_\infty, \mathbb{C})$ (3.5.2). It is explained in (3.5.4) why this filtration has different 'centres' on subspaces $H^n(X_\infty, \mathbb{C})_{\neq 1}$ and $H^n(X_\infty, \mathbb{C})_{\lambda=1}$. Varchenko proposed the idea of introducing the Hodge filtration $F^\cdot$ on $H^n(X_\infty, \mathbb{C})$ directly (by means of integral asymptotics), without using an embedding in a projective family. Following this idea, Scherk and Steenbrink [Sc-S] introduced the filtration $F^\cdot$ in a different way. We begin by introducing the filtration $F^\cdot = F^\cdot_{SS}$ on $H^n(X_\infty, \mathbb{C})$ according to Scherk–Steenbrink because it gives a genetically more naturally motivated construction of the limit Hodge filtration.

7.1 *Motivation of Scherk–Steenbrink's construction of the Hodge filtration*

The sequence of steps leading to the construction of filtration $F^\cdot_{SS}$ is as follows.

(1) The sheaf $\mathscr{H} = \underline{H} \otimes_{S'} \mathscr{O}_{S'} = \mathscr{H}^n_{DR}(X'/S')$ has a canonical extension $\mathscr{L}$ on S, and there is a canonical isomorphism (6.4.8)

$$\psi\colon H = H^n(X_\infty, \mathbb{C}) \rightrightarrows \mathscr{L}/t\mathscr{L} = \bigoplus_{-1<\alpha\leqslant 0} C_\alpha.$$

$$A = \sum A_\lambda \mapsto s[A] = \sum s[A, \alpha],\ s[A, \alpha] = t^\alpha \cdot t^N A(t),$$

of the space $H^n(X_\infty, \mathbb{C})$ considered as a space of multivalued horizontal sections of the fibration $\underline{H}$ and the zero fibre of the sheaf $\mathscr{L}$.

(2) The Hodge filtration $F^\cdot$ on $H^n(Y_\infty, \mathbb{C})$ according to Schmid (see §2.2) (coinciding with the Hodge filtration according to Steenbrink) is the limit Hodge filtration on the cohomology of the projective varieties $H^n(Y_t, \mathbb{C})$ (2.6.6). Let a bar denote an object for $Y \to S$ analogous to the corresponding object for $X \to S$, $\overline{\mathscr{H}} = \overline{\underline{H}} \otimes_{\mathbb{C}_{S'}} \mathscr{O}_{S'} = \mathscr{H}^n_{\mathrm{DR}}(Y'/S')$ and so on. Then the limit Hodge filtration $\overline{F}^\cdot$ on $\overline{H} = H^n(Y_\infty, \mathbb{C})$ is a filtration induced by the filtration on the zero fibre $\overline{\mathscr{L}}/t\overline{\mathscr{L}}$ of the canonical extension of the sheaf $\overline{\mathscr{H}}$ on S, though after a preliminary lift of the base $\tilde{S} \to S$ 'killing' the semisimple part of monodromy T_s. The filtration on $\overline{\mathscr{L}}$ is obtained by an extension of the Hodge filtration $\overline{\mathscr{F}}^\cdot$ on $\overline{\mathscr{H}}$, where $\overline{\mathscr{F}}^\cdot$ is the variation of the Hodge filtration on $H^n(Y_t, \mathbb{C})$. The consequences of the necessity of a base lift $\tilde{S} \to S$ are discussed below in (5).

(3) One can calculate the cohomology of a hypersurface by means of a complex of principal parts of differential forms on the ambient manifold with poles along this hypersurface (4.3.5). The sheaf $\overline{\mathscr{H}}$ is identified with the sheaf $\mathscr{H}_{Y|S'}$, $\mathscr{H}_Y = \mathbb{R}^{n+1}\overline{p}_*(DR^\cdot_{\overline{Z}/S}(B_{[\overline{\Gamma}]\overline{Z}}))$, where $\overline{\Gamma} \subset \overline{Z} = Y \times S$ is the graph of the morphism $Y \to S$ (5.1.4). The Hodge filtration is identified with the pole order filtration. We obtain an extension of the Hodge filtration $\overline{\mathscr{F}}^\cdot$ on $\overline{\mathscr{H}}$ to a filtration on the sheaf $\mathscr{H}_Y$ and in particular on the canonical extension $\mathscr{L}_Y = \overline{\mathscr{L}}$ of the sheaf $\overline{\mathscr{H}}$ to S.

(4) We can repeat the construction in (3) for the morphism $f\colon X \to S$ (§5). We can use this to introduce the limit Hodge filtration on the zero fibre of the canonical extension $\mathscr{L}$ of the sheaf $\mathscr{H}$ from S' to S, without using an embedding into the projective family. Thus, we have to consider the differential system (5.3.12)

$$\mathscr{H}_X = \mathscr{H}^{n+1}(p_*(DR^\cdot_{Z/S}(B_{[\Gamma]Z})) = \mathscr{H}^{n+1}(f_*\Omega^\cdot_X[D]),$$

where

$$\Omega^\cdot_X[D] \simeq \Omega^\cdot_X\left[\frac{1}{f(x)-t}\right],\ \omega D^i \mapsto \partial_t^i\left(\left[\frac{\omega}{f(x)-t}\right]\right) = \left[\frac{i!\omega}{(f(x)-t)^{i+1}}\right],$$

to take its Hodge filtration $F^\cdot$ (5.3.13) induced by the pole order filtration, and to consider the induced filtration on the canonical extension $\mathscr{L} \subset \mathscr{H}_X$ of the sheaf $\mathscr{H}$ on S'.

We have an inclusion of the Brieskorn lattice $\mathscr{H}^{(0)} \subset \mathscr{H}_X$ onto the n-th term of the filtration $F^n\mathscr{H}_X$, $\omega \mapsto [\omega/(f(x)-t)]$ (5.4.1). Moreover, the remaining terms of the Hodge filtration are obtained from $\mathscr{H}^{(0)} = F^n\mathscr{H}_X$ by applying the operator ∂_t, $F^{n-k}\mathscr{H}_X = \partial_t^k F^n\mathscr{H}_X = \partial_t^k\mathscr{H}^{(0)}$ (5.4.3).

Thus, the Hodge filtration is defined by the inclusion of the Brieskorn lattice $\mathscr{H}^{(0)} \subset \mathscr{H}_X$.

Consider a filtration on $\mathscr{L} \subset \mathscr{H}_X$ induced by $F^\cdot$ on $\mathscr{H}_X$,

$$F^p\mathscr{L} = F^p\mathscr{H}_X \cap \mathscr{L} = \partial_t^{n-p}\mathscr{H}^{(0)} \cap \mathscr{L} = \partial_t^{n-p}(V^{>n-p-1}\mathscr{H}^{(0)}).$$

The last equality is valid because $\mathscr{L} = V^{>-1}$ and the operator ∂_t takes V^α to $V^{\alpha-1}$. By definition $F^p\mathscr{L}$ is generated by elements of the form $\partial_t^{n-p}s[\omega]$, where $s[\omega] = \sum_{\alpha\geqslant\alpha(\omega)>-1}s[\omega, \alpha] \in \mathscr{H}^{(0)}$ are elements of order $\alpha(\omega) > n-p-1$. On $\mathscr{L}/t\mathscr{L} = \oplus_{-1<\alpha\leqslant 0}C_\alpha$ the induced filtration is of the form

$$F^p(\mathscr{L}/t\mathscr{L}) = \partial_t^{n-p}(V^{>n-p-1}\mathscr{H}^{(0)}/V^{>n-p}\mathscr{H}^{(0)}).$$

This is the desired limit Hodge filtration for the unipotent monodromy $T = T_{\mathrm{u}}$. Scherk and Steenbrink defined it in this way in the initial version of the paper [Sc-S]. However, in general, it is still not the desired filtration [Sa2, Ph2] because we have not yet dealt with the base lift $\tilde{S} \to S$.

(5) The necessity of having a base lift $\tilde{S} \to S$ 'kill' T_{s} leads to the following. In the above definition instead of using the elements $\partial_t^{n-p}s[\omega]$ to generate $F^p\mathscr{L}$ we have to take their leading parts $\partial_t^{n-p}s[\omega, \alpha(\omega)]$. Let us explain this.

Without the base lift the filtration term $F^p\mathscr{L}$ is generated by elements of the form $\omega = \sum_{\alpha>-1}t^\alpha t^N A_\alpha^\omega$ such that $\partial_t^{-(n-p)}\omega \in \mathscr{H}^{(0)}$, and for the filtration induced on $H = \oplus_\lambda H_\lambda \overset{\psi}{\rightrightarrows} \oplus_{-1<\alpha\leqslant 0}C_\alpha$ the subspace F^pH is generated by the elements $\sum_{-1<\alpha\leqslant 0}A_\alpha^\omega$.

According to Schmid's theorem we have to do the following. We have to take a covering π: $\tilde{S} \to S$, $t = \tilde{t}^m$ 'killing' the semisimple part of the monodromy, $T_{\mathrm{s}}^m = id$, and making it unipotent, $\tilde{T} = T^m = T_{\mathrm{u}}^m = \tilde{T}_{\mathrm{u}}$ (a tilde denotes the corresponding object on $\tilde{S}$). We have to take the inverse image $\tilde{\mathscr{F}}^\cdot$ on $\tilde{\mathscr{H}} = \pi^*\mathscr{H}$ of the Hodge filtration $\mathscr{F}^\cdot$ on $\mathscr{H}$ and to consider a filtration on $\tilde{\mathscr{L}}$ induced by the filtration on $\mathscr{L}$,

$$F^p\tilde{\mathscr{L}} = \pi^*(F^p\mathscr{L}) \cap \tilde{\mathscr{L}},$$

where $\pi^*(F^p\mathscr{L}) = (F^p\mathscr{L}) \otimes_{\mathscr{O}_S} \mathscr{O}_{\tilde{S}}$. Then, finally, we have to take the filtration $F^\cdot$ induced on the zero fibre $\tilde{\mathscr{L}}/t\tilde{\mathscr{L}} \simeq H$. This is the limit Hodge filtration according to Schmid. Since the monodromy $\tilde{T}$ is unipotent, we have that in the decomposition $\tilde{\mathscr{L}} \simeq \oplus_{\alpha>-1}\tilde{C}_\alpha$ all α are integers, $\tilde{\mathscr{L}} = \oplus_{k\in\mathbb{N}_0}C_k$, and the mapping $\tilde{\mathscr{L}} \to \tilde{\mathscr{L}}/t\tilde{\mathscr{L}}$ reduces to $\tilde{\omega} = \sum_{k\geqslant 0}\tilde{\omega}_k \mapsto \tilde{\omega}_0$. The filtration $\pi^*(F^p\mathscr{L}) = (F^p\mathscr{L}) \otimes_{\mathscr{O}_S} \mathscr{O}_{\tilde{S}}$ is generated by elements from $\mathbb{C}\{\tilde{t}\}\pi^*\omega = \mathbb{C}\{\tilde{t}\}\cdot\sum_{\alpha>-1}\tilde{t}^{m\alpha}\cdot\tilde{t}^{mN}A_\alpha^\omega$. To obtain $F^p\tilde{\mathscr{L}} = \pi^*(F^p\mathscr{L}) \cap \tilde{\mathscr{L}}$ we have to take elements of degree $\geqslant 0$ in $\tilde{t}$. We see that the subspace F^p of $\tilde{\mathscr{L}}/\tilde{t}\tilde{\mathscr{L}} \simeq H$ is generated by elements $A^{\omega}_{\alpha(\omega)}$, i.e. by first summands in $\sum_{-1<\alpha\leqslant 0} A^{\omega}_{\alpha}$.

(6) Lastly the final note: instead of operating in the context of the Gauss–Manin differential system $\mathscr{H}_X$ we may operate in the more usual classical context of the meromorphic connection $\mathscr{M} = \mathscr{H}^{(0)} \otimes_{\mathscr{O}_S} \mathscr{K}$. This follows from the fact that the canonical lattice $\mathscr{L}$ lies in $\mathscr{H}_X$ as well as in $\mathscr{M}$, $\mathscr{L} \subset \mathscr{H}_X$ and $\mathscr{L} \subset \mathscr{M}$. We'll talk about the relation between $\mathscr{H}_X$ and $\mathscr{M}$ in more detail in the supplement to this section in (7.7.9)–(7.7.11). The sheaves $\mathscr{L}$, $\mathscr{H}_X$ and $\mathscr{M}$ are topological invariants of a singularity defined by the Jordan structure of T on H. Analytic invariants of a singularity f (moduli) are reflected in the inclusion of the Brieskorn lattice to the canonical lattice, $\mathscr{H}^{(0)} \subset \mathscr{L}$, which defines among other things the MHS on the vanishing cohomology.

The above explains the following definition:

7.2 The definition of the limit Hodge filtration F_{SS} according to Scherk–Steenbrink

Let $\mathscr{M} = \hat{\oplus} C_\alpha$ be the meromorphic connection of a singularity f, $\mathscr{L} = V^{>-1} = \hat{\oplus}_{\alpha>-1} C_\alpha \subset \mathscr{M}$ be the canonical lattice, and $\mathscr{H}^{(0)}$ be the Brieskorn lattice. By (6.5.6) we have $\mathscr{H}^{(0)} \subset \mathscr{L}$.

(7.2.1) *Definition* The *Hodge filtration* $F^{\cdot} = F^{\cdot}_{\mathrm{SS}}$ *according to Scherk–Steenbrink* is defined on the zero fibre $\mathscr{L}/t\mathscr{L} = \oplus_{-1<\alpha\leqslant 0} C_\alpha$ on each C_α separately,

(7.2.2) $$F^p(\mathscr{L}/t\mathscr{L}) \underset{\mathrm{df}}{=} \bigoplus_{-1<\alpha\leqslant 0} F^p C_\alpha,$$

where

(7.2.3) $$F^p C_\alpha = \partial_t^{n-p} Gr_V^{\alpha+n-p} \mathscr{H}^{(0)},$$

(7.2.4) $Gr^{\beta}_{V}\mathscr{H}^{(0)} = V^{\beta} \cap \mathscr{H}^{(0)}/V^{>\beta} \subset Gr^{\beta}_{V}\mathscr{M} = C_\beta$ is a subspace generated by the leading parts of geometric sections $s[\omega] \in \mathscr{H}^{(0)}$ of order β.

Since $\mathscr{H}^{(0)} \subset V^{>-1}$, we have $Gr^{\beta}_{V}\mathscr{H}^{(0)} = 0$ for $\beta \leqslant -1$ and hence:

(7.2.5) $$F^p C_\alpha = 0 \text{ for } p>n,\ F^p(\mathscr{L}/t\mathscr{L}) = 0 \text{ for } p>n.$$

If $\beta > -1$, then there is an isomorphism ∂_t^{-1}: $C_\beta \rightrightarrows C_{\beta+1}$, inducing inclusions

$$Gr^{\beta}_{V}\mathscr{H}^{(0)} \overset{\partial_t^{-1}}{\to} Gr^{\beta+1}_{V}\mathscr{H}^{(0)}$$

(recall that ∂_t^{-1} takes $\mathscr{H}^{(0)}$ into itself and $\partial_t^{-1}(\mathscr{H}^{(0)}) = \mathscr{H}^{(-1)} \subset \mathscr{H}^{(0)}$ by (I.5.3.4)).

(7.2.6) We can schematically visualize the inclusion $\mathscr{H}^{(0)} \subset \mathscr{M}$, more exactly $Gr_V \mathscr{H}^{(0)} \subset Gr \mathscr{M}$, and the Hodge filtration $F^{\cdot}$ on C_α, $-1 < \alpha \leqslant 0$, by means of the following diagram [He1] (see picture below). Let us picture $\mathscr{L} = \hat{\oplus}_{\alpha > -1} C_\alpha$ as an infinite set of columns C_β of height

(7.2.7) $$\dim C_\beta = \dim H_\lambda = \mu_\lambda, \quad \text{where } \lambda = \mathrm{e}^{-2\pi i \beta}.$$

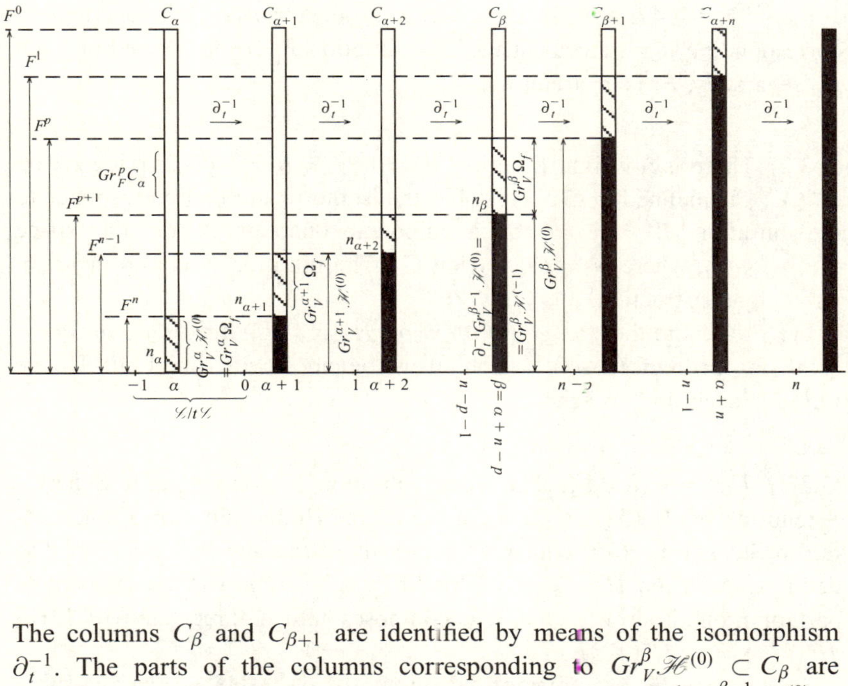

The columns C_β and $C_{\beta+1}$ are identified by means of the isomorphism ∂_t^{-1}. The parts of the columns corresponding to $Gr_V^\beta \mathscr{H}^{(0)} \subset C_\beta$ are hatched and the parts corresponding to $\partial_t^{-1} Gr_V^{\beta-1} \mathscr{H}^{(0)} = Gr_V^\beta \mathscr{H}^{(-1)} \subset Gr_V^\beta(\mathscr{H}^{(0)})$ are shaded. The 'figure' $Gr_V^{\cdot} \mathscr{H}^{(-1)}$ is obtained from $Gr_V^{\cdot}(\mathscr{H}^{(0)})$ by a shift of 1 to the right. Part of $Gr_V^{\cdot}(\mathscr{H}^{(0)})$ corresponding to one eigenvalue λ of the monodromy is shown in the picture above, and as an example the whole of $Gr_V^{\cdot}(\mathscr{H}^{(0)})$ is shown for the singularity $\tilde{E}_7 = X_9 = T_{2,4,4}$ in picture on the next page.

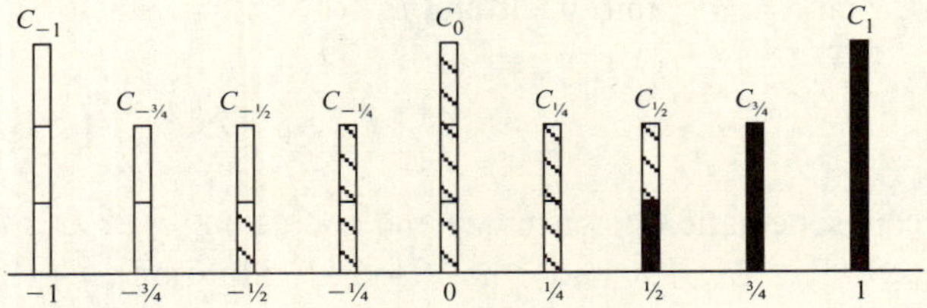

7.3 The Scherk–Steenbrink theorem

(7.3.1) The filtration $F^{\cdot}_{SS}$ on $Gr^{\cdot}_V(\mathscr{L}/t\mathscr{L}) = \oplus_{-1<\alpha\leqslant 0} C_\alpha$ is carried to $H^n(X_\infty, \mathbb{C})$ by means of the isomorphism ψ: $H^n(X_\infty, \mathbb{C}) \rightrightarrows \mathscr{L}/t\mathscr{L} \simeq Gr^{\cdot}_V(\mathscr{L}/t\mathscr{L})$. By construction, ψ: $H^n(X_\infty, \mathbb{C})_\lambda \xrightarrow{\sim} C_\alpha$, where $\lambda = e^{-2\pi i\alpha}$, $-1 < \alpha \leqslant 0$, $H^n(X_\infty, \mathbb{C})_\lambda$ is a root subspace of the operator T_s corresponding to an eigenvalue λ. The filtration $F^{\cdot}$ being defined on each C_α separately, $F^{\cdot}$ is invariant wrt T_s.

(7.3.2) There is a weight filtration $W = W(N)$, $N = -(1/2\pi i)\log T_u$, on $H^n(X_\infty, \mathbb{C})$ defined in (3.5.2). Under the isomorphism ψ it corresponds to the filtration $W = W(N)$ of a nilpotent operator N on the space $\oplus_{-1<\alpha\leqslant 0} C_\alpha$, where $N = t\partial_t - \alpha$ on C_α. The filtration W is also invariant wrt T_s by construction.

Let $F^{\cdot}_S$ denote the *Hodge filtration on* $H^n(X_\infty, \mathbb{C})$ *according to Steenbrink* in theorem (3.4.4), which also coincides with the limit Hodge filtration according to Schmid ([S3]).

(7.3.3) *Theorem* ([Sc-S]) The Hodge filtration $F^{\cdot}_{SS}$ according to Scherk–Steenbrink on $H^n(X_\infty, \mathbb{C})$ coincides with the Hodge filtration $F^{\cdot}_S$ according to Steenbrink. Consequently (3.4.4), the filtrations $W_{\cdot}$ and $F^{\cdot} = F^{\cdot}_{SS}$ define an MHS on $H^n(X_\infty, \mathbb{C})$. This MHS is invariant wrt the semisimple part of monodromy T_s, and it decomposes into a direct sum of MHS $H^n(X_\infty) = \oplus_\lambda H^n(X_\infty)_\lambda$.

Let us consider examples of calculation of the MHS on the vanishing cohomology.

(7.3.4) *The MHS in the case of quasihomogeneous singularities*
Let $f(x) \in \mathbb{C}[x_0, \ldots, x_n]$ be a quasihomogeneous singularity of degree 1 with weights $\operatorname{wt} x_i = w_i$. We use notations from (I.5.5.1–I.5.5.3). The

differential forms $\omega_m = x^m \,\mathrm{d}x$, $m \in A$, $|A| = \mu$, represent a basis of the $\mathcal{O}_S$-module $\mathcal{H}^{(0)} = \oplus_{m\in A}\mathcal{O}_S\omega_m$. We use the same symbol to denote differential forms, geometric sections defined by them, elements of the space $\mathcal{L}/t\mathcal{L} = \oplus_{-1<\alpha\leqslant 0} = C$, etc. In (I.5.5.4) we showed that $\omega_m \in \mathcal{H}^{(0)}$ are eigenvectors of the operator $t\partial_t$,

$$\omega_m \in C_{a(m)-1}.$$

Thus the nilpotent operator N is zero and the weight filtration is as follows:

$$0 \subset W_{n-1} \subset W_n \subset W_{n+1} = C,\ W_n = \bigoplus_{-1<\alpha<0} C_\alpha,\ W_{n+1}/W_n = C_0.$$

The Hodge filtration $F^\cdot = F_{SS}$, $C = F^0 \supset F^1 \supset \ldots \supset F^n \supset F^{n+1} = 0$, is defined as follows. Let us take the partition of the set of indices (= the set of exponents of monomials) A into subsets, $A = \cup A_p$, where $A_p = \{m \in A,\ n - p - 1 < a(m) - 1 \leqslant n - p\}$. We have that if $m \in A_p$, then $\eta_m = \partial_t^{n-p}\omega_m \in C_\alpha$, where $-1 < \alpha \leqslant 0$. The elements η_m generate a basis of the lattice $\mathcal{L}$, $\mathcal{L} = \oplus_{m\in A}\mathcal{O}_S\eta_m$, and correspondingly of the space $C = \oplus_{m\in A}\mathbb{C}\eta_m$. Then by definition

$$F^pC = \bigoplus_{m\in A_{\geqslant p}} \mathbb{C}\eta_m, \quad \text{where } A_{\geqslant p} = A_p \cup A_{p+1} \cup \ldots \cup A_n.$$

(7.3.5) *The MHS in the case of singularities $T_{p,q,r}$ ([Sc-S])* The unimodal singularities of series $T_{p,q,r}$ are defined by equations

$$f(x, y, z) = x^p + y^q + z^r + axyz,\ a \neq 0,$$

where $p^{-1} + q^{-1} + r^{-1} < 1$ ([AGV]). These are simplest not (semi)quasi-homogeneous singularities. For them $\mu = p + q + r - 1$, and the monomials M_i

$$1, \qquad xyz, \qquad x^k (0 < k < p), \qquad y^k (0 < k < q), \qquad z^k (0 < k < r),$$

form a basis of the Jacobian algebra Q_f. The differential forms $\omega_i = M_i\omega$, $1 \leqslant i \leqslant \mu$, where $\omega = \mathrm{d}x \wedge \mathrm{d}y \wedge \mathrm{d}z$, form a basis of the vector space $\Omega_f = \mathcal{H}^{(0)}/\mathcal{H}^{(-1)}$. Applying the method developed by Scherk [Sc1], one finds that a basis of the canonical lattice $\mathcal{L}$ consists of the forms

$$\omega, \qquad t\partial_t\omega, \qquad \partial_t(x^k\omega)\ (0 < k < p), \qquad \partial_t(y^k\omega)\ (0 < k < q),$$
$$\partial_t(z^k\omega)\ (0 < k < r),$$

(by (III.3.4.14) we can replace $t\partial_t\omega$ by $\partial_t(xyz\omega)$). Moreover, except for a few small values of p, q and r, the operator $t\partial_t$ on $C = \mathcal{L}/t\mathcal{L}$ has the following form relative to this basis:

$$t\partial_t(t\partial_t\omega) = 0;\ t\partial_t(\partial_t x^k\omega) = \frac{k-p}{p}\partial_t x^k\omega \text{ for } 0 < k < p;$$

$$t\partial_t(\partial_t y^k\omega) = \frac{k-q}{q}\partial_t y^k\omega \text{ for } 0<k<q;$$

$$t\partial_t(\partial_t z^k\omega) = \frac{k-r}{r}\partial_t z^k\omega \text{ for } 0<k<r.$$

Thus, this basis is a Jordan basis for the operator $t\partial_t$. The space $C = \oplus_{-1<\alpha\leqslant 0} C_\alpha = C_{\neq 0} \oplus C_0$, where the space C_0 is generated by vectors ω and $t\partial_t\omega$. The weight filtration W on C is the weight filtration $W(N)$ of the nilpotent operator N with center $n=2$ on $C_{\neq 0}$ and with center $n=3$ on C_0. On $C_{\neq 0}$ the operator $N=0$ and hence $W_1C_{\neq 0}=0$, $W_2C_{\neq 0}=C_{\neq 0}$. On C_0 the operator $N\neq 0$ and $N^2=0$, and we have $W_1C_0=0$, $W_2C_0=W_3C_0=\mathbb{C}t\partial_t\omega$, $W_4C_0=C_0$. Therefore the weight filtration is as follows: $0=W_1\subset W_2=W_3\subset W_4=C$, where $W_2=C_{\neq 0}\oplus\mathbb{C}t\partial_t\omega$ is generated by all base vectors different from ω. By definition of the Hodge filtration we have $C=F^1\supset F^2\supset F^3=0$, where F^2 is the subspace generated by ω.

7.4 Varchenko's theorem about the operator of multiplication by f in Ω_f

By the definition of the filtration $F^\cdot$ in (7.2.1) we have

$$\begin{aligned}
(7.4.1)\qquad Gr_F^p C_\alpha &= F^pC_\alpha/F^{p+1}C_\alpha \\
&= \partial_t^{n-p}Gr_V^{\alpha+n-p}\mathscr{H}^{(0)}/\partial_t^{n-p-1}Gr_V^{\alpha+n-p-1}\mathscr{H}^{(0)} \\
&= \partial_t^{n-p}Gr_V^{\alpha+n-p}\mathscr{H}^{(0)}/\partial_t^{n-p}Gr_V^{\alpha+n-p}\mathscr{H}^{(-1)} \\
&\simeq Gr_V^{\alpha+n-p}\mathscr{H}^{(0)}/Gr_V^{\alpha+n-p}\mathscr{H}^{(-1)} \\
&\simeq Gr_V^{\alpha+n-p}\Omega_f,
\end{aligned}$$

where $\Omega_f=\mathscr{H}^{(0)}/\mathscr{H}^{(-1)}=\Omega_{X,0}^{n+1}/\mathrm{df}\wedge\Omega_{X,0}^n$, and $V^\cdot$ also denotes the filtration induced by $V^\cdot$ on Ω_f. Thus,

$$(7.4.2)\qquad \partial_t^{n-p}\colon Gr_V^\beta\Omega_f \xrightarrow{\sim} Gr_F^pC_\alpha,\ \text{ where } \beta=\alpha+n-p,$$

and

$$(7.4.3)\qquad \bigoplus_{\alpha,p}\partial_t^{n-p}\colon Gr_V^\cdot\Omega_f \xrightarrow{\sim} Gr_F^\cdot\Big(\bigoplus_{1<\alpha\leqslant 0}C_\alpha\Big)=Gr_F^\cdot H^n(X_\infty).$$

Moreover, since the operator N is equal to $t\partial_t-\alpha$ on C_α, the mapping induced by it, $Gr_F^\cdot N\colon Gr_F^pC_\alpha\to Gr_F^{p-1}C_\alpha$, coincides with $t\partial_t$, and we have a commutative diagram

$$(7.4.4)\qquad \begin{array}{ccc}
Gr_V^\beta\Omega_f & \xrightarrow{t} & Gr_V^{\beta+1}\Omega_f \\
\partial_t^{n-p}\downarrow\wr & & \downarrow\wr\ \partial_t^{n-p+1} \\
Gr_F^pC_\alpha & \xrightarrow{GrN} & Gr_F^{p-1}C_\alpha
\end{array},\quad \beta=\alpha+n-p.$$

However, under the identification of the $\mathbb{C}\{t\}$-module Ω_f with $\Omega^{n+1}_{X,0}/df \wedge \Omega^n_{X,0}$ the operator of multiplication by t reduces to the operator of multiplication by $f(x)$. This implies the following theorem [V4, Sc-S]:

(7.4.5) *Varchenko's theorem* The operator of multiplication by $f(x)$ on $Gr^{\cdot}_V\Omega_f$ and the operator $N = -(1/2\pi\mathrm{i})\log T_\mathrm{u}$ on $H^n(X_\infty) \simeq \oplus_{-1<\alpha\leqslant 0} C_\alpha$ have the same Jordan normal form.

7.5 The definition of the limit Hodge filtration $F^{\cdot}$ on $H^n(X_\infty)$ according to Varchenko

The filtration $F^{\cdot}_{\mathrm{SS}}$ on $H^n(X_\infty, \mathbb{C}) \overset{\psi}{\rightarrow} \oplus_{-1<\alpha\leqslant 0} C_\alpha$ was defined by Scherk and Steenbrink after Varchenko had defined the asymptotic Hodge filtration $F^{\cdot}_{\mathrm{Va}}$. Varchenko defines this filtration $F^{\cdot}_{\mathrm{Va}}$ on fibres of the cohomological fibration $\underline{H}_t = H^n(X_t, \mathbb{C})$ or the corresponding filtration $\mathscr{F}^{\cdot}_{\mathrm{Va}}$ on the sheaf $\mathscr{H} = \underline{H} \otimes \mathscr{O}_{S'}$ in such a way that $\mathscr{H}^p_{\mathrm{Va}}$ is generated by leading parts $\omega_1 = s[\omega, \alpha(\omega)]$ of geometric sections $s[\omega] \in \mathscr{H}^{(0)}$ of order $\alpha(\omega) \leqslant n - p$. In terms of the canonical extension $\mathscr{L} = V^{>-1}$ of the sheaf $\mathscr{H}$ this means that on $\mathscr{L}$ the induced filtration $\mathscr{F}^{\cdot}_{\mathrm{Va}}$ is generated by sections $t^k\omega_1$, where k is such that $-1 < \alpha(t^k\omega_1) \leqslant 0$. In other words:

(7.5.1) *Definition* The *Hodge filtration* $F^{\cdot} = F^{\cdot}_{\mathrm{Va}}$ *according to Varchenko* is defined on the zero fibre $\mathscr{L}/t\mathscr{L} = \oplus_{-1<\alpha\leqslant 0} C_\alpha$ and on each C_α separately

(7.5.2) $$F^p_{\mathrm{Va}}(\mathscr{L}/t\mathscr{L}) = \bigoplus_{-1<\alpha\leqslant 0} F^p_{\mathrm{Va}} C_\alpha$$

and

(7.5.3) $$F^p_{\mathrm{Va}} C_\alpha = t^{-(n-p)} Gr_V^{\alpha+n-p} \mathscr{H}^{(0)}.$$

7.6 Comparison of the filtrations F_{SS} and F_{Va}

From formulae (7.2.3) and (7.5.3) it follows that to get filtrations $F^{\cdot}_{\mathrm{SS}}$ and $F^{\cdot}_{\mathrm{Va}}$ on $\mathscr{L}/t\mathscr{L}$ we have to take the same geometric sections

$$s[\omega] \in \mathscr{H}^{(0)},\ n - p - 1 < \alpha(\omega) \leqslant n - p,$$

take their leading parts

$$s[\omega, \beta] = t^\beta t^N A^\omega,\ \alpha(\omega) = \beta = \alpha + n - p,\ -1 < \alpha \leqslant 0,$$

and then carry them into $\mathscr{L}/t\mathscr{L} = \oplus_{-1<\alpha\leqslant 0} C_\alpha$ firstly by ∂_t^{n-p} and secondly by dividing by t^{n-p}. It follows from (6.4.2) that $\partial_t\colon C_\beta \to C_{\beta-1}$ takes $s[A, \beta]$ to

(7.6.1) $$\partial_t s[A,\beta] = \beta s[A,\beta-1] + s[NA,\beta-1] = s[(\beta E+N)A,\beta-1].$$

Hence,

(7.6.2) $$\partial_t^{n-p} s[A,\beta] = s[((\alpha+1)E+N) \times ((\alpha+2)E+N)\ldots(\beta E+N)A,\alpha] = ((\alpha+1)E+N)\ldots(\beta E+N)A\cdot t^{-(n-p)}s[A,\beta]$$

and, consequently,

(7.6.3) $$F^p_{\mathrm{SS}}C_\alpha = ((\alpha+1)E+N)\ldots(\beta E+N)F^p_{\mathrm{Va}}C_\alpha.$$

This formula shows that

(7.6.4) $$F^p_{\mathrm{SS}}(\mathscr{L}/t\mathscr{L}) = F^p_{\mathrm{Va}}(\mathscr{L}/t\mathscr{L}) \bmod N,$$

or

$$F^p_{\mathrm{SS}}H^n(X_\infty) = F^p_{\mathrm{Va}}H^n(X_\infty) \bmod N\cdot H^n(X_\infty).$$

Since $N(W_k)\subset W_{k-2}$, we obtain:

(7.6.5) *Corollary* The filtrations F^p_{SS} and F^p_{Va} on $Gr^W_k H^n(X_\infty) = W_k/W_{k-1}$ coincide.

This corollary and theorem (7.3.3) imply Varchenko's theorem [V3]:

(7.6.6) *Theorem* The filtrations $W_\cdot = W(N)$ from (3.5.2) and $F^\cdot = F^\cdot_{\mathrm{Va}}$ define an MHS on the vanishing cohomology $H^n(X_\infty,\mathbb{C})$.

We emphasize once more that the idea of defining the Hodge filtration $F^\cdot = F^\cdot_{\mathrm{Va}}$ by means of the Brieskorn lattice $\mathscr{H}^{(0)}$ (and the proof of this theorem) are due to A. N. Varchenko.

7.7 Supplement on the connection between the Gauss–Manin differential system $\mathscr{H}_X$ and its meromorphic connection $\mathscr{M}$

In this review we have tried to avoid appealing to the theory of D-modules. Nevertheless we want to give some feeling of how the sheaves $\mathscr{H}_X$ and $\mathscr{M}$ are related. Let $S\subset\mathbb{C}$ be a disk centered at the point $t=0$, $D = D_{S,0} = \mathbb{C}\{t\}[\partial_t]$ be the ring of differential operators in one variable, and D_S be the sheaf of differential operators on S. If M is a D-module then an operation of localization $M_{(t)} = M\otimes_{\mathscr{O}}\mathscr{K} = M[t^{-1}]$ is defined. Here $\mathscr{K} = \mathbb{C}\{t\}[t^{-1}]$ is the field of meromorphic functions – the field of fractions of the ring $\mathscr{O} = \mathscr{O}_{S,0} = \mathbb{C}\{t\}$. $M_{(t)}$ is naturally a D-module, where the action of ∂_t on $M_{(t)}$ is defined by the rule of differentiation of fractions. One can find in [Sa7, Sa8] a survey of the theory of D-modules finite over D, with its localization $M_{(t)}$ being a regular meromorphic connection in the sense (I.7.7.2) (i.e. the theory of regular holonomic D-modules). Let $M_{\mathrm{rh}}(D)$ be the category of such D-modules. D-modules $\mathscr{H}_X = M$ and $\mathscr{M} = M_{(t)} \in M_{\mathrm{rh}}(D)$ (see (5.4.5)), and by (5.4.2) the operator of multiplying by ∂_t is invertible on $\mathscr{H}_X$, and the operator of multiplying by t is invertible on $\mathscr{M}$. The main fact is that any such D-module is a direct sum of indecomposable D-modules

(7.7.1) $$\mathscr{M}^{\alpha,q} = D/D\cdot(t\partial_t - \alpha)^q = D/D\cdot[\partial_t t - (\alpha+1)]^q.$$

One can assume that up to isomorphism (see the lemma about shift in (I.7.7.6)): if $\alpha \notin \mathbb{Z}$, then $M = \mathscr{M}^{\alpha,q}$, $\alpha \in \Lambda'$, where $\Lambda' \subset \mathbb{C}$, a set of representatives of non-zero classes of $\mathbb{C}/\mathbb{Z}$; and if $\alpha \in \mathbb{Z}$, then either $\alpha = 0$ (for $\alpha \geqslant 0$) or $\alpha = -1$ (for $\alpha < 0$), i.e. we have:

(7.7.2)
$$\text{either } M = \mathscr{M}^{0,q} = D/D\cdot(t\partial_t)^q, \quad \text{or } M = \mathscr{M}^{-1,q} = D/D\cdot(\partial_t t)^q.$$

Moreover, for $\alpha \in \Lambda'$ both operators t and ∂_t are invertible on the D-module $\mathscr{M}^{\alpha,q}$, ∂_t is (and t isn't) invertible on $\mathscr{M}^{0,q}$, and t is (and ∂_t isn't) invertible on $\mathscr{M}^{1,q}$.

To understand better what happens when we pass from $\mathscr{H}_X$ to $\mathscr{M}$, i.e. from M to its localization $M_{(t)}$, we consider a D-module $M = \mathscr{M}^{\alpha,q}$ and for simplicity we'll assume that $q = 1$. In general there is a homomorphism of a D-module M to its localization $M_{(t)}$ and we have an exact sequence

(7.7.3) $$0 \to {}^{\tau}M \to M \xrightarrow{\varphi} M_{(t)} \to M^* \to 0,$$

where ${}^{\tau}M = \operatorname{Ker}\varphi$ is an $\mathscr{O}_{S,0}$-torsion of M, and $M^* = \operatorname{Coker}\varphi$.

The following simple lemma plays an important role in clarifying of the structure of D-modules.

(7.7.4) *Lemma* If $Q = q_0(t) + q_1(t)\partial_t + \ldots + q_m(t)\partial_t^m \in D$ is a differential operator of order m, then for any $\alpha \in \mathbb{C}$ it can be presented in the following form

$$Q = d_m\partial_t^m + d_{m-1}\partial_t^{m-1} + \ldots + d_1\partial_t + c(t) + P\cdot(t\partial_t - \alpha),$$

where $d_k \in \mathbb{C}$ are constants, $c(t) = \sum_{j=0}^{\infty} c_j t^j \in \mathscr{O}_{S,0}$ is a holomorphic function, and $P \in D$ a differential operator.

Proof This easily follows from the commutator rule

(7.7.5) $$[\partial_t, t] = 1, \quad \text{i.e. } \partial_t t - t\partial_t = 1.$$ ■

Thus, let $M = \mathscr{M}^{\alpha,1} = D/D\cdot(t\partial_t - \alpha) = D\cdot u$, where u is a generator satisfying the relation $(t\partial_t - \alpha)u = 0$. Let us consider a D-module $\mathscr{K}\cdot e$, which is a vector space of dimension $q = 1$ over the field $\mathscr{K} = \mathbb{C}\{t\}[t^{-1}]$ with a generator e, on which ∂_t acts such that $(t\partial_t - \alpha)e = 0$, i.e.

$$(7.7.6) \qquad \partial_t e = \frac{\alpha}{t} e \text{ and then } \partial_t^m e = \frac{\alpha(\alpha-1)\ldots(\alpha-m+1)}{t^m} e$$

(i.e. we can assume that $e = t^\alpha$ and ∂_t is the differentiation in t). Then it is easy to see that $\mathscr{K}\cdot e = M_{(t)}$ is the localization and the homomorphism $\varphi\colon M \to M_{(t)} = \mathscr{K}\cdot e, \quad u \mapsto e,$ takes $Q\cdot u = (d_m\partial_t^m + \ldots + d_1\partial_t + c(t))u \bmod D(t\partial_t - \alpha)$ to the Laurent series

$$(7.7.7) \qquad \left(\sum_{j=1}^m d_j \frac{\alpha(\alpha-1)\ldots(\alpha-j+1)}{t^j} + c(t)\right) e.$$

From this formula we obtain:

(7.7.8) *Corollaries*

(1) If $\alpha \notin \mathbb{Z}$, then φ is an isomorphism, i.e. M coincides with its localization $M_{(t)}$ and, moreover, ∂_t is an isomorphism, $\partial_t\colon C_\beta \rightrightarrows C_{\beta-1}$, where $M_{(t)} = \hat{\oplus} C_\beta$, $C_\beta = \mathbb{C}\cdot t^k e$ is a root subspace of the operator $t\partial_t$ on M with the eigenvalue $\beta = \alpha + k$.

(2) If $\alpha = -1$ ($\alpha \in -\mathbb{N}$), $M = \mathscr{M}^{-1,1}$, then again φ is an isomorphism, but ∂_t is not invertible because $C_\beta \overset{\partial_t}{\rightrightarrows} C_{\beta-1}$ is an isomorphism for $\beta \neq 0$, and $\partial_t\colon C_0 \to C_{-1}$ is a zero map.

(3) If $\alpha = 0$ ($\alpha \in \mathbb{N}_0$), $M = \mathscr{M}^{0,1}$, then φ is not an isomorphism: $\operatorname{Ker}\varphi = \{(d_m\partial_t^m + \ldots + d_1\partial_t)u\}$, and $\operatorname{Im}\varphi = \{c(t)e\} = \mathbb{C}\{t\}e \subset \mathscr{K}\cdot e$ and $\operatorname{Coker}\varphi = \mathbb{C}\{t\}[t^{-1}]/\mathbb{C}\{t\} = M^*$ is a space of principle parts of meromorphic functions.

(7.7.9) Let us consider this third case $M = \mathscr{M}^{0,1} = D/D\cdot t\partial_t$ in more detail. By (7.7.4) elements $Q\cdot u \in M$ can be written in a form $(d_m\partial_t^m + \ldots + d_1\partial_t + c(t))u$. Such a representation corresponds to a root decomposition of the operator $t\partial_t$ on M:

$$M = \hat{\oplus} M_\alpha = \ldots \oplus M_{-1} \oplus M_0 \oplus M_1 \ldots \ni \ldots + d_1\partial_t + c_0 + c_1 t + \ldots$$

∂_t is an isomorphism: the central place is

$$\partial_t\colon M_0 \rightrightarrows M_1,\ c_0 \mapsto c_0\partial_t.$$

Conversely, the operator t is not an isomorphism: since $t\partial_t = 0$, we have that $t\colon M_{-1} \to M_0$ is a zero map. Consequently, $\hat{\oplus}_{\alpha\in-\mathbb{N}} M_\alpha = \operatorname{Ker}\varphi$ is a $\mathbb{C}\{t\}$-torsion.

Let $M_{(t)} = \hat{\oplus} C_\alpha \ni \sum c_k t^k$, $k \geqslant k_0$, be a root decomposition of $M_{(t)}$.

We see that in going from $M = \hat{\oplus} M_k$ to $M_{(t)} = \hat{\oplus} C_k$ in (7.7.3) we 'cut off the tail' $\oplus_{k<0} M_k = M^\tau$ in M and 'change' it by $\oplus_{k<0} C_k = M^*$. Thus, $(\mathscr{M}^{0,1})_{(t)} = \mathscr{M}^{-1,1}$.

Note that the canonical lattice $\mathscr{L} = V^{>-1} = \hat{\oplus}_{k\geqslant 0} M_k = \hat{\oplus}_{k\geqslant 0} C_k$ lies in M as well as in $M_{(t)}$:

(7.7.10) $$\mathscr{L} \subset M \text{ and } \mathscr{L} \subset M_{(t)}.$$

(7.7.11) Therefore, if for a D-module $M = \oplus \mathscr{M}^{\alpha,q}$ the operator ∂_t is invertible, then the transition from M to $M_{(t)}$ consists in changing the summands $\mathscr{M}^{0,q}$ by $\mathscr{M}^{-1,q}$. (Conversely, if the operator t is invertible, $M = M_{(t)}$, then the transition to a D-module with invertible operator ∂_t (microlocalization) consists in exchanging $\mathscr{M}^{-1,q}$ and $\mathscr{M}^{0,q}$).

8 Spectrum of a hypersurface singularity

The spectrum *Sp(f)* of an isolated singularity $f: (\mathbb{C}^{n+1}, 0) \to (\mathbb{C}, 0)$ is a set of μ rational numbers $\alpha_1, \ldots, \alpha_\mu$, which is its most important discrete invariant. It codes the relation between the semisimple part of the monodromy T and the asymptotic Hodge filtration $F^\cdot$ of the MHS on $H^n(X_\infty)$. The numbers $\alpha_j = -(1/2\pi\mathrm{i}) \log \lambda_j$ are the logarithms of the eigenvalues of monodromy $\lambda_j = \mathrm{e}^{-2\pi\mathrm{i}\alpha_j}$, and the choice of values (branches) of logarithm is defined by the filtration $F^\cdot$. More detailed information about the MHS on $H^n(X_\infty)$ is given by a set *Spp(f)* of spectral pairs (α_j, l_j), which, in addition to the Hodge filtration, takes into account the weight filtration $W_.$ on $H^n(X_\infty)$. Giving the *Spp(f)* is equivalent to giving the Hodge numbers $h_\lambda^{p,q}$ of the MHS, which are the main set of discrete invariants of a singularity. The initial definition of the spectrum for isolated singularities [S3] was extended by Steenbrink to the case of non-isolated singularities [S6].

8.1 The definition of the spectrum of an isolated singularity

Let $F^\cdot$, $H = F^0 \supset \ldots \supset F^p \supset F^{p+1} \supset \ldots$, be the Hodge filtration on the vanishing cohomology $H = H^n(X_\infty)$ of an isolated singularity f. The filtration $F^\cdot$ is invariant wrt the action of the semisimple part of monodromy T_s (7.3.3). Hence, T_s acts on $Gr_F^p H = F^p/F^{p+1}$ and $Gr_F^p H = \oplus_\lambda (Gr_F^p)_\lambda$, where $(Gr_F^p)_\lambda = (Gr_F^p H)_\lambda = Gr_F^p H_\lambda$ is the eigensubspace corresponding to λ. Let

(8.1.1) $$\mu^p = \dim Gr_F^p H,\ \mu_\lambda^p = \dim (Gr_F^p)_\lambda.$$

Then $\sum_p \mu^p = \mu$ is the Milnor number, $\sum_\lambda \mu_\lambda^p = \mu^p$, and $\sum_p \mu_\lambda^p = \mu_\lambda$ is the multiplicity of an eigenvalue λ, $\mu_\lambda = \dim H_\lambda$. We can consider the set of eigenvalues of a monodromy $\lambda_1, \ldots, \lambda_\mu$ as a 'divisor', i.e. as an element of a free abelian group $\mathbb{Z}^{(\mathbb{C})}$ with generators (λ), $\lambda \in \mathbb{C}$,

$$(8.1.2) \qquad (\lambda_1) + \ldots + (\lambda_\mu) = \sum_\lambda \mu_\lambda \cdot (\lambda) = \sum_{\lambda, p} \mu_\lambda^p \cdot (\lambda).$$

Thus, all eigenvalues λ are distributed to levels p according to the filtration $F^\cdot$.

Now to each eigenvalue λ we juxtapose its logarithm $\alpha = -(1/2\pi\mathrm{i}) \log \lambda$. Since λ is a root of unity, α is a rational number defined modulo an integer. We normalize α (choose a value of the logarithm) according to the level p of λ wrt $F^\cdot$ by the condition

$$(8.1.3) \qquad \alpha = -\frac{1}{2\pi\mathrm{i}} \log \lambda, \; n - p - 1 < \alpha \leqslant n - p$$

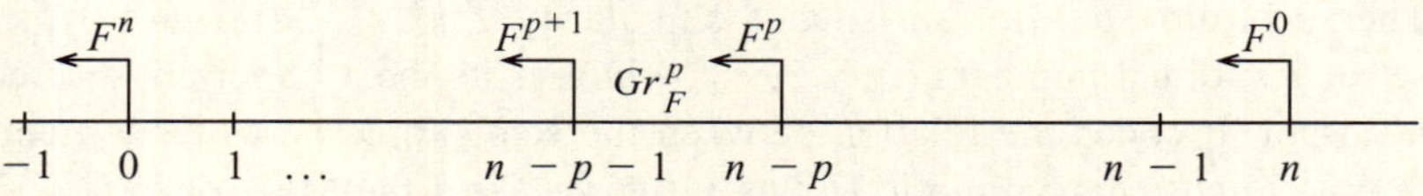

We obtain an element of the group $\mathbb{Z}^{(\mathbb{Q})}$

$$(8.1.4) \qquad Sp(f) = (\alpha_1) + \ldots + (\alpha_\mu) = \sum_\alpha n_\alpha \cdot (\alpha)$$

which is called the *spectrum* of the singularity f. The numbers α_j are called *spectral numbers*, and $n_\alpha = \mu_\lambda^p$ the *spectral multiplicities*.

By (7.4.2) we also have

$$(8.1.4') \qquad n_\beta = \dim Gr_F^p C_\alpha = \dim Gr_V^\beta \Omega_f,$$

where $\beta = \alpha + n - p$, $-1 < \alpha \leqslant 0$, and

$$(8.1.4'') \qquad \dim Gr_V^\beta \mathscr{H}^{(0)} = n_\alpha + n_{\alpha+1} + \ldots + n_\beta = f_\lambda^p$$

(see picture, p. 107).

(8.1.5) The choice of logarithm values (normalization) (8.1.3) is motivated by the definition of the filtration $F^\cdot$ in the previous section, (7.2.3), : F^p is generated by leading parts of geometric sections of differential forms $\omega \in \mathscr{H}^{(0)}$ with orders $\alpha(\omega) \in (n-p-1, n-p]$. Thus, if $\omega_1, \ldots, \omega_{\mu^p} \in \mathscr{H}^{(0)}$ are such that $n - p - 1 < \alpha(\omega_j) \leqslant n - p$ and $s[\omega_j, \alpha(\omega_j)]$ form a basis of F^p/F^{p+1}, then the orders $\alpha(\omega_1), \ldots, \alpha(\omega_{\mu^p})$ are the part of the spectrum belonging to $(n-p-1, n-p]$.

The spectrum defined as a set of μ rational numbers, or as a 'divisor' $Sp(f) = \sum n_\alpha \cdot (\alpha) \in \mathbb{Z}^{(\mathbb{Q})}$ can also be considered, following M. Saito [Sa9], as a fractional Laurent polynomial

$$Sp(f) = t^{\alpha_1} + \ldots + t^{\alpha_\mu} = \sum_\alpha n_\alpha t^\alpha \in \mathbb{Z}[t^{1/m}, t^{-1/m}], \tag{8.1.6}$$

where m is a common denominator of the numbers α_j, $T_s^m = id$.

(8.1.7) *Remark* M. Saito normalized logarithms in a different way: the spectral number α corresponding to an eigenvalue λ of level p satisfies the condition

$$\alpha = -\frac{1}{2\pi i} \log \lambda, \; n - p < \alpha \leqslant n - p + 1. \tag{8.1.8}$$

If we denote by $\widetilde{Sp}(f)$ the spectrum of a singularity according to Saito, then $\widetilde{Sp}$ is obtained from Sp by 'a shift by 1 to the right' or

$$\widetilde{Sp}(f) = t \cdot Sp(f) = t \cdot \sum_\alpha n_\alpha t^\alpha. \tag{8.1.9}$$

The spectral numbers contained in $\widetilde{Sp}(f)$, i.e. numbers $\alpha + 1$, Saito calls the *exponents* of f.

8.2 The spectral pairs $Spp(f)$

The spectral pairs, or characteristic pairs as Steenbrink called them in [S3], give more detailed information than the spectrum does. They take into account a distribution of eigenvalues of monodromy λ not only wrt the Hodge filtration $F^\cdot$, but also wrt the weight filtration $W_.$ Since both filtrations $F^\cdot$ and $W_.$ on the space $H = H^n(X_\infty, \mathbb{C})$ are invariant wrt the semisimple part of monodromy T_s, T_s acts on $Gr_F^p Gr_{p+q}^W H$ and $Gr_F^p Gr_{p+q}^W H = \oplus_\lambda Gr_F^p Gr_{p+q}^W H_\lambda$. In (3.5.5) we denoted Hodge numbers by $h_\lambda^{p,q} = \dim Gr_F^p Gr_{p+q}^W H_\lambda$.

(8.2.1) *Definition* Let us juxtapose these $h_\lambda^{p,q}$ eigenvalues λ and the set of $n_{\alpha,1} = h_\lambda^{p,q}$ *spectral pairs* (α, l), where $\alpha = -(1/2\pi i) \log \lambda$ is a *spectral number* normalized by the level p by formula $n - p - 1 < \alpha \leqslant n - p$ (8.1.3), and l is a *weight number*

$$l = \begin{cases} p + q, & \text{if } \lambda \neq 1, \\ p + q - 1, & \text{if } \lambda = 1. \end{cases} \tag{8.2.2}$$

As in the case of the spectrum we can consider the set of spectral pairs $Spp(f)$ as an element of an abelian group with generators $(\alpha, 1) \in \mathbb{Q} \times \mathbb{Z}$,

(8.2.3) $$Spp(f) = \sum n_{\alpha,1}\cdot(\alpha, 1).$$

Obviously, giving $Spp(f)$ is equivalent to giving the set of all Hodge numbers $h_\lambda^{p,q}$, and their spectral multiplicities are

(8.2.4) $$n_\alpha = \sum_l n_{\alpha,l}.$$

8.3 Properties of the spectrum

The symmetry of Hodge numbers in §3 implies the following symmetries of multiplicities of spectral pairs of an isolated singularity f: $(\mathbb{C}^{n+1}, 0) \to (\mathbb{C}, 0)$.

(8.3.1) *Proposition*

(i) $n_{\alpha,l} = n_{2n-1-l-\alpha,l}$,
(ii) $n_{\alpha,l} = n_{\alpha-n+l,2n-l}$,
(iii) $n_{\alpha,l} = n_{n-1-\alpha,2n-l}$.

Proof The proof of the first formula follows immediately from the symmetry (3.5.6) $h_\lambda^{p,q} = h_{\bar\lambda}^{q,p}$, and that of the second from (3.5.7) $h_{\neq 1}^{p,q} = h_{\neq 1}^{n-q,n-p}$ and (3.5.8) $h_1^{p,q} = h_1^{n+1-q,n+1-p}$. The second number of a spectral pair is defined by formula (8.2.2) in such a way that for both types of second symmetry of $h_\lambda^{p,q}$ the symmetry for spectral pairs is given by the same formula. The third formula in (8.3.1) follows from the two first. Similarly, any two of these formulae imply the third. ∎

(8.3.2) *Corollary* (symmetry of the spectrum) The spectrum $Sp(f)$ of an isolated singularity f: $(\mathbb{C}^{n+1}, 0) \to (\mathbb{C}, 0)$ is symmetric wrt the point $(n-1)/2$,

(8.3.3) $$n_\alpha = n_{n-1-\alpha}.$$

This follows immediately from (8.2.4) and formula (iii) in (8.3.1).

(8.3.4) *Corollary* (range of the spectrum) The spectrum (as a set of rational numbers) is contained in the interval $(-1, n)$

(8.3.5) $$Sp(f) \subset (-1, n).$$

Moreover,

(8.3.6) $$F^0 H^n(X_\infty, \mathbb{C}) = H^n(X_\infty, \mathbb{C})$$

and

(8.3.7) $$V^{>-1} \supset \mathscr{H}^{(0)} \supset V^{n-1},$$

where $\mathscr{H}^{(0)}$ is the Brieskorn lattice and $V^{\cdot}$ is the root filtration.

Proof By Malgrange's theorem (I.6.2.1) it follows from (6.5.6) that $\mathscr{H}^{(0)} \subset V^{>-1}$, and hence $Gr^{\beta}_{V}\mathscr{H}^{(0)} = 0$ for $\beta \leqslant -1$ and, consequently, $F^{n+1} = 0$. Hence, $n_\alpha = 0$ for $\alpha \leqslant -1$, i.e. $Sp(f) \subset (-1, \infty)$. The symmetry (8.3.3) implies that $Sp(f) \subset (-1, n)$. It follows that $Gr^{p}_{F} H_\lambda = 0$ for $p < 0$ (and, moreover, $Gr^{p}_{F} H_\lambda = 0$ for $p < 1$, if $\lambda = 1$), i.e. $F^0 H = H$. Furthermore, $Gr^{\beta}_{V}\mathscr{H}^{(0)} = C_\beta$ for $\beta \geqslant n-1$ (otherwise there exists a spectral number $\geqslant n$), i.e. leading parts of geometrical sections generate C_β for $\beta \geqslant n-1$. It follows that $\mathscr{H}^{(0)} \supset V^{n-1}$. ■

8.4 The spectra of a quasihomogeneous and a semiquasihomogeneous singularity

Let $f(x) = \sum c_m x^m$, $m = (m_0, \ldots, m_n)$, $x^m = x_0^{m_0} \cdot \ldots \cdot x_n^{m_n}$, be a quasihomogeneous singularity of degree 1 with variables of weights $\operatorname{wt} x_j = w_j$. The (quasi)degree of monomials x^m

(8.4.1) $$\nu(x^m) = \sum_{j=0}^{n} w_j m_j$$

defines the order function (valuation)

$$\nu\colon \mathscr{O}_{X,0} = \mathbb{C}\{x_0, \ldots, x_n\} \to \mathbb{Q}_{\geqslant 0} \cup \{\infty\},$$

where for a series $g(x) = \sum b_m x^m$ we set

(8.4.2) $$\nu(g) = \min\{\nu(x^m)\colon b_m \neq 0\}.$$

The inclusion $\Omega^{n+1}_{X,0} \subset \mathscr{O}_{X,0}\cdot(\mathrm{d}x_0/x_0) \wedge \ldots \wedge (\mathrm{d}x_n/x_n)$ defines an order function on the space of germs of differential forms $\Omega^{n+1}_{X,0}$, which we also denote by $\nu\colon \Omega^{n+1}_{X,0} \to \mathbb{Q}_{>-1} \cup \{\infty\}$, and where for a form $\omega = g(x)\,\mathrm{d}x_0 \wedge \ldots \wedge \mathrm{d}x_n$, $g \in \mathbb{C}\{x_0, \ldots, x_n\}$, we set (with a shift of -1)

(8.4.3) $$\nu(\omega) = \nu(g\cdot x_0 \cdot \ldots \cdot x_n) - 1.$$

In particular, for a monomial $\omega_m = x^m\,\mathrm{d}x$

(8.4.4) $$\nu(\omega_m) = \sum_{j=0}^{n} w_j\cdot(m_j + 1) - 1 \underset{\mathrm{df}}{=} a(m) - 1.$$

The order function ν defines a decreasing filtration on $\Omega^{n+1}_{X,0}$. We want to transfer the order function and this filtration to the quotient $\mathscr{H}^{(0)} = \Omega^{n+1}_{X,0}/\mathrm{d}f \wedge \mathrm{d}\Omega^{n-1}_{X,0}$. In what follows we'll meet this situation of coming to a quotient space more than once. Therefore we make the following general remark.

(8.4.5) On the order function and the filtration on a quotient. Assume that on a group (module, ring etc.) M there is an order function α: $M \to \mathbb{Q}_{\geqslant 0} \cup \{\infty\}$ satisfying the condition

$$\alpha(m_1 + m_2) \geqslant \min\{\alpha(m_1), \alpha(m_2)\}, \forall\, m_1, m_2 \in M,$$

where the equality holds if $\alpha(m_1) \neq \alpha(m_2)$.

With an order function α one associates a decreasing $\mathbb{Q}$-filtration $V^{\cdot}$,

$$V^{\beta}M = \{m \in M\colon \alpha(m) \geqslant \beta\},$$

with $V^{>\beta}M$ defined analogously. The joined object $Gr^{\cdot}_V M$ has components $Gr^{\beta}_V M = V^{\beta}M/V^{>\beta}M$. The order function α may be recovered by the filtration V.

$$\alpha(m) = \max\{\beta\colon m \in V^{\beta}\}.$$

Now let $N \subset M$ be a subgroup and $\overline{M} = M/N$ its quotient group. We can define a function $\overline{\alpha}$: $\overline{M} \to \mathbb{Q}_{\geqslant 0} \cup \{\infty\}$ on $\overline{M}$,

$$\overline{\alpha}([m]) = \max\{\alpha(m+n),\ n \in N\}.$$

Then we can consider two filtrations on $\overline{M}$:

(i) a filtration $\overline{V}^{\cdot}$ associated with $\overline{\alpha}$,

$$\overline{V}^{\beta}(\overline{M}) = \{[\overline{m}], \overline{\alpha}([m]) \geqslant \beta\} \text{ and}$$

(ii) a quotient filtration $\overline{\overline{V}}^{\cdot}$,

$$\overline{\overline{V}}^{\beta}(\overline{M}) = (V^{\beta}M + N)/N \subset M/N = \overline{M}.$$

It is easy to verify the following proposition.

Proposition. $\overline{V}^{\cdot} = \overline{\overline{V}}^{\cdot}$ Therefore in the following we denote by $\overline{\alpha}$ and $\overline{V}^{\cdot}$ the order function and the filtration on a quotient.

We now return to the Newton order function ν on $\Omega^{n+1}_{X,0}$. Let $\overline{\nu}$: $\mathscr{H}^{(0)} \to \mathbb{Q}_{\geqslant 0} \cup \{\infty\}$ be the corresponding order function on the Brieskorn lattice $\mathscr{H}^{(0)}$. We showed in (I.5.5.4) that for monomial forms $\omega_m = x^m\,\mathrm{d}x$ we have $t\partial_t[\omega_m] = (\alpha(m) - 1)[\omega_m] = \nu(\omega_m)[\omega_m]$, i.e. $[\omega_m] \in C_{\nu(\omega)}$ and

(8.4.6) $$\alpha([\omega_m]) = \nu(\omega_m) = \overline{\nu}([\omega_m]),$$

where α is the order function defined by the root decomposition $\mathscr{M} = \hat{\oplus} C_\alpha$ in (6.3.1). It follows that

(8.4.7) $$\alpha(\omega) = \overline{\nu}(\omega) \text{ for all } \omega \in \mathscr{H}^{(0)}.$$

This implies the following proposition:

(8.4.8) *Proposition* If $\{x^m\}$, $m \in A$, $|A| = \mu$, is a monomial basis of the Jacobian algebra $Q_f = \mathcal{O}_{X,0}/J_f$, then the forms $\omega_m = x^m\,dx$ form a basis of the $\mathcal{O}_{S,0}$-module $\mathcal{H}^{(0)}$, and their orders

$$\alpha(\omega_m) = \nu(\omega_m) = a(m) - 1$$

form the set of spectral numbers,

(8.4.9) $$Sp(f) = \sum_{m\in A}(\nu(\omega_m)).$$

For the spectrum as a fractional Laurent polynomial we have the elegant formula:

(8.4.10) $$\widetilde{Sp}(f) = \prod_{j=0}^{n} \frac{t^{w_j} - t}{1 - t^{w_j}}$$

(see [S3] and (8.5.8)).

(8.4.11) *Example* Let $f(x, y) = x^3 + y^4$ be the singularity of type E_6. Then $\mathrm{wt}\,x = \frac{1}{3}$, $\mathrm{wt}\,y = \frac{1}{4}$. The following monomials x^m form a basis of Q_f, and for the values of $\nu(\omega_m) = a(m) - 1$ we have

x^m:	1	y	x	y^2	xy	xy^2
$\nu(\omega_m)$:	$-\frac{5}{12}$	$-\frac{2}{12}$	$-\frac{1}{12}$	$\frac{1}{12}$	$\frac{2}{12}$	$\frac{5}{12}$.

By the formula (8.4.10) we have

$$\widetilde{Sp}(f) = (t^{1/3} - t)(t^{1/4} - t)(t^{1/3} - 1)^{-1}(t^{1/4} - 1)^{-1}$$
$$= t(t^{-5/12} + t^{-2/12} + t^{-1/12} + t^{1/12} + t^{2/12} + t^{5/12}).$$

As shown by Steenbrink [S2], $Gr_k^W H^n(X_\infty) = 0$ for $k \neq n, n+1$ and

(8.4.12)

$$\{\omega_m : n - p - 1 < \nu(\omega_m) < n - p\} \text{ is a basis of } Gr_F^p Gr_n^W H^n(X_\infty),$$

$$\{\omega_m : \nu(\omega_m) = n - p\} \text{ is a basis of } Gr_F^p Gr_{n+1}^W H^n(X_\infty).$$

In particular, for quasihomogeneous singularities all second numbers l of spectral pairs are equal to n.

A singularity $f(x)$ is called *semiquasihomogeneous* with weights $w_0, \ldots, w_n$, if $f(x) = f_0(x) + g(x)$, where $f_0(x)$ is an isolated quasihomogeneous singularity of degree $\deg f_0 = \nu(f_0) = 1$, and $\nu(g) > 1$. A semiquasihomogeneous singularity is a μ-const deformation of its quasihomogeneous part $f_0(x)$. By Varchenko's theorem (see (8.9.11) below) the spectrum is constant in μ-const deformations. Thus, for a semiquasihomogeneous singularity $f(x)$ the spectra $Sp(f)$ and $Sp(f_0)$ coincide.

8.5 Calculation of the spectrum of an isolated singularity in terms of a Newton diagram

(8.5.1) Let $f = \sum c_m x^m \in \mathscr{O} = \mathscr{O}_{\mathbb{C}^{n+1},0} = \mathbb{C}\{x_0, \ldots, x_n\}$ be a germ of an isolated singularity, $\operatorname{supp} f = \{m \in \mathbb{N}_0^{n+1}: c_m \neq 0\}$ be its support, $\Gamma_+(f)$ be the Newton polyhedron, i.e. a convex hull of $\bigcup_{m \in supp\, f}(m + \mathbb{R}_+^{n+1}) \subset \mathbb{R}^{n+1}$, and $\Gamma(f)$ be the Newton boundary (or the Newton diagram), i.e. a union of compact faces σ of $\Gamma_+(f)$. We assume that f is Γ-non-degenerate (non-degenerate with respect to $\Gamma(f)$), i.e. for each compact face σ of $\Gamma(f)$ the polynomials $\partial f_\sigma / \partial x_0, \ldots, \partial f_\sigma / \partial x_n$ have no common zeroes on $(\mathbb{C} \backslash \{0\})^{n+1}$. Here $f_\sigma = \sum_{m \in \operatorname{cone}(0,\sigma)} c_m x^m$, $\operatorname{cone}(0, \sigma)$ is a cone over the face σ with a vertex at the point 0. Moreover, we assume that Γ is convenient, i.e. Γ intersects all coordinate axes or $f(x)$ contains monomials $x_j^{m_j}$, $j = 0, \ldots, n$. This doesn't affect the generality, because f has an isolated singularity and for large m_i the germs f and $f + \sum_{j=0}^n x_j^{m_j}$ are R-equivalent.

(8.5.2) Let us consider a homogeneous function $h: \mathbb{R}_+^{n+1} \to \mathbb{R}_+$, $h(\lambda a) = \lambda h(a)$, such that $h(\Gamma) = 1$. Γ defines a decreasing filtration on the ring $\mathscr{O} = \mathscr{O}_{\mathbb{C}^{n+1},0}$

(8.5.3) $$\mathscr{N}^{\cdot a}\mathscr{O} = \{g(x) \in \mathscr{O}: h(\operatorname{supp} g) \geqslant a\},$$

called the *Newton filtration*, and correspondingly $\mathscr{N}^{\cdot > a}\mathscr{O}$ and $Gr^a_{\mathscr{N}} \cdot = \mathscr{N}^{\cdot a}/\mathscr{N}^{\cdot > a}$. The filtration $\mathscr{N}^{\cdot\cdot}$ defines an order function ν: $\mathscr{O} \to \mathbb{Q}_{\geqslant 0} \cup \{\infty\}$, where the *Newton order* or the Newton degree of an element $g \in \mathscr{O}$ is given by

(8.5.4) $$\nu(g) = \max\{a | g \in \mathscr{N}^{\cdot a}\}.$$

The Newton filtration $\mathscr{N}^{\cdot\cdot}$ induces an order function $\overline{\nu}$ of a quotient filtration on the Jacobian algebra $Q^f = \mathscr{O}/J_f$.

The filtration $\mathscr{N}^{\cdot\cdot}$ and the order ν on ν induce the Newton filtration and the Newton order on $\Omega^{n+1}_{X,0}$, where ν: $\Omega^{n+1}_{X,0} \to \mathbb{Q}_{>-1} \cup \{\infty\}$ is defined by

(8.5.5) $$\nu(\omega) = \nu(g \cdot x_0 \ldots x_n) - 1 \text{ for } \omega = g(x)\,\mathrm{d}x_0 \wedge \ldots \wedge \mathrm{d}x_n.$$

Correspondingly one defines the quotient filtration $\mathscr{N}^{\cdot\cdot}$ and the Newton order $\overline{\nu}$: $\mathscr{H}^{(0)} \to \mathbb{Q}_{>-1} \cup \{\infty\}$ on the quotient $\mathscr{H}^{(0)} = \Omega^{n+1}_{X,0}/\mathrm{d}f \wedge \mathrm{d}\Omega^{n-1}_{X,0}$

(8.5.6) $$\overline{\nu}([\omega]) = \max\{\nu(\eta): [\eta] = [\omega]\},$$
$$\mathscr{N}^{\cdot a}\mathscr{H}^{(0)} = \{[\omega] \in \mathscr{H}^{(0)}: \overline{\nu}([\omega]) \geqslant a\},$$

and analogously for $\Omega_f = \mathscr{H}^{(0)}/\mathscr{H}^{(-1)} = \Omega^{n+1}_{X,0}/\mathrm{d}f \wedge \Omega^n_{X,0}$.

With a filtration $F^\cdot$ on the vector space H there is associated the Poincaré polynomial

(8.5.7) $$p_{H,F^\cdot}(t) = \sum g_{\alpha} \cdot t^{\alpha},$$

where $g_\alpha = \dim F^\alpha / F^{>\alpha}$.

M. Saito [Sa6] proved Steenbrinks's conjecture [S3] on the calculation of the spectrum in terms of the Newton filtration

(8.5.8) *Theorem* For a Γ-non-degenerate function f with an isolated singularity the root filtration $V^\cdot$ and the Newton filtration $\mathcal{N}^\cdot$ on the Brieskorn lattice $\mathcal{H}^{(0)}$ coincide and therefore

(8.5.9) $$\mathrm{Sp}(f) = \sum_{i=1}^{\mu} t^{\alpha_i} = p_{\Omega_f, \mathcal{V}^\cdot}(t) = p_{\Omega_f, \mathcal{N}^\cdot}(t).$$

Varchenko and Khovanskii [V-Kh] gave an elementary proof of this theorem.

(8.5.10) Steenbrink [S3] obtained, using results of Kouchnirenko [Ko], the following formula for the calculation of $p_{\Omega_j, \mathcal{N}^\cdot}(t)$ in terms of the Newton diagram. Let σ be a face of $\Gamma(f)$. We define A_c to be the $\mathbb{C}$-algebra generated by the monomials x^m for $m \in \mathrm{cone}\,(0, \sigma)$. The Newton filtration on $\mathcal{O}$ induces a filtration on $A_\sigma \subset \mathcal{O}$. Let $p_{A_\sigma}(t)$ be the Poincaré series of A_σ associated with this filtration. Then

(8.5.11)

$$\tilde{p}_{\Omega_f, \mathcal{N}^\cdot}(t) \stackrel{\mathcal{D}f}{=} t p_{\Omega_f, \mathcal{N}^\cdot}(t) = \sum_{\sigma} (-1)^{n-\dim \sigma} (1-t)^{k(\sigma)} p_{A_\sigma}(t) + (-1)^{n+1},$$

where the summation is taken over all faces σ of $\Gamma(f)$, and $k(\sigma)$ is the dimension of the minimal coordinate plane containing σ.

The function h from (8.5.2) is linear on $\mathrm{cone}\,(0, \sigma)$, and if this cone is 'basis', i.e. it is generated by a part of a basis of the lattice of monomials, then A_σ is isomorphic to a polynomial algebra, $A_\sigma = \mathbb{C}[y_1, \ldots, y_k]$, graded by the condition $\mathrm{wt}\, y_i = w_i = \nu(y_i)$, $i = 1, \ldots, k$. In this case the Poincaré series is well known ([AGV]),

(8.5.12) $$p_{A_\sigma}(t) = 1/(1 - t^{w_1}) \ldots (1 - t)^{w_k},$$

(8.5.13) *Example* Let $f(x, y) = x^r - x^2y^2 + ay^s$, $a \neq 0$, $r, s \geqslant 4$, r or $s > 4$. This is the singularity of type $T_{2,r,s}$ [AGV]. In this case $\Gamma(f)$ consists of two segments σ_{13} and σ_{23} joining points σ_1 and σ_3, and σ_2 and σ_3 where $\sigma_1 = (r, 0)$, $\sigma_2 = (0, s)$, $\sigma_3 = (2, 2)$. There are five faces in $\Gamma(f)$: $\sigma_1, \sigma_2, \sigma_3, \sigma_{13}$ and σ_{23}. For them the number $k(\sigma)$ equals 1, 1, 2, 2 and 2, respectively. The cone $\mathrm{cone}\,(0, \sigma_{13})$ (respectively cone$(0, \sigma_{23})$) is generated by $m_1 = (1, 0)$ and $m_2 = (1, 1)$, and it is 'basis'. Since $\nu(x) = 1/r$, $\nu(xy) = \frac{1}{2}$, we have by (8.5.12) that $p_{A_{\sigma_{13}}} = (1 - t^{1/2})^{-1}(1 - t^{1/r})^{-1}$. The cones cone$(0, \sigma_1)$, cone$(0, \sigma_2)$ and cone$(0, \sigma_3)$ are generated by $(1, 0)$, $(0, 1)$ and $(1, 1)$, and also are 'basis'. We have $p_{A_{\sigma_1}} = (1 - t^{1/r})^{-1}$, $p_{A_{\sigma_2}} = (1 - t^{1/s})^{-1}$, $p_{A_{\sigma_3}} = (1 - t^{1/2})^{-1}$. Theorem (8.5.8) and formula (8.5.11) give that $\widetilde{Sp}(f)$ is equal to

$$
\begin{aligned}
\tilde{p}_{\Omega_f,\mathcal{A}}(t) = {} & (1-t)^2(1-t^{1/2})^{-1}(1-t^{1/r})^{-1} \\
& + (1-t)^2(1-t^{1/2})^{-1}(1-t^{1/s})^{-1} \\
& - (1-t)(1-t^{1/r})^{-1} - (1-t)(1-t^{1/s})^{-1} \\
& - (1-t)^2(1-t^{1/2})^{-1} + 1 \\
= {} & t^{1/2}(t^{(r-1)/r} + \ldots + t^{1/r}) + t^{1/2}(t^{(s-1)/s} + \ldots + t^{1/s}) \\
& + t^{3/2} + t + t^{1/2}.
\end{aligned}
$$

(8.5.14) *Example* Let $f(x, y, z) = x^p + y^q + z^r + axyz$, $a \neq 0$, $p^{-1} + q^{-1} + r^{-1} < 1$, be a singularity of type $T_{p,q,r}$. In this case $\Gamma(f)$ consists of three triangles σ_{124}, σ_{234}, σ_{314} defined by three of the four points $\sigma_1(p, 0, 0)$, $\sigma_2(0, q, 0)$, $\sigma_3(0, 0, r)$ and $\sigma_4(1, 1, 1)$. There are 13 faces in $\Gamma(f)$: $\sigma_1, \sigma_2, \sigma_3, \sigma_4, \sigma_{12}, \sigma_{23}, \sigma_{31}, \sigma_{14}, \sigma_{24}, \sigma_{34}, \sigma_{124}, \sigma_{234}, \sigma_{314}$. For σ_4 and for the last six of these faces $k(\sigma) = 3$. The cones over all the faces are 'basis'. Hence, by (8.5.12) we have

$$
\begin{aligned}
p_{A_{\sigma_1}} &= (1-t^{1/p})^{-1},\ p_{A_{\sigma_{12}}} = (1-t^{1/p})^{-1}(1-t^{1/q})^{-1}, \\
p_{A_{\sigma_{14}}} &= (1-t)^{-1}(1-t^{1/p})^{-1}, \\
p_{A_{\sigma_{124}}} &= (1-t)^{-1}(1-t^{1/p})^{-1}(1-t^{1/q})^{-1} \text{ etc.}
\end{aligned}
$$

Calculating $\tilde{p}_{\Omega_f,\mathcal{A}}(t)$ using the formula (8.5.11), we obtain by the theorem (8.5.8)

$$
\begin{aligned}
Sp(f) = {} & 1 + t + (t^{(p-1)/p} + \ldots + t^{1/p}) + (t^{(q-1)/q} + \ldots + t^{1/q}) \\
& + (t^{(r-1)/r} + \ldots + t^{1/r}).
\end{aligned}
$$

(8.5.15) *Example* Let $f(x, y, z) = x^8 + y^8 + z^8 + x^2y^2z^2$.

The same calculations as in (8.5.14) will give

$$\tilde{S}p(f) = t(t^{-1/2} + 3t^{-3/8} + 6t^{-2/8} + 9t^{-1/8} + 13 + 18t^{1/8} + 21t^{2/8} + 24t^{3/8}$$
$$+\ 25t^{1/2} + 24t^{5/8} + 21t^{6/8} + 18t^{7/8} + 13t$$
$$+\ 9t^{9/8} + 6t^{10/8} + 3t^{11/8} + t^{3/2}).$$

In particular, we obtain that $\mu = 215$ and the characteristic polynomial of the monodromy is $(t^8 - 1)^{27}(t - 1)^{-1}$.

See also Danilov's paper [Da] about the calculation of the Hodge numbers $h_\lambda^{p,q}$ in terms of the Newton diagram.

(8.5.16) The calculation of the negative part of spectrum. On the Brieskorn lattice $\mathscr{H}^{(0)} = \Omega^{n+1}_{X,0}/\mathrm{d}f \wedge \mathrm{d}\Omega^{n-1}_{X,0}$ as on a quotient, the Newton order $\overline{\nu}$ is defined by formula (8.5.6)

$$\overline{\nu}([\omega]) = \max\{\nu(\omega + \eta), \eta \in \mathrm{d}f \wedge \mathrm{d}\Omega^{n-1}_{X,0}\}.$$

The function ν satisfies the property $\nu(\omega + \eta) \geqslant \min\{\nu(\omega), \nu(\eta)\}$ and $\nu(\omega + \eta) = \nu(\omega)$, if $\nu(\omega) \leqslant \nu(\eta)$. Hence for ω such that $\nu(\omega) < \min\{\nu(\eta), \eta \in \mathrm{d}f \wedge \mathrm{d}\Omega^{n-1}_{X,0}\}$, we have

(8.5.17) $$\overline{\nu}([\omega]) = \nu(\omega).$$

Let

(8.5.18)
$$s \underset{\mathrm{df}}{=} \min\{\nu(\omega), \omega \in \mathrm{d}f \wedge \mathrm{d}\Omega^{n}_{X,0}\}$$
$$= \min\left\{\nu(g(x)x_0, \ldots, x_n) - 1,\ g \in J_f = \left(\frac{\partial f}{\partial x_0}, \ldots, \frac{\partial f}{\partial x_n}\right)\right\}$$
$$= \min\left\{\nu\left(\frac{\partial f}{\partial x_j}\cdot x_0\cdot \ldots \cdot x_n\right) - 1,\ j = 0, \ldots, n\right\}.$$

It is obvious that $s > 0$ (because $\nu(\partial f/\partial x_j \cdot x_j) = \nu(f) = 1$) and that $s \leqslant \min\{\nu(\omega), \omega \in \mathrm{d}f \wedge \mathrm{d}\Omega^{n-1}_{X,0}\}$. We then obtain:

(8.5.19) *Proposition* If $\nu(\omega) < s$, then

$$\alpha([\omega]) = \overline{\nu}([\omega]) = \nu(\omega).$$

(8.5.20) Let $A \subset \mathbb{N}_0^{n+1}$ be a set of exponents m of monomials x^m for which

$$\nu(x^m x_0\cdot \ldots \cdot x_n) - 1 < s$$

and correspondingly $A_0 \subset A$ be a set of exponents of the *underdiagram monomials*

$$A_0 = \{m: \nu(x^m x_0 \cdot \ \ldots \ \cdot x_n) - 1 \leqslant s\}.$$

A_0 is a set of those m for which $m + (1, \ldots, 1)$ doesn't belong to the interior of the Newton polyhedron.

Then for the forms $\omega_m = x^m\,\mathrm{d}x$, $m \in A$, we have

$$\alpha(\omega_m) = \nu(\omega_m) = \nu(x^m x_0 \cdot \ \ldots \ \cdot x_n) - 1.$$

Now note that the images of monomials x^m, $m \in A$, are linear independent in the Jacobian algebra Q_f and hence forms ω_m, $m \in A$, are linear independent $\bmod \mathrm{d}f \wedge \Omega^n_{X,0}$, i.e. they are linear independent in $\Omega_f = \mathscr{H}^{(0)}/\mathscr{H}^{(-1)}$. Indeed, by definition of the set A for any linear combination $\sum_{m\in A} c_m x^m$ we have $\nu((\sum_{m\in A} c_m x^m)x_0 \ldots x_n) - 1 < s$ and $\sum_{m\in A} c_m x^m\,\mathrm{d}x = \sum_{m\in A} c_m \omega_m \notin \mathrm{d}f \wedge \Omega^n_{X,0}$ by the definition of s.

Moreover, any form $\omega = g(x)\,\mathrm{d}x$, $g(x) = \sum_m c_m x^m = \sum_{m\in A} c_m x^m + \sum_{m\notin A} c_m x^m$, can be written in the form

$$\omega = \sum_{m\in A} c_m x^m + \eta,$$

where $\nu(\eta) \geqslant s$ and, consequently, $\alpha([\eta]) = \overline{\nu}([\eta]) \geqslant \nu(\eta) \geqslant s$. We obtain:

(8.5.21) *Corollary* The images of forms ω_m, $m \in A$, form a basis in the space $\oplus_{-1<\alpha<s} Gr^{\alpha}_{V}\mathscr{H}^{(0)}$, and also in $\oplus_{-1<\alpha<s} Gr^{\alpha}_{V}\Omega_f$, and their orders $\alpha(\omega)$, $m \in A$, are exactly the spectral numbers in the interval $(-1, s)$. In particular, $\alpha(\omega_m)$, $m \in A_0$, is exactly the part of spectrum in $(-1, 0]$ and hence

$$\sum_{-1<\alpha<0} n_\alpha = \dim F^n H^n(X_\infty) = \#(A_0). \tag{8.5.22}$$

Moreover, the least spectral number is given by

$$\alpha_{\min} = \nu(x_0 \ldots x_n) - 1. \tag{8.5.23}$$

(8.5.24) Corollary (8.5.21) enables us to calculate the spectrum completely in the cases of curves ($n = 1$) and surfaces ($n = 2$):

(1) If $n = 1$, $f\colon (\mathbb{C}^2, 0) \to (\mathbb{C}, 0)$, then $\mathrm{Sp}(f) \subset (-1, 1)$ and $\mathrm{Sp}(f)$ is symmetrical wrt the point 0. Knowing the part of $\mathrm{Sp}(f)$ in $(-1, 0]$ we know $\mathrm{Sp}(f)$ completely by symmetry.

(2) If $n = 2$, $f\colon (\mathbb{C}^3, 0) \to (\mathbb{C}, 0)$, then $\mathrm{Sp}(f) \subset (-1, 2)$ and $\mathrm{Sp}(f)$ is symmetrical wrt the point $\frac{1}{2}$. Knowing the spectrum in the interval $(-1, 0]$, we find it in $[1, 2)$ by symmetry. For the other eigenvalues λ of the monodromy the numbers $\alpha = -(1/2\pi \mathrm{i}) \log \lambda$ give the part of spectrum in the interval $(0, 1)$.

8.6 Calculation of the geometric genus of a hypersurface singularity in terms of the spectrum

Recall that the geometric genus p_g of a normal isolated singularity $(Y, 0)$ of dimension n is defined by means of its resolution $\pi\colon \overline{Y} \to Y$.

(8.6.1) $$p_g = \dim R^{n-1}\pi_*\mathcal{O}_{\overline{Y}},$$

if $n \geqslant 2$. For curves ($n = 1$) $p_g = \delta$ is the δ-invariant of the singularity, $\delta = \dim \pi_*\mathcal{O}_{\overline{Y}}/\mathcal{O}_Y$.

(8.6.2) *M. Saito's theorem* [Sa3] For an isolated hypersurface singularity $(Y, 0) \subset (\mathbb{C}^{n+1}, 0)$ with equation $f(x) = 0$

$$p_g = \dim F^n H^n(X_\infty, \mathbb{C}) = \sum_{-1<\alpha\leqslant 0} n_\alpha,$$

i.e. the genus is equal to the degree of the non-positive part of spectrum.

8.7 Spectrum of the join of isolated singularities

Let $f(x) \in \mathbb{C}\{x_0, \ldots, x_n\}$ and $g(y) \in \mathbb{C}\{y_0, \ldots, y_m\}$ be isolated singularities of $n + 1$ and $m + 1$ variables, respectively.

(8.7.1) *Definition* The *join* (or the direct sum) of singularities f and g is the singularity $f \oplus g$ in $n + m + 2$ variables defined by the function $f(x) + g(y) \in \mathbb{C}\{x_0, \ldots, x_n; y_0, \ldots, y_m\}$.

This singularity is also isolated. Let $X^f \to S$, X^f_∞, $\mathcal{H}^{(0)}_f$ etc. be the Milnor fibration, the canonical fibre, the Brieskorn lattice etc. for the singularity f (and analogously for g and $f \oplus g$).

Singularities of the form $f \oplus g$ and the results about them are called Sebastiani–Thom singularities and 'Sebastiani–Thom' results, respectively because Sebastiani and Thom [Se-T] proved that the Milnor fibre $X^{f\oplus g}_\infty$ has the homotopy type of the join of the Milnor fibres of f and g, $X^{f\oplus g}_\infty \simeq X^f_\infty * X^g_\infty$, and in particular

(8.7.2) $$H^{n+m+1}(X^{f\oplus g}_\infty, \mathbb{C}) \simeq H^n(X^f_\infty, \mathbb{C}) \otimes H^m(X^g_\infty, \mathbb{C})$$

and $\mu^{f\oplus g} = \mu^f \cdot \mu^g$. They also calculated the monodromy of the singularity $f \oplus g$, $T^{f\oplus g} = T^f \otimes T^g$ (see also [AGV]).

(8.7.3) *Theorem* If $\{\alpha_i\}$ and $\{\beta_i\}$ are the spectra of isolated singularities $f(x)$ and $g(x)$, then

(8.7.4) $$\{\alpha_i + \beta_i + 1\},\ i = 1, \ldots, \mu^f,\ j = 1, \ldots, \mu^g,$$

is the spectrum of the singularity $f \oplus g$, or

(8.7.5) $$\mathrm{Sp}(f \oplus g) = t\cdot\mathrm{Sp}(f)\cdot\mathrm{Sp}(g),$$

if we consider the spectrum as an element of $\mathbb{Z}[t^{1/m}, t^{-1/m}]$.

This formula is of a more symmetrical form $\widetilde{\mathrm{Sp}}(f \oplus g) = \widetilde{\mathrm{Sp}}(f)\cdot\widetilde{\mathrm{Sp}}(g)$ for the spectrum $\widetilde{\mathrm{Sp}}(f) = t\mathrm{Sp}(f)$ according to M. Saito. This theorem was conjectured by Steenbrink [S3] and proved by Varchenko [V3] and Scherk and Steenbrink [Sc-S] for the case of isolated singularities. M. Saito proved it for the general case [Sa9].

(8.7.6) The spectrum of the zero-dimensional singularity A_1, $g(y) = y^2$, consists of one number $-1/2$, $\mathrm{Sp}(g) = (-1/2)$ or $\mathrm{Sp}(g) = t^{-1/2}$. We obtain the useful consequence:

(8.7.7) *Corollary* If a singularity $f(x)$ has spectrum $\{\alpha_i\}$, then the singularity $f(x) + y^2$ has spectrum $\{\alpha_i + 1/2\}$, i.e. adding a square of a new variable leads to a shift of the spectrum by $1/2$.

Actually Scherk and Steenbrink [Sc-S] expressed the limit MHS on $H^{f\oplus g} = H^{n+m+1}(X_\infty^{f\oplus g}, \mathbb{C})$ in terms of the MHS on $H^f = H^n(X_\infty^f, \mathbb{C})$ and $H^g = H^m(X_\infty^g, \mathbb{C})$. They proved the following theorem:

(8.7.8) *Theorem* Let $(H^f, W_., F^\cdot)$ and $(H^g, W_., F^\cdot)$ be limit MHSs on H^f and H^g and let $H^f = \oplus_\lambda H_\lambda^f = \oplus_{-1<\alpha\leqslant 0} H_\alpha^f$, $H^g = \oplus_{-1<\beta\leqslant 0} H_\beta^g$ be root decompositions of the monodromy operators on H^f and H^g. Then the MHS on $H^{f\oplus g} = H = H^f \otimes_{\mathbb{C}} H^g$ is the join of the MHS on H^f and H^g in the following sense:

(i) the weight filtration $W_.$ on H

$$W_k H = \left(\sum_{\substack{\alpha=\beta=0\\ i+j=k}} W_i H_\alpha^f \otimes W_j H_\beta^g \right) \oplus \left(\bigoplus_{\alpha+\beta=-1} \sum_{i+j=k-2} W_i H_\alpha^f \otimes W_j H_\beta^g \right)$$

$$\oplus \left(\bigoplus_{\substack{-1<\alpha,\beta\leqslant 0\\ \alpha+\beta\neq -1,\ \text{either } \alpha\neq 0 \text{ or } \beta\neq 0}} \sum_{i+j=k-1} W_i H_\alpha^f \otimes W_j H_\beta^g \right),$$

(ii) the Hodge filtration $F^\cdot$ on H

$$F^p H = \left(\bigoplus_{\substack{-1<\alpha,\beta\leqslant 0\\ \alpha+\beta>-1}} \sum_{i+j=p} F^i H_\alpha^f \otimes F^j H_\beta^g \right)$$

$$\oplus\left(\bigoplus_{\substack{-1<\alpha,\beta\leqslant 0\\ \alpha+\beta\leqslant -1}}\sum_{i+j=p-1} F^i H^f_\alpha \otimes F^j H^g_\beta\right).$$

As a consequence of this theorem we obtain the formula expressing Hodge numbers $h_\lambda^{p,q}$ for the singularity $f \oplus g$ by means of Hodge numbers for the singularities f and g.

8.8 Spectra of simple, uni- and bimodal singularities

The methods of calculation of spectra considered in (8.4.8), (8.5.8), (8.5.10) and (8.7.3) enable us to calculate spectra for a wide class of singularities. In particular, there are tables of spectra of simple, uni- and bimodal singularities [Go]. We give such a table in (8.8.1) for the case of simple and unimodal singularities f: $(\mathbb{C}^3, 0) \to (\mathbb{C}, 0)$, $n = 2$. In this case the spectra are contained in the interval $(-1, 2)$ and are symmetric wrt the point $1/2$. Table (8.8.1) show the singularity notation of according to Arnold and the products $(N\alpha_1, \ldots, N\alpha_\mu)$ of spectra $\{\alpha_i\}$ and the number N.

(8.8.1) Table

Type of singularity	N	$(N\alpha_1, \ldots, N\alpha_u)$
A_μ	$\mu+1$	$(1, 2, \ldots, \mu)$
D_μ	$2\mu-2$	$(1, 3, \ldots, 2\mu-3, \mu-2)$
E_6	12	$(1, 4, 5, 7, 8, 11)$
E_7	18	$(1, 5, 7, 9, 11, 13, 17)$
E_8	30	$(1, 7, 11, 13, 17, 19, 23, 29)$
$T_{p,q,r}$	pqr	$(0, pqr, k\cdot qr, l\cdot pr, m\cdot pq)$ $0<k<p, 0<l<q, 0<m<r$
E_{12}	42	$(-1, 5, 11, 13, 17, 19, 23, 25, 29, 31, 37, 43)$
E_{13}	30	$(-1, 3, 7, 9, 11, 13, 15, 17, 19, 21, 23, 27, 31)$
E_{14}	24	$(-1, 2, 5, 7, 8, 10, 11, 13, 14, 16, 17, 19, 22, 25)$
Z_{11}	30	$(-1, 5, 7, 11, 13, 15, 17, 19, 23, 25, 31)$
Z_{12}	22	$(-1, 3, 5, 7, 9, 11, 11, 13, 15, 17, 19, 23)$
Z_{13}	18	$(-1, 2, 4, 5, 7, 8, 9, 10, 11, 13, 14, 16, 19)$
Q_{10}	24	$(-1, 5, 7, 8, 11, 13, 16, 17, 19, 25)$
Q_{11}	18	$(-1, 3, 5, 6, 7, 9, 11, 12, 13, 15, 19)$
Q_{12}	15	$(-1, 2, 4, 5, 5, 7, 8, 10, 10, 11, 13, 16)$
W_{12}	20	$(-1, 3, 4, 7, 8, 9, 11, 12, 13, 16, 17, 21)$
W_{13}	16	$(-1, 2, 3, 5, 6, 7, 8, 9, 10, 11, 13, 14, 17)$
S_{11}	16	$(-1, 3, 4, 5, 7, 8, 9, 11, 12, 13, 17)$
S_{12}	13	$(-1, 2, 3, 4, 5, 6, 7, 8, 9, 10, 11, 14)$
U_{12}	12	$(-1, 2, 3, 3, 5, 6, 6, 7, 9, 9, 10, 13)$

This table contains, in particular, the spectra of simple elliptic singularities $\tilde{E}_6 = P_8 = T_{3,3,3}$, $\tilde{E}_7 = X_9 = T_{2,4,4}$ and $\tilde{E}_8 = J_{10} = T_{2,3,6}$.

Now we proceed to the investigation of variations of Brieskorn lattices and MHSs in families of hypersurface singularities. The next chapter is devoted to continuous invariants (moduli), but here we'll consider the behaviour of discrete invariants under deformations.

8.9 Semicontinuity of the spectrum. Stability of the spectrum for μ-const deformations

Let $f\colon (\mathbb{C}^{n+1}, 0) \to (\mathbb{C}, 0)$ be a germ of a holomorphic function. Assume that 0 is an isolated singularity. Let $F_1\colon \mathbb{C}^{n+1} \times Y \to \mathbb{C}$ be a deformation of f, i.e. a family of functions $f_y = F_1(., y)\colon \mathbb{C}^{n+1} \times \{y\} \to \mathbb{C}$ parametrized by Y. Let

(8.9.1) $$F\colon \mathscr{X} \to S \times Y,\ F(x, y) = (F_1(x, y), y)$$

be good representative of the deformation of the corresponding hypersurface singularity $(f^{-1}(0), 0) = (X, 0)$, where $S = \{t \in \mathbb{C}\colon |t| < \delta\}$, $Y = \{y \in \mathbb{C}^m\colon |y| < \delta'\}$, $\mathscr{X} = \{(x, y) \in \mathbb{C}^{n+1} \times Y\colon\ |x| < \varepsilon,\ |F_1(x, y)| < \delta\}$, $0 < \delta,\ \delta' \ll 1$. It appears uncertain how the spectra of singularities vary under deformations.

(8.9.2) Arnol'd was the first to discover the importance of the spectrum for deformation theory [Ar]. He conjectured that the spectrum behaves semicontinuously under deformation in the following sense: if a singularity P adjoins a (simpler) singularity P' with $\mu' < \mu$ and $\alpha_1 \leqslant \alpha_2 \leqslant \ldots \leqslant \alpha_\mu$, $\alpha'_1 \leqslant \ldots \leqslant \alpha'_{\mu'}$ are their ordered spectra, then $\alpha_k \leqslant \alpha'_k$. From this (and from the symmetry of the spectrum) it follows that when $\mu' = \mu - 1$ the spectrum of P' 'separates' the spectrum of P: $\alpha_1 \leqslant \alpha'_1 \leqslant \alpha_2 \leqslant \alpha'_2 \leqslant \ldots \leqslant \alpha'_{\mu-1} \leqslant \alpha_\mu$. We formulate the property of the semicontinuity of the spectra in terms of the notion of the semicontinuity domain [S5].

(8.9.3) *Definition* The *degree of the spectrum* $Sp(f) = \sum n_\alpha \cdot (\alpha)$ of a singularity f *relative to a subset* $B \subset \mathbb{R}$ is the number

$$\deg_B \mathrm{Sp}(f) = \sum_{\alpha \in B} n_\alpha$$

of spectral numbers contained in B.

In particular, if $B = \mathbb{R}$, then

(8.9.4) $$\deg_B Sp(f) = \sum n_\alpha = u \text{ is the Milnor number,}$$

and if B is a half-open interval of the form $B = (n-p-1,\ n-p]$, then

(8.9.5) $$\deg_B Sp(f) = \sum_{n-p-1<\alpha\leqslant n-p} n_\alpha = \dim Gr_F^p H^n(X_\infty, \mathbb{C}).$$

When $p = n$, $B = (-1, 0]$, we obtain by M. Saito's theorem (8.6.2) the geometric genus of the singularity: $\deg_{(-1,0]} \mathrm{Sp}(f) = p_g$. It is known that the geometric genus is semicontinuous under deformations of the singularity [E]. This is a particular case of the semicontinuity of the spectrum.

Assume that there are several singular points $x_1, \ldots, x_r$ in the fibre $F^{-1}(z) = \mathscr{X}_z$, $z = (t, y)$, of a deformation $F: \mathscr{X} \to Z = S \times Y$ of a singularity $(X, 0)$. Denote by $\mathrm{Sp}(f_z) = \sum_{i=1}^r \mathrm{Sp}(f_y, x_i)$ the sum of all the spectra of the function $f_y: \mathbb{C}^{n+1} \times \{y\} \to \mathbb{C}$ at the critical points $x_1, \ldots, x_r$ having the same critical value t. We define the degree

(8.9.6) $$\deg_B Sp(f_z) = \sum_{i=1}^r \deg_B Sp(f_y, x_i).$$

(8.9.7) *Definition* A subset $B \subset \mathbb{R}$ is called a *semicontinuity domain* for deformations of isolated singularities, if for every deformation F, the function $z \mapsto \deg_B Sp(f_z)$ is upper semicontinuous on Z,

$$\deg_B Sp(f) \geqslant \deg_B Sp(f_z)$$

on some neighborhood of the point $0 \in Z$, $f = f_0$.

(8.9.8) *Theorem* [S5] Every half-interval $B = (a, a+1]$ of length 1 is a semicontinuity domain for deformations of isolated hypersurface singularities.

The theorem is first proved for the case $B = (n-p-1,\ n-p]$, i.e. the semicontinuity of the function $\dim Gr_F^p H^n(X_\infty, \mathbb{C})$ is proved. Then the theorem is proved for any a by means of Varchenko's trick [V8] using the relation between the spectra of the functions $f(x)$ and $f(x) + z^q$,

(8.9.9) $$Sp(f(x) + z^q) = \{\alpha + k/q \mid \alpha \in \mathrm{Sp}(f),\ k = 1, \ldots, q-1\},$$

(see (8.7.3) and (8.8.1)).

(8.9.10) Arnol'd conjectured that any half-line $B = (-\infty, a]$ is a semicontinuity domain. Theorem (8.9.8) implies the validity of this conjecture. Varchenko's conjecture which is stronger than theorem (8.9.8), is that any open interval $B = (a, a+1)$ is a semicontinuity domain. He proved this [V8] for low-weight deformations of quasi-homogeneous singularities.

Theorem (8.9.8) also implies the low semicontinuity of the smallest spectrum number $\alpha_{\min}$ (or, equivalently, of the complex singularity index), which was conjectured by Malgrange [M1]. Finally theorem (8.9.8) implies a new proof of the following theorem of Varenchenko [V5]:

(8.9.11) *Theorem* The spectrum is constant under a deformation of isolated hypersurface singularities with constant Milnor number (= under a μ-const deformation).

Indeed, $\mu = \sum n_\alpha$ remains constant and $\deg_B Sp(f) = \sum_{\alpha \in B} n_\alpha$, $B = (a, a+1]$ and $\deg_B Sp(f)$ can only jump. Hence, $\deg_B Sp(f)$ remains constant for any a. This is only possible if the spectrum remains constant.

8.10 Spectrum of a non-isolated singularity

The spectrum $Sp(f)$ of an isolated singularity $f\colon (\mathbb{C}^{n+1}, 0) \to (\mathbb{C}, 0)$ was defined in (8.1.4) by the following data. Given a vector space $H = H^n(X_\infty, \mathbb{C})$ of dimension μ, an operator $\gamma = T_s$ acts on H with eigenvalues $\lambda_1, \ldots, \lambda_\mu$ which are roots of unity, and there is a decreasing γ-stable filtration $F^\cdot$ (the limit Hodge filtration). The operator γ also acts on $Gr_F^\cdot(H) = \oplus_p Gr_F^p(H) = \oplus_p F^p/F^{p+1}$. By definition the spectrum $Sp(f)$ is associated with the triple $(H, \gamma, F^\cdot)$, $Sp(f) = Sp(H, \gamma, F^\cdot)$, and it is a set of logarithms of the eigenvalues γ_j, $\alpha_j = -(1/2\pi\mathrm{i})\log\lambda_j$, $j = 1, \ldots, \mu$, which are normalized according to the distribution of λ_j in the decomposition $\oplus_p F^p/F^{p+1}$ by condition (8.1.3). The choice of the normalization (8.1.3) is motivated by the definition of the filtration $F^\cdot$ on H. In addition this choice depends on n. We can consider a more natural normalization:

(8.10.1) $$\alpha = \frac{1}{2\pi\mathrm{i}}\log\lambda \text{ and } [\alpha] = p, \text{ i.e. } p \leqslant \alpha < p+1,$$

if an eigenvalue λ corresponds to F^p/F^{p+1}.

Let $Sp' = Sp'(H, \gamma, F^\cdot)$ be the spectrum obtained according to normalization (8.10.1). Just as in (8.1.4) and (8.1.6) we can think of Sp' in three ways:

(8.10.2) as a set of spectral numbers $Sp' = \{\alpha_1, \ldots, \alpha_\mu\}$;
as a 'divisor' $Sp' = \sum_\alpha n_\alpha \cdot (\alpha) \in \mathbb{Z}^{(\mathbb{Q})}$;
as a Laurent polynomial $Sp' = \sum_\alpha n_\alpha t^\alpha \in \mathbb{Z}[t^{1/m}, t^{-1/m}]$.

The way in which the spectrum Sp' is considered is always obvious from the context.

In the group $\mathbb{Z}^{(\mathbb{Q})}$, respectively in the ring $\mathbb{Z}[t^{1/m}, t^{-1/m}]$, there is an involution ι defined by $(\alpha) \mapsto (-\alpha)$, respectively, $t^{\alpha} \mapsto t^{-\alpha}$. Under the change $\alpha \mapsto n - \alpha$ the condition $p \leqslant \alpha < p + 1$ (8.10.1) transforms into condition (10.1.3) or equivalently $n - p - 1 < n - \alpha \leqslant n - p$. Thus, if we define

(8.10.3) $$Sp_n(H, \gamma, F^{\cdot}) = t^n \iota(Sp'(H, \gamma, F^{\cdot})),$$

we obtain that for an isolated singularity

$$Sp(f) = Sp_n(H, \gamma, F^{\cdot})$$

where $H = H^n(X_\infty, \mathbb{C})$, $\gamma = T_s$, and $F^{\cdot}$ is the limit Hodge filtration. We recall that $\widetilde{Sp}(f) = tSp(f)$ is the spectrum according to M. Saito (8.1.9).

Now let f: $(\mathbb{C}^{n+1}, 0) \to (\mathbb{C}, 0)$ be any, not necessarily isolated, singularity, f: $X \to S$ be the Milnor fibration, which is defined in the same way as in the case of isolated singularities, and X_∞ be the canonical fibre. Then there is an MHS [N] on the cohomology $H^*(X_\infty)$ which is invariant wrt the semisimple part of the monodromy $\gamma = T_s$.

The *spectrum* of a singularity f is defined [S6] by the formula

(8.10.4) $$Sp(f) = \sum_{k=0}^{n} (-1)^{n-k} Sp_n(\tilde{H}(X_\infty), T_s, F^{\cdot})$$
$$= Sp_n(\tilde{H}^n(X_\infty)) - Sp_n(\tilde{H}^{n-1}(X_\infty)) + \ldots$$

where $\tilde{H}^k(X_\infty)$ are reduced cohomologies.

In the case of isolated singularities this definition coincides with the previous one.

(8.10.5) *Examples*

(1) A_∞: f: $(\mathbb{C}^3, 0) \to (\mathbb{C}, 0)$, $f(x, y, z) = xy$.

In this case $Sp(f) = -(1)$ or $Sp(f) = -t$ since the fibre X_∞ is homeomorphic to the affine variety $xy = 1$ in $\mathbb{C}^3$, i.e. to $\mathbb{C}^* \times \mathbb{C}$, and $T = \mathrm{id}$.

(2) $T_{\infty,\infty,\infty}$. f: $(\mathbb{C}^3, 0) \to (\mathbb{C}, 0)$, $f(x, y, z) = xyz$.

In this case $X_\infty \simeq \{xyz = 1\} \simeq \mathbb{C}^* \times \mathbb{C}^*$, $T = id$ and the MHS on $H^i(X_\infty)$ is pure of type (i, i), $i = 0, 1, 2$. Hence we obtain that $Sp(f) = (0) - 2{\cdot}(1)$.

These examples show that unlike what happens with isolated singularities the spectrum $Sp(f)$ need not be symmetric wrt the point $(n - 1)/2$.

(8.10.6) *Remark* The Sebastiani–Thom formula (8.7.5) for the join of singularities, which for the spectrum according M. Saito $\widetilde{Sp}(f) = t{\cdot}Sp(f)$ reads $\widetilde{Sp}(f \oplus g) = \widetilde{Sp}(f){\cdot}\widetilde{Sp}(g)$, was also proved by M. Saito [Sa9] in the case of non-isolated singularities. This formula also remains valid in the case of a zero function in k variables, if for such a function we set $\widetilde{Sp}(g) = (-1)^k t^k$. In particular, for $k = 1$ we obtain

(8.10.7) $$\mathrm{Sp}(f \oplus 0) = -t\mathrm{Sp}(f),$$

i.e. if f: $(\mathbb{C}^n, 0) \to (\mathbb{C}, 0)$ has the spectrum $Sp(f) = \sum n_\alpha \cdot (\alpha)$, then the function g: $(\mathbb{C}^{n+1}, 0) \to (\mathbb{C}, 0)$, $g(x_0, x_1, \ldots, x_n) = f(x_1, \ldots, x_n)$ has the spectrum $\mathrm{Sp}(g) = -\sum n_\alpha \cdot (\alpha + 1)$.

8.11 Relation between the spectrum of a singularity with a one-dimensional critical set and spectra of isolated singularities of its Iomdin series

Let f: $(\mathbb{C}^{n+1}, 0) \to (\mathbb{C}, 0)$ be a singularity with a one-dimensional critical locus $\Sigma = Sing\, f = Sing\, f^{-1}(0)$. Let $\Sigma = \Sigma_1 \cup \ldots \cup \Sigma_r$ be decomposition into the union of irreducible components. Let m_i be the multiplicity of the one-dimensional singularity $(\Sigma_i, 0)$. There are two ways to relate f to a deformation of isolated singularities. Firstly, f can be considered as a limit (degeneration) of one-dimensional families f_ε, $\varepsilon \to 0$, of isolated singularities as follows. Let l be a general linear form (coordinate) on $\mathbb{C}^{n+1}$. We consider a deformation

(8.11.1) $$f_\varepsilon = f + \varepsilon \cdot 1^k$$

of the singularity f. Then for all k sufficiently large and $0 < |\varepsilon| \ll 1$ the functions f_ε have an isolated singularity at 0. We call the series (8.11.1) of isolated singularities f_k (with variable k) the *Yomdin series*, since Yomdin [Y] obtained the relation for Milnor numbers of singularities f and f_k

(8.11.2) $$\mu(f_k) = \mu(f) + k \cdot m(\Sigma),$$

where $\mu(f) = \mu_n(f) - \mu_{n-1}(f)$ is the Milnor number of the non-isolated singularity f which, by (8.10.4) is equal to the degree of its spectrum $\mu(f) = \deg Sp(f)$ (as in the case of isolated singularities), and $m(\Sigma)$ is the multiplicity of the singularity $(\Sigma, 0)$. Steenbrink [S6] generalized this result and obtained a formula relating the spectra of singularities f and f_k. Steenbrink proved this formula in different specific cases, and M. Saito [Sa9] proved it in the general case.

(8.11.3) The second way to relate f to deformations of isolated singularities is to consider f as itself a deformation $f = f_y$ of isolated singularities f_y of one less dimension. Let us take a general linear form l on $\mathbb{C}^{n+1}$ to be one of the coordinate functions, which we denote by y. Let $(x_1, \ldots, x_n, y)$ be coordinates on $\mathbb{C}^{n+1}$, $f(x_1, \ldots, x_n, y) =$

$f_y(x_1, \ldots, x_n)$. We can consider a function $f\colon \mathbb{C}^{n+1} \to \mathbb{C}$ as a family of functions $f_y\colon \mathbb{C}^n \times \{y\} \to \mathbb{C}$. The hyperplane $\mathbb{C}^n_y = \mathbb{C}^n \times \{y\}$ intersects the curve Σ transversely in r points $P_i = \Sigma_i \cap \mathbb{C}^n_y$, $i = 1, \ldots, r$. For $y \neq 0$ we get r isolated hypersurface singularities:

(8.11.4) $$g_i\colon (\mathbb{C}^n_y, P_i) \to (\mathbb{C}, 0),\ g_i = f_{y,i} = f_{|\mathbb{C}^n}.$$

The μ-class of the singularity g_i depends only on f (for general l) and is called the *transverse type* of the singularity f *along* Σ_i.

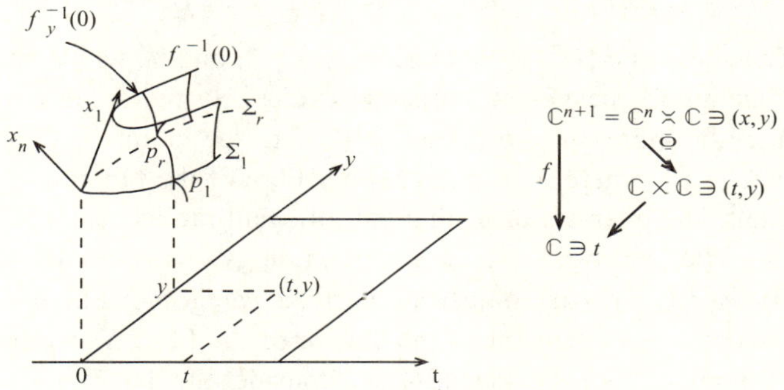

Let $\Sigma_i^* = \Sigma_i \backslash \{0\}$. We can (locally) consider $f\colon \mathbb{C}^{n+1} \backslash (\mathbb{C}^n \times \{0\}) \to (\mathbb{C}, 0)$ as r μ-const one-parameter deformations $f_y\colon \bigcup_{i=1}^r (\mathbb{C}^n_y, P_i) \to (\mathbb{C}, 0)$ parametrized by the curves Σ_i^*, $i = 1, \ldots, r$ (or by $\mathbb{C}\backslash\{0\} \ni y$).

In the (8.11.5)–(8.11.8), we will make a small digression about one important general question: Deligne's method of vanishing cycles.

Deligne's sheaves of vanishing cycles

(8.11.5) Let $f\colon X \to S$ be a morphism of an analytic space X onto the unit disk and let its restriction $f\colon X' \to S'$ be topologically locally trivial over $S' = S' \backslash \{0\}$ and let there exist a retraction $r\colon X \to X_0$ onto the zero fibre, compatible with the trivialization. By Thom's isotopy theorem any proper morphism f is of such a topological type in a neighborhood of the zero fibre (see [Di3], Ch. 1). The situation is the same for the case in which f is the Milnor fibration of a germ of a holomorphic function. Let $X_t = f^{-1}(t)$ and let $r_t\colon X_t \to X_0$ be the restriction of the retraction onto the fibre X_t. The local triviality of f over S' defines the monodromy transformation $h\colon X_t \to X_t$. From the topological point of view the map $f\colon X \to S$ is reconstructed by the quadruple (X_t, X_0, h, r_t): first we get the fibration $f^1\colon X^1 \to S^1$ over the circle S^1 gluing $X_t \times [0, 1]$ by means of h, and then we get X as a cone $\operatorname{cone}(X^1 \xrightarrow{r^1} X_0)$ of the map r^1 defined by the retraction r_t. The map $f\colon X \to S$ corresponds to the map of cones $\mathrm{cone}\,(X^1 \xrightarrow{r^1} X_0) \to \mathrm{cone}\,(S^1 \to pt) = S$.

(8.11.6) Deligne's method of vanishing cycles consists of obtaining information about the difference between the cohomologies of X_0 and X_t by means of the Leray spectral sequence of the map $r_t\colon X_t \to X_0$, which for the sheaf $F = \mathbb{C}_{X_t}$ on X_t reads as follows:

$$E_2^{p,q} = H^p(X_0, R^q(r_t)_*\mathbb{C}_{X_t}) \Rightarrow H^{p+q}(X_t, \mathbb{C}).$$

The sheaves $\varphi_f^q(\mathbb{C}) = R^q(r_t)_*\mathbb{C}_X$ (or ψ_f^q) on X_0, which appear here, are just Deligne's sheaves of vanishing cycles. By the definition of the direct image the fibre of the sheaf $\varphi_f^q(\mathbb{C})$ at the point $x \in X_0$ is equal to $H^q(X_{t,x}, \mathbb{C})$, where $X_{t,x}$ is the Milnor fibre of the function f at the point x (we take the intersection with a ball of small radius centered at the point x etc.). The sheaves ψ_f^q are cohomology sheaves of the complex $\psi^\cdot_f = \mathbb{R}(r_t)_*\mathbb{C}_X$ (the object of derived category). Finally to make the construction independent of the choice of the fibre X_t we have to pass to the canonical fibre X_∞ (see (2.4.5)): consider the diagram

(8.11.7)
$$\begin{array}{ccccc} X_0 \underset{i}{\subset} X & \overset{k}{\leftarrow} & X_\infty = X \underset{S}{\times} U = X' \underset{S'}{\times} U \\ f\downarrow & & \downarrow \\ S & \overset{e}{\leftarrow} & U, \end{array}$$

where $U \xrightarrow{e} S'$ is the universal covering of the punctured disk S'. The canonical fibre X_∞ has the homotopical type of fibres X_t since U is simply connected. The space X has the homotopical type of X_0 by virtue of the retraction $r\colon X \to X_0$. So we can consider the map $k\colon X_\infty \to X$ instead of the map $r_t\colon X_t \to X_0$. We now come to the final definition of the sheaves of vanishing cycles.

(8.11.8) *Definition* The *sheaves of nearby cycles* $\psi_f^q(\mathbb{C}_X)$ are the cohomology sheaves of the complex

$$\psi_f(\mathbb{C}_X) = i^*\mathbb{R}k_*(k^*\mathbb{C}_X).$$

This complex of sheaves $\psi_f(\mathbb{C}_X)$ on X_0 (or more exactly, the object of the derived category D_{X_0}) is also called the sheaf of nearby cycles. The *sheaves of vanishing cycles* $\varphi_f^q\mathbb{C}_X$ are the cohomology sheaves of the complex

$$\varphi_f(\mathbb{C}_X) = \mathrm{cone}\,(\mathbb{C}_{X_0} \to \psi_f(\mathbb{C}_X)),$$

the cone over the natural morphism $\mathbb{C}_{X_0} = i^*\mathbb{C}_X \to \psi_f(\mathbb{C}_X)$.

Fibres of the sheaves $\psi_f^q\mathbb{C}_X$ at points $x \in X_0$ are the vanishing cohomology, i.e. the cohomology $H^q(X_{t,x}, \mathbb{C})$ of the Milnor fibre at the point x, and fibres of the sheaves $\varphi_f^q\mathbb{C}_X$ are the reduced cohomology $\tilde{H}^q(X_{t,x}, \mathbb{C})$.

According to the ideology of the 'new' homological algebra we can change $\mathbb{C}_X$ by an arbitrary sheaf $\mathcal{F}$ or a complex of sheaves on X and obtain functors of nearby and vanishing cycles $\psi_f, \varphi_f: D_X \to D_{X_0}$. These functors are used in the construction of perverse sheaves on X out of sheaves on X_0 and X' by 'gluing' [G-M].

We now return to singularities with a one-dimensional critical locus.

(8.11.9) *Vertical and horizontal monodromies* Let $f: X \to S$ be the Milnor fibration of a singularity $f: (\mathbb{C}^{n+1}, 0) \to (\mathbb{C}, 0)$. There is a sheaf of vanishing cycles $\varphi_f(\mathbb{C}_X)$ on X_0. This is a complex of sheaves with the constructable cohomology $\varphi_f^q(\mathbb{C}_X) = \mathcal{H}^q(\varphi_f(\mathbb{C}_X))$, where $\varphi_f^q(\mathbb{C}_X) = 0$ on $X_0 \backslash \Sigma$ for all q and $\varphi_f^q(\mathbb{C}_X) = 0$ on $\Sigma^* = \Sigma\backslash\{0\} = \Sigma_1^* \cup \ldots \cup \Sigma_r^*$ for $q \neq n-1$, and the fibre $\varphi_f^{n-1}(\mathbb{C}_X)_{P_i}$ at a point $P_i = \Sigma_i \cap \mathbb{C}_y$ is $\tilde{H}^{n-1}(X(g_i), \mathbb{C})$, the vanishing cohomology of the isolated singularity $g_i: (\mathbb{C}_y^n, P_i) \to (\mathbb{C}, 0)$. Thus on $\Sigma_i^* = \Sigma_i\backslash\{0\}$ there are local systems $\underline{H}_i = \varphi_f^{n-1}(\mathbb{C}_{X_i})_{|\Sigma^*}$ with fibres $\tilde{H}^{n-1}(X(g_i), \mathbb{C})$ at points $P_i \in \Sigma_i^*$.

On the space $\tilde{H}^{n-1}(X(g_i), \mathbb{C})$ there are two monodromy transformations: the monodromy T_i of the singularity $g_i: (\mathbb{C}_y^n, P_i) \to (\mathbb{C}, 0)$ corresponding to a circuit around the origin in the complex plane $\mathbb{C}$ with coordinate t, which is called the *horizontal monodromy*; and the monodromy τ_i of the local system $\underline{H}_i$ on the curve $\Sigma_i^* \subset X_0 = f^{-1}(0)$ corresponding to a circuit around the origin in the complex plane $\mathbb{C}$ with coordinate y $(t = 0)$. The monodromies T_i and τ_i, $i = 1, \ldots, r$, commute with each other (since the group $\pi_1(\mathbb{C}^* \times \mathbb{C}^*) = \mathbb{Z} \oplus \mathbb{Z}$ is abelian).

(8.11.10) Let us consider one of the branches Σ_i, $i = 1, \ldots, r$. Let μ_i be the Milnor number of the singularity g_i, $Sp(g_i) = \sum_{j=1}^{\mu_i} t^{\alpha_{ij}}$ be the spectrum of the singularity g_i, and $\lambda_{i,j} = e^{-2\pi\sqrt{-1}\alpha_{ij}}$ be the eigenvalues of T_i. Let $\kappa_{i,1}, \ldots, \kappa_{i,\mu_i}$ be the eigenvalues of τ_i. Since T_i and τ_i commute, we can make the eigenvalues $\lambda_{i,j}$ and $\kappa_{i,j}$ consistent, i.e. set them in one to one correspondence in the following way. Choose a common eigenbasis $e_{i,1}, \ldots, e_{i,\mu}$ of the commuting semisimple operators $(T_i)_s$ and $(\tau_i)_s$, and associate to the eigenvalues $\lambda_{i,j}$ the eigenvalues $\kappa_{i,j}$ of the operator $(\tau_i)_s$ corresponding to the vector $e_{i,j}$. Finally take a number $\beta_{i,j}$ such that

$$\kappa_{i,j} = e^{-2\pi\sqrt{-1}\beta_{i,j}}, \ \beta_{ij} \in (0, 1].$$

(8.11.11) *Theorem* If $f: (\mathbb{C}^{n+1}, 0) \to (\mathbb{C}, 0)$ is a singularity with a one-dimensional critical locus $\Sigma = \bigcup_{i=1}^{r} \Sigma_i$ and $f_k = f + \varepsilon{\cdot}l^k$ is the isolated singularity of its Yomdin series, then for sufficiently large $k \geqslant k_0$ the spectra of f and f_k are related by the formula

$$Sp(f_k) - Sp(f) = \sum_{i,j} t^{\alpha_{i,j}+\beta_{i,j}/m_i k} . \frac{1-t}{1-t^{1/m_i k}}, \tag{8.11.12}$$

where m_i is the multiplicity of the branch $(\Sigma_i, 0)$ and α_{ij} and β_{ij} are defined in (8.11.10) by μ-const deformations of isolated singularities along $\Sigma_i \backslash \{0\}$, $i = 1, \ldots, r$, which are defined by the function f. The number k_0 is determined by the discriminant of the map defined by (f, l).

This theorem was conjectured by Steenbrink [S6] and proved by him for $n = 1$ and for some other cases. M. Saito [Sa9] proved the conjecture in general. He reformulated and proved the conjecture in the more general context of his theory of mixed Hodge modules. D. Siersma [Si], proved by topological methods a formula relating ζ-functions of the singularities f and f_k, which can be considered as 'exp $(2\pi$i[formula (8.11.12)])'.

III

The period map of a μ-const deformation of an isolated hypersurface singularity associated with Brieskorn lattices and MHSs

Until now we studied each singularity $f: (\mathbb{C}^{n+1}, 0) \to (\mathbb{C}, 0)$ individually. From now on we'll be concerned with families of singularities. Accordingly to the general ideas of moduli spaces and period maps for non-singular compact varieties in algebraic geometry, we have to consider the set of all algebraic (or analytic) structures on the same underlying topological variety. The period map is connected with the 'linearization' of these objects: we replace the variety by its cohomology space, and the algebraic variety structures by the spaces of differential forms on algebraic varieties and HSs defined by them. In our case the underlying topological object connected with a singularity f is its Milnor fibration $f: X \to S$ and the corresponding linear objects, *viz.* the cohomological Milnor fibration $\underline{H} \to S'$ of rank $\mu = \dim H^n(X_t, \mathbb{C})$ equal to the Milnor number, the topological Gauss–Manin connection ∇ on $\underline{H}$, the monodromy transformation T on the cohomology, the meromorphic connection $\mathscr{M}$, the type of which is defined by the Jordan structure of the monodromy T, and so on. The complex structure of f is reflected in the Brieskorn lattice $\mathscr{H}^{(0)}$ and in its embedding in $\mathscr{M}$, and in the corresponding MHS defined by this embedding.

1 Gluing of Milnor fibrations and meromorphic connections of a μ-const deformation of a singularity

Let $f: (\mathbb{C}^{n+1}, 0) \to (\mathbb{C}, 0)$ be an isolated singularity. In chapters I and II we associated to each such individual singularity the following objects: the Milnor fibration $f: X \to S$; the corresponding cohomological (homological) fibration $\underline{H} = \bigcup_{t \in S'} H^n(X_t, \mathbb{C})$ on S'; the vector space of multivalued horizontal sections H of the fibration $\underline{H}$, which can also be considered as the canonical fibre $\underline{H}_\infty = H^n(X_\infty, \mathbb{C})$; the sheaf of sections

$\mathscr{H} = \underline{H} \otimes_{\mathbb{C}_{S'}} \mathscr{O}_{S'} \simeq \mathscr{H}^n_{\mathrm{DR}}(X'/S')$ of the fibration $\underline{H}$ with connection ∇ defined by this fibration; the De Rham cohomology sheaf $\mathscr{H}^{(-2)} = \mathscr{H}^n_{\mathrm{DR}}(X/S)$ which is a natural extension of $\mathscr{H}$ on S; the Brieskorn lattice $\mathscr{H}^{(0)}$ which is another natural extension of $\mathscr{H}$ defined by the geometric sections $s[\omega]$, where ω runs through the set of $(n+1)$-forms on X; the meromorphic connection on S, $\mathscr{M} = \mathscr{H}^{(-2)} \otimes_{\mathscr{O}_S} \mathscr{O}_S[t^{-1}] = \mathscr{H}^{(0)} \otimes_{\mathscr{O}_S} \mathscr{O}_S[t^{-1}]$, $\mathscr{M}$ is a sheaf on S such that $\mathscr{M}_{|S'} = \mathscr{H}$ and its fibre at the point $0 \in S$ (denoted again by $\mathscr{M}$) is the localization of the fibre $\mathscr{H}^{(0)}_0$; $\mathscr{M}$ has the root decomposition $\mathscr{M} = \hat{\oplus}_\alpha C_\alpha$, where C_α is isomorphic to the subspace $H_\lambda \subset H$ corresponding to the eigenvalue $\lambda = \mathrm{e}^{-2\pi i\alpha}$, and the subspace $C_\alpha \subset \mathscr{M}$ consists of the (single-valued) sections $s[A, \alpha] = t^\alpha \cdot t^N A(t)$ of the sheaf $\mathscr{M}$, where $A = A(t) \in H$, and $N = -(1/2\pi \mathrm{i}) \log T_{\mathrm{u}}$; and finally, the canonical lattice (extension) $\mathscr{L} \subset \mathscr{M}$, $\mathscr{L} = V^{>-1}\mathscr{M} = \hat{\oplus}_{\alpha>-1} C_\alpha$. Now we wish to consider the corresponding objects for the singularities of a μ-const deformation of the singularity f and to glue them to form a unified family.

1.1 Milnor fibrations

Let

(1.1.1) $$F\colon (\mathbb{C}^{n+1} \times \mathbb{C}^m, 0) \to (\mathbb{C} \times \mathbb{C}^m, 0),$$

$t = F_1(x, y)$, $y = y$, be a μ-const deformation of the singularity f with a smooth base space $\mathbb{C}^m \ni y = (y_1, \ldots, y_m)$. To simplify the notation, we'll often write $t = F(x, y)$ instead of $F_1(x, y)$, if this does not lead to an ambiguity. Thus, we have an m-parametric family of isolated singularities of functions

(1.1.2) $$f_y\colon (\mathbb{C}^{n+1} \times \{y\}, 0) \to (\mathbb{C}, 0),\ f_y(x) = F(x, y),$$

with constant Milnor number μ for y in some neighborhood $D \subset \mathbb{C}^m$ of the point 0, where $f_0(x) = F(x, 0) = f(x)$.

(1.1.3) For every fixed $y \in D$ the singularity f_y has the Milnor fibration $f_y\colon X(y) \to S(y) = S$, the corresponding cohomological fibration $\underline{H}(y)$ on S' and so on. The letter y in parenthesis means that we are considering the corresponding object for an individual singularity f_y for fixed y. We want to show, following Varchenko [V1, V5], how we can glue $X(y)$, $\underline{H}(y)$ etc. to form a uniform fibration over $S' \times D$. The problem lies in the fact that when constructing the Milnor fibrations $X(y) \to S(y)$ of singularities f_y one chooses the radii $\delta = \delta(y)$ and $\varepsilon = \varepsilon(y)$ for each y separately. Recall the notation: $B = B_\varepsilon = \{x \in \mathbb{C}^{n+1}: |x| < \varepsilon\}$, $S = S_\delta = \{t \in \mathbb{C}: |t| < \delta\}$ and $D = D_\eta = \{y \in \mathbb{C}^m: |y| < \eta\}$.

Choose $\varepsilon > 0$ such that the spheres $\partial B_{\varepsilon'}$ intersect the hypersurface $f(x) = F(x, 0) = 0$ transversely for $\varepsilon' \in (0, \varepsilon]$ and that the point 0 is the unique singular point of the function $f(x)$ in B_ε. Choose $\eta > 0$ and $\delta > 0$ such that: (1) for any $y_0 \in D_\eta$ the function $f_{y_0}(x)$ has no critical points in B_ε distinct from 0; (2) for any $y_0 \in D_\eta$ and any $t \in S_\delta$ the sphere ∂B_ε intersects the hypersurface $f_{y_0}(x) = t$ transversely. Put $\mathscr{X} = F^{-1}(S \times D) \cap (B \times D)$ and consider the map

(1.1.4) $$F: \mathscr{X} \to S \times D.$$

Then over $S' \times D$ we have a smooth locally trivial fibration F': $\mathscr{X}' \to S' \times D$. We get the family of fibrations:

(1.1.5)
$$F(y) = f_y: \mathscr{X}(y) \to S \times \{y\}, \text{ where } \mathscr{X}(y) = F^{-1}(S \times \{y\}),\ y \in D.$$

To construct the Milnor fibration $X(y) \to S_{\delta(y)}$ of a singularity f_y: $\mathbb{C}^{n+1} \times \{y\}, 0) \to (\mathbb{C}, 0)$, we have to take $0 < \delta(y) \ll \varepsilon(y) \ll 1$ and to consider the restriction of F onto the subset $F^{-1}(S_{\delta(y)} \times \{y\}) \cap (B_{\varepsilon(y)} \times \{y\}) \cap \mathscr{X}(y) \subset \mathscr{X}(y)$. The next theorem follows from the paper of Lê and Ramanujam [L-R].

(1.1.6) *Theorem* The embeddings of fibres $X(y)_t \subset \mathscr{X}(y)_t$ are homotopic equivalences. ■

Thus, from the homotopic (homological) point of view we can consider F in (1.1.4) as a family of Milnor fibrations $X(y) \to S$ of singularities f_y.

1.2 Cohomological fibration

The locally trivial fibration F': $\mathscr{X}' \to S' \times D$ defines a flat cohomological fibration over $S' \times D$,

(1.2.1) $$\underline{H} = \bigcup_{(t,y) \in S' \times D} H^n(\mathscr{X}_{(t,y)}, \mathbb{C}).$$

where $\mathscr{X}_{(t,y)} = F^{-1}(t, y)$ is the fibre of the mapping F. Since $\mathscr{X}_{(t,y)} = \mathscr{X}(y)_t$ is also the fibre of the fibration $\mathscr{X}(y) = F^{-1}(S \times \{y\})$ at the point t, we have from (1.1.6) that the restriction of $\underline{H}$ to $S' \times \{y\}$,

(1.2.2) $$\underline{H}_{|S' \times \{y\}} = \underline{H}(y)$$

is the Milnor fibration of the singularity f_y. One can say the same about the dual homological fibration

$$\underline{H}_* = \bigcup_{(t,y)\in S'\times D} \underline{H}_n(\mathscr{X}_{(t,y)}, \mathbb{C}).$$

Denote by H, respectively $H(y)$, the vector space of multivalued horizontal sections of the fibration $\underline{H}$, respectively $\underline{H}(y)$. Obviously, H and $H(y)$ are canonically isomorphic: the isomorphism $H(y) \xrightarrow{\sim} H$ is realized by extending sections from $S' \times \{y\}$ to $S' \times D$. The vector space H is also identified with the space of single-valued horizontal sections of the fibration $(e \times \mathrm{id})^{-1}\underline{H}$, where $U \xrightarrow{e} S'$ is the universal covering of S', respectively $U \times D \xrightarrow{e\times id} S' \times D$, and with the cohomology space $H^n(\mathscr{X}_\infty, \mathbb{C})$ of the canonical fibre $\mathscr{X}_\infty = \mathscr{X} \times_{S\times D} (U \times D)$ of the mapping F.

Let

(1.2.3) $$\mathscr{H} = \underline{H} \otimes_{\mathbb{C}_{S'\times D}} \mathscr{O}_{S'\times D} = R^n F_* \mathbb{C}_{\mathscr{X}'} \otimes \mathscr{O}_{S'\times D}$$

be the sheaf of holomorphic sections of the fibration $\underline{H}$, and ∇ be the integrable connection on $\mathscr{H}$ defined by the local system $\underline{H}$, i.e. the Gauss–Manin connection. Obviously, the restriction of $\mathscr{H}$ to $S' \times \{y\}$, $\mathscr{H} \otimes \mathscr{O}_{S'\times\{y\}} = \mathscr{H}(y)$, is the sheaf of holomorphic sections of the fibration $\underline{H}(y)$. By the relative Poincaré Lemma we have a resolution $0 \to F^{-1}\mathscr{O}_{S'\times D} \to \Omega^{\cdot}_{\mathscr{X}'/S'\times D}$, and the sheaf $\mathscr{H}$ is identified with the sheaf of relative De Rham cohomology (chapter I, §3)

$$\mathscr{H} = \mathscr{H}^n_{\mathrm{DR}}(\mathscr{X}'/S' \times D) = \mathbb{R}^n F'_*(\Omega^{\cdot}_{\mathscr{X}'/S'\times D}) = \mathscr{H}^n(F'_*(\Omega^{\cdot}_{\mathscr{X}'/S'\times D}).$$

1.3 Canonical extension of the sheaf $\mathscr{H}$ and the meromorphic connection

Recall ((II.6.3.5), (II.6.4.5)) that in the case of an individual singularity $f(x)$ we constructed the canonical extension $\mathscr{L}$ of the sheaf $\mathscr{H}$ from S' to the whole disk S by means of the trivialization of the sheaf $\mathscr{H}$ by elementary sections $s[A, \alpha] = \psi_\alpha(A) = t^\alpha \cdot t^N A$ corresponding to multivalued horizontal sections $A = A(t) \in H_\lambda$, $\alpha = -(1/2\pi\mathrm{i}) \log \lambda$. We proceed in the same way for the sheaf $\mathscr{H}$ in the case of a family of singularities.

Now let H be the space of multivalued horizontal sections of the sheaf $\underline{H}$ (1.2.1) over $S' \times D$. For an eigenvalue λ of the monodromy T, a section $A = A(t, y) \in H_\lambda$ and a value β of logarithm $-(1/2\pi\mathrm{i}) \log \lambda$, we define the elementary section

(1.3.1) $$s[A, \beta](t, y) = t^\beta \cdot t^N A(t, y).$$

This is a single-valued holomorphic section of the fibration $\underline{H}$, i.e. $s[A, \beta]$

is a section of the sheaf $\mathscr{H}$. Since A is a horizontal section, it is clear (II.6.4.2) that $s[A, \beta]$ satisfies the conditions

$$(1.3.2)\qquad (t\partial_t - \beta)^{n+1}\cdot s[A, \beta] = 0,\quad \partial_j\cdot s[A, \beta] = 0,\ j = 1, \ldots, m,$$

where $\partial_t = \nabla_{\mathrm{d}/\mathrm{d}t}$, $\partial_j = \nabla_{\partial/\partial y_j}$.

(1.3.3) *Definition* Choose bases in the spaces H_λ and let $A_1, \ldots, A_\mu$ be the corresponding basis in the vector space $H = \oplus H_\lambda$. If we choose the values α_j of logarithms $\alpha = -(1/2\pi\mathrm{i})\log\lambda$ of the eigenvalues corresponding to the vectors A_j, we obtain sections $s[A_j, \alpha_j]$ defining a trivialization of the sheaf $\mathscr{H}$ over $S' \times D$. If α_j are chosen such that condition $-1 < \alpha_j \leqslant 0$ holds, then the extension of the sheaf $\mathscr{H}$ corresponding to this trivialization is called the *canonical extension* $\mathscr{L}$ of the sheaf $\mathscr{H}$ to $S \times D$,

$$\mathscr{L} = \bigoplus_{j=1}^{\mu} \mathscr{O}_{S\times D}\cdot s[A_j, \alpha_j],\ -1 < \alpha \leqslant 0.$$

Let

$$(1.3.4)\qquad \mathscr{M} = \mathscr{L} \otimes \mathscr{O}_{S\times D}[t^{-1}]$$

be the localization of the sheaf $\mathscr{L}$ along the subvariety $\{0\} \times D \subset S \times D$ defined by the equation $t = 0$. It is obvious that the restrictions of $\mathscr{L}$ and $\mathscr{M}$ to $S \times \{y\}$,

$$\mathscr{L}(y) = \mathscr{L} \otimes \mathscr{O}_{S\times\{y\}},\ \mathscr{M}(y) = \mathscr{M} \otimes \mathscr{O}_{S\times\{y\}},$$

are the canonical lattice and the meromorphic Gauss–Manin connection of the singularity $f_y\colon (\mathbb{C}^{n+1}, 0) \to (\mathbb{C}, 0)$.

For any of the logarithms $\beta = -(1/2\pi\mathrm{i})\log\lambda$ of an eigenvalue of the monodromy we denote by $\underline{C}_\beta$ the constant sheaf on $S \times D$ generated by sections $s[A, \beta]$, $A \in H_\lambda$, and let C_β be the sheaf of holomorphic sections of the fibration $\underline{C}_{\beta|\{0\}\times D}$; C_β is a free $\mathscr{O}_D$-module of rank $\mu_\lambda = \dim H_\lambda$. We can also consider the sheaf C_β as the sheaf of germs of sections of the sheaf $\mathscr{M}$ of the form $s[A(t, y), \beta] = t^\beta\cdot t^N A(t, y)$, where $A(t, y)$ depends holomorphically on y and for a fixed y $A(t, y)$ is a horizontal section of $\underline{H}(y)$.

Any section of the sheaf $\mathscr{M}$ can be written in the form

$$(1.3.5)\qquad s = \sum_{j=1}^{\mu} g_j(t, y)\, s[A_j, \alpha_j],$$

where $g_j(t, y)$ are meromorphic functions with poles along $\{0\} \times D$. Decomposing $g_j(t, y)$ into series in t, $g_j(t, y) = \sum_k g_{j,k}(t, y)t^k$, and taking into account that $t^k s[A, \alpha] = s[A, \alpha + k]$, we get a decomposition of the form

$$s = \sum_\beta s[A_\beta(y), \beta] = \sum_\beta s_\beta. \tag{1.3.6}$$

Thus, we have a decomposition

$$\mathscr{M} = \hat{\oplus}_\beta C_\beta, \tag{1.3.7}$$

where β runs the arithmetic progressions $-(1/2\pi\mathrm{i}) \log \lambda$. The components s_β in (1.3.6) are the (holomorphic) sections of the sheaves C_β. The restriction of (1.3.7) onto $S \times \{y\}$ gives the decomposition (II.6.2.5) $\mathscr{M}(y) = \hat{\oplus}_\beta C_\beta(y)$ of the meromorphic connection of the singularity f_y into the sum of root subspaces of the operator $t\partial_t$. As in the case of individual singularities f_y we often do not distinguish between $\mathscr{M}$ and the restriction $\mathscr{M}_{|\{0\}\times D}$. There is a filtration $V^\cdot$ (II.6.3.2) in $\mathscr{M}$, and $\mathscr{L} = V^{>-1}\mathscr{M} = \hat{\oplus}_{\beta>-1} C_\beta$.

2 Differentiation of geometric sections and their root components wrt a parameter

When studying individual singularities in (I.5.1.5) we associated with each $(n+1)$-form ω the section $s[\omega]$ of the sheaf $\mathscr{H}$, the geometric section of $\mathscr{H}$. The geometric sections generate the Brieskorn lattice $\mathscr{H}^{(0)} \subset \mathscr{M}$. Now we want to extend this to families of singularities and to learn to differentiate the obtained geometric sections wrt the parameter.

2.1 Geometric sections and their root components

Let f_y be a μ-const deformation of a singularity $f\colon (\mathbb{C}^{n+1}, 0) \to (\mathbb{C}, 0)$ as in §1 and let $\omega \in \Gamma(\mathscr{X}^\cdot, \Omega^{n+1}_{\mathscr{X}/S\times D})$ be a relative $(n+1)$-form represented by the differential form

$$\omega = g(x, y)\,\mathrm{d}x_0 \wedge \ldots \wedge \mathrm{d}x_n, \tag{2.1.1}$$

where $g(x, y)$ is a holomorphic function.

As in (I.5.1.5) for a fixed $y \in D$ the restriction ω_y of the form ω to $\mathscr{X}^\cdot(y)$ defines a section $s[\omega_y]$ of the fibration $\underline{H}(y)$ over $S' \times \{y\}$: the form $(\omega/\mathrm{d}f_y)(t) = \mathrm{Res}_{\mathscr{X}^\cdot(y)_t}(\omega/(f_y - t)$ defines the cohomology class $s[\omega_y](t)$ in the fibre $\mathscr{X}^\cdot(y)_t = \mathscr{X}_{(t,y)}$. Varying y we obtain a section $s[\omega]$ of the fibration $\underline{H}$:

$$s[\omega](t, y) = \left[\frac{\omega}{\mathrm{d}f_y} \bigg|_{\mathscr{X}(t,y)} \right] \in H^n(\mathscr{X}_{(t,y)}, \mathbb{C}). \tag{2.1.2}$$

In the same way as in (I.4.2.7) one can verify that this is a holomorphic section of the fibration $\underline{H}$, i.e. $s[\omega]$ is a section $\mathscr{H}$. We call $s[\omega]$ the *geometric section* of the form ω.

We now consider the root components. By (II.6.5.6) for a fixed $y \in D$, $s[\omega_y] \in \mathscr{L}(y)$ and the root decomposition (II.6.2.6) for $s[\omega_y]$ is written in the form

$$s[\omega](y) = \sum_{\beta > -1} s(\omega, \beta)(y), \tag{2.1.3}$$

where the root components $s(\omega, \beta) \in C_\beta(y)$. By varying y we obtain a section $s(\omega, \beta)$ of the fibration $\underline{C}_\beta$ on D.

(2.1.4) *Remark* $s[\omega]$ and $s(\omega, \beta)$ can be considered in two senses. On the one hand $s[\omega](y)$ as a function of y only is a section of $\mathscr{M}_{|\{0\}\times D} = \hat{\oplus}_\beta C_\beta$. On the other hand, these sections are considered as germs of sections of the sheaf $\mathscr{H} = \mathscr{M}_{|S'\times D}$, $S' \times D = S \times D \backslash \{0\} \times D$, i.e. $s[\omega]$, as a function in t and y, is a section of $\mathscr{H}$ of the form

$$s[\omega](t, y) = \sum_\beta t^\beta t^N s(\omega, \beta)(y).$$

(2.1.5) Finally, we note that the sections $s(\omega, \beta)$ of the fibrations $\underline{C}_\beta$ are holomorphic, i.e. the $s(\omega, \beta)$ are sections of the sheaf C_β. Indeed, if we present $s[\omega]$ as a section of the sheaf $\mathscr{L}$ in the form (1.3.5), then the functions $g_j(t, y)$ are holomorphic functions on $S' \times D$ because $s[\omega]$ are sections of $\mathscr{H}$ and $s[A_j, \alpha_j]$ is a basis of sections of this sheaf. On the other hand, the functions $g_j(t, y)$ are holomorphic on the whole of $S \times \{y\}$ for every fixed y because $s[A_j, \alpha_j]$ is a basis of $\mathscr{L}(y)$ and $s[\omega_y] \in \mathscr{H}^{(0)}(y) \subset \mathscr{L}(y)$. Now by Hartog's theorem it follows that the functions $g_j(t, y)$ are holomorphic on the whole of $S \times D$. Decomposing $g_j(t, y)$ into series in t as in (1.3.6) we obtain that $s(\omega, \beta)$ are holomorphic functions.

We now want to learn to differentiate $s[\omega](y)$ and $s(\omega, \beta)(y)$ wrt the parameter y.

2.2 Formulae for derivatives of geometric sections and their root components wrt a parameter

To differentiate $s[\omega](y)$ along the parameter y_j we have to consider $s[\omega]$ as a section $s[\omega](t, y)$ of the sheaf $\mathscr{H}$, take its covariance derivative $\partial_{y_j} = \nabla_{\partial/\partial y_j}$ and then again consider $\partial_{y_j}(s[\omega](t, y)$ as a section of $\mathscr{M}_{|\{0\}\times D}$.

Consider the following function on $S' \times D$

$$I_{\omega,\sigma}(t, y) = \langle s[\omega](t, y), \sigma(t, y)\rangle = \int_{\sigma(t,y)} \frac{\omega}{\mathrm{d}f_{y}}_{|\mathscr{X}_{(t,y)}}, \tag{2.2.1}$$

where $\sigma(t, y)$ is a flat section of the homological fibration. Since $\partial_{y_j}\sigma(t, y) = 0$, it is sufficient by (I.2.7.3), in the same way as for calculating $\partial_t = \nabla_{\mathrm{d}/\mathrm{d}t}$ in (I.4.3.1), to find

$$\frac{\partial}{\partial y_j} I_{\omega,\sigma}(t, y) = \langle \partial_{y_j} s[\omega](t, y), \sigma(t, y)\rangle. \tag{2.2.2}$$

In the same way as in (I.4.3.1), to calculate the derivative we can use Leray's residue theorem (I.4.2.6) to present the function $I_{\omega,\sigma}(t, y)$,

$$\int_{\sigma(t,y)} \frac{\omega_y}{\mathrm{d}f_y}_{|\mathscr{X}_{(t,y)}} = \frac{1}{2\pi\mathrm{i}} \int_{\partial(\sigma(t,y))} \frac{\omega_y}{f_y(x) - t}, \tag{2.2.3}$$

where ∂ is the Leray coboundary and $\omega_y = \mathrm{d}f_y \wedge \eta$, $\eta_{|\mathscr{X}_{(t,y)}} = (\omega_y/\mathrm{d}f_y)_{|\mathscr{X}_{(t,y)}}$. We can assume that the cycle $\partial(\sigma(t, y))$ is the same in a neighborhood of the point (t, y) and we can apply the standard theorem of analysis when differentiating an integral wrt a parameter.

The derivative of the integrand on the right-hand side of (2.2.3) is

$$\frac{\partial}{\partial y_j}\left(\frac{\omega_y}{f_y(x) - t}\right) = \frac{\partial}{\partial y_j}\left(\frac{\omega}{F(x, y) - t}\right) = \frac{\partial \omega}{\partial y_j} \cdot \frac{1}{F - t} - \frac{\partial F}{\partial y_j} \cdot \frac{\omega}{(F - t)^2}.$$

We can transform the second summand using that $\omega_y = \mathrm{d}f_y \wedge \eta$ and

$$\mathrm{d}\left(\frac{\dfrac{\partial F}{\partial y_j}\cdot\eta}{f_y(x) - t}\right) = \frac{\mathrm{d}\left(\dfrac{\partial F}{\partial y_j}\eta\right)}{f_y(x) - t} - \frac{\dfrac{\partial F}{\partial y_j}\cdot \mathrm{d}f_y \wedge \eta}{(f_y(x) - t)^2}.$$

From this

$$\frac{\partial}{\partial y_j}\left(\frac{\omega_y}{f_y(x) - t}\right) = \frac{\dfrac{\partial \omega}{\partial y_j}}{f_y(x) - t} - \frac{\mathrm{d}\left(\dfrac{\partial F}{\partial y_j}\eta\right)}{f_y(x) - t} + \mathrm{d}\left(\frac{\dfrac{\partial F}{\partial y_j}\cdot\eta}{f_y(x) - t}\right).$$

Integrating along $\partial(\sigma(t, y))$, we obtain by (2.2.1) and (2.2.3):

$$\langle \partial_{y_j} s[\omega](t, y), \sigma(t, y)\rangle = \left\langle s\left[\frac{\partial\omega}{\partial y_j}\right](t, y), \sigma(t, y)\right\rangle - \left\langle s\left[\mathrm{d}\left(\frac{\partial F}{\partial y_j}\eta\right)\right], \sigma(t, y)\right\rangle.$$

It remains to note that by (I.5.2.1) $\partial_t((\partial F/\partial y_j)\omega) = \partial_t(\mathrm{d}f_y \wedge (\partial F/\partial y_j)\eta) = \mathrm{d}(\partial F/\partial y_j)\eta)$, i.e. $s[\mathrm{d}((\partial F/\partial y_j)\eta)] = \partial_t s[(\partial F/\partial y_j)\omega]$, giving

(2.2.4) $$\partial_{y_j} s[\omega](t, y) = s\left[\frac{\partial\omega}{\partial y_j}\right](t, y) - \partial_t s\left[\frac{\partial F}{\partial y_j}\omega\right](t, y).$$

Finally, we obtain:

(2.2.5) *Theorem* If a function $F(x, y)$ defines a μ-const deformation of the singularity $f\colon (\mathbb{C}^{n+1}, 0) \to (\mathbb{C}, 0)$, $y \in D \subset \mathbb{C}^m$ and $\omega = g(x, y)\,\mathrm{d}x_0 \wedge \ldots \wedge \mathrm{d}x_n$ is a holomorphic $(n+1)$-form, then for the geometric section $s[\omega](y) \in \mathscr{M}$ we have

(2.2.6) $$\partial_{y_j} s[\omega](y) = s\left[\frac{\partial\omega}{\partial y_j}\right](y) - \partial_t s\left[\frac{\partial F}{\partial y_j}\omega\right](y).$$

For root components $s(\omega, \alpha)(y)$ this gives

(2.2.7) $$\partial_{y_j} s(\omega, \alpha)(y) = s\left(\frac{\partial\omega}{\partial y_j}, \alpha\right)(y) - \partial_t s\left(\frac{\partial F}{\partial y_j}\omega, \alpha + 1\right)(y).$$

Formula (2.2.7) is obtained by comparing root components in (2.2.6) since $\partial_t\colon C_\alpha \to C_{\alpha-1}$.

(2.2.8) *Corollary* If a form $\omega = g(x, y)\,\mathrm{d}x_0 \wedge \ldots \wedge \mathrm{d}x_n$ does not depend on y, then

(2.2.9) $$\partial_y s[\omega] = -\partial_t s\left[\frac{\partial F}{\partial y}\omega\right],$$

(2.2.10) $$\partial_{y_j} s(\omega, \alpha) = -\partial_t s\left(\frac{\partial F}{\partial y_j}\omega, \alpha + 1\right),$$

i.e. to find ∂_{y_j}, we ought to multiply ω by $-\partial F/\partial y_j$, and then to apply ∂_t.

Since the operators $\partial y_1, \ldots, \partial y_m$ and ∂_t commute, we can iterate formulae (2.2.6) and (2.2.7) to find higher derivatives, and use them for decomposition into Taylor series. As an example, examined by Hertling [He], we consider the following.

2.3 Decomposition of the root components of geometric sections into Taylor series for upper diagonal deformations of quasihomogeneous singularities

(2.3.1) Let $f\colon (\mathbb{C}^{n+1}, 0) \to (\mathbb{C}, 0)$ be a quasihomogeneous singularity defined by the polynomial $f(x_0, \ldots, x_n)$ of degree $\deg f = 1$ with weights $\operatorname{wt} x_i = w_i$. In this case by proposition (I.5.5.4) the section $s[\omega]$ defined by the monomial $(n+1)$-form

$$\omega = x^k\,\mathrm{d}x = x_0^{k_0} \cdot \ldots \cdot x_n^{k_n}\,\mathrm{d}x_0 \wedge \ldots \wedge \mathrm{d}x_n$$

is homogeneous, i.e. it contains only one root component

(2.3.2) $$s[\omega] = s(w, \nu(\omega)) \in C_{\nu(\omega)},$$

where

(2.3.3) $$\mathscr{V}(\omega) = a(k) - 1 = \sum_{i=0}^{n} w_i(k_i + 1) - 1.$$

In this case the order function α defined by the roof decomposition (II.6.3.1) coincides with the Newton order ν (II.8.4.4), $\alpha(s[\omega]) = \nu(\omega)$.

For later use it is useful to reformulate (2.3.2) in the following form:

(2.3.4) *Lemma* If $\omega = x^k\,\mathrm{d}x$, and $M = x^l = x_0^{l_0} \cdot \ldots x_n^{l_n}$ is a monomial of weighted degree $\nu(M) = w_0 l_0 + \ldots + w_n l_n =: b(M) + 1$, then $s[M\omega] = s(M\omega, \nu(M\omega))$, where

$$\nu(M\omega) = \nu(M) + \nu(\omega) = b(M) + 1 + \nu(\omega).$$

(2.3.5) We now consider a μ-const deformation of the singularity f. Let $M_j = x^{l_j} = x_0^{l_{0,j}} \cdot \ldots x_n^{l_{n,j}}$, $j = 1, \ldots, m$, be monomials of weighted degree $\nu(M_j) > 1$. Then

$$b_j = b(M_j) = \nu(M_j) - 1 > 0.$$

Consider the m-parametric deformation

(2.3.6) $$f_y(x) = F(x, y) = f(x) + \sum_{j=1}^{m} x^{l_j} \cdot y_j$$

of the singularity $f_0(x) = f(x)$, which is, as is well known [AGV], a μ-const deformation. Consider the geometric section $s[\omega](y)$ induced by a monomial form $\omega = x^k\,\mathrm{d}x$ on $D = \mathbb{C}^m$, and consider its root components $s(\omega, \alpha)(y)$, which are sections of sheaves C_α on D. Let us find derivatives of sections $s(\omega, \alpha)(y)$ and their decompositions into Taylor series at the point $y = 0$.

Formula (2.2.10) applied to the deformation (2.3.6) gives

$$\partial_{y_j} s(\omega, \alpha)(y) = -\partial_t s(z^{l_j}\omega, \alpha+1)(y),$$

and iterating we obtain

(2.3.7) $$\partial^i_y s(\omega, \alpha)(y) = (-\partial_t)^{|i|} s(M^i\omega, \alpha + |i|)(y),$$

where the standard notation is used for

$$i = (i_1, \ldots, i_m) \in \mathbb{N}_0^m,\ |i| = i_1 + \ldots + i_m,\ i! = i_1! \ldots i_m!,$$
$$y^i = y_1^{i_1} \cdot \ldots y_m^{i_m},\ \partial^i_y = \partial^{i_1}_{y_1} \cdot \ldots \partial^{i_m}_{y_m},\ M^i = M_1^{i_1} \cdot \ldots \cdot M_m^{i_m}.$$

The decomposition of $s(\omega, \alpha)$ into Taylor series at the point $y = 0$ is of the form

$$s(\omega, \alpha) = \sum_{i \in \mathbb{N}_0^m} (\partial^i_y s(\omega, \alpha)(0) \frac{y^i}{i!},$$

or applying (2.3.7),

(2.3.8) $$s(\omega, \alpha)(y) = \sum_{i \in \mathbb{N}_0^m} (-\partial_t)^{|i|} s(M^i\omega, \alpha - |i|)(0) \cdot \frac{y^i}{i!}.$$

By (2.3.4) it follows that $s[M^i\omega](0) = s(M^i\omega, \nu(M^i\omega(0)))$, where $\nu(M^i\omega(0)) = |i| + b_1 i_1 + \ldots + b_m i_m + \nu(\omega)$, $b_j = b(M_j) > 0$. Hence, $s(M^i\omega, \alpha + |i|)(0) \neq 0$ only if

(2.3.9) $$\alpha = b_1 i_1 + \ldots + b_m i_m + \nu(\omega).$$

Finally, we obtain:

(2.3.10) *Proposition* For the deformation (2.3.6) of a quasihomogeneous singularity and for a monomial form $\omega = x^k\, dx$ the root components of the geometric section $s[\omega]$ have a decomposition into Taylor series of the form

(2.3.11) $$s(\omega, \alpha)(y) = \sum_i (-\partial_t)^{|i|} s(M^i\omega](0) \cdot \frac{y^i}{i!},$$

where the summation is taken only over the multiindices $i \in \mathbb{N}_0^m$ satisfying (2.3.9).

In fact the Taylor series (2.3.11) contains only a finite number of non-zero terms. Indeed, the set of $i \in \mathbb{N}_0^m$ satisfying (2.3.9), i.e. the set of integer points of the hyperplane $b_1 z_1 + \ldots + b_m z_m = \alpha - \nu(\omega)$ in the positive 'octant' is finite because all $b_j > 0$. In particular:

(2.3.12): if $\alpha < \nu(\omega)$, then $s(\omega, \alpha)(y) \equiv 0$; and

(2.3.13): if $\alpha = \nu(\omega)$, then

$$s(\omega, \nu(\omega))(y) = \text{const} = s[\omega](0).$$

2.4 The sheaves $Gr_V^\beta \mathscr{H}^{(0)}$

We have considered how the geometric sections vary under μ-const deformation of a singularity. We recall (II.7.2.1) that for an individual singularity f_y the Hodge filtration of the MHS on $H^n(X(y)_\infty, \mathbb{C})$ is defined by means of $Gr_V^\beta \mathscr{H}^{(0)}(y)$. So now we consider how the Brieskorn lattices $\mathscr{H}^{(0)}(y)$ and their subfactors $Gr_V^\beta \mathscr{H}^{(0)}(y)$ generated by the main parts of geometric sections vary under the variation of y.

(2.4.1) The geometric section $s[\omega]$ of the sheaf $\mathscr{L} \subset \mathscr{M}_{|\{0\}\times D}$ on D was defined for a relative $(n+1)$-form ω in (2.1.1). It has the root decomposition (2.1.3)

$$s[\omega] = \sum_{\beta > -1} s(\omega, \beta),$$

where $s(\omega, \beta)$ is the holomorphic section of the filbration $\underline{C}_\beta$ (2.1.5). Let $\mathscr{H}^{(0)} \subset \mathscr{L}$ be the subsheaf generated by geometric sections. The fibre $\mathscr{H}^{(0)}(y) = \mathscr{H}^{(0)} \otimes \mathscr{O}_{S\times\{y\}}$ is the Brieskorn lattice of the singularity f_y.

(2.4.2) *Definition* The *order of a geometric section* $s[\omega]$ is

$$\alpha(\omega) = \min\{\beta | s(\omega, \beta) \neq 0\}.$$

One should distinguish $\alpha(\omega)$ from the orders of geometric sections $s[\omega](y)$ at points y

(2.4.3) $$\alpha(s[\omega](y)) = \min\{\beta | s(\omega, \beta)(y) \neq 0\}.$$

For a geometric section $s[\omega] = s(\omega, \alpha(\omega)) + \dots$ the main part $s(\omega, \alpha(\omega)) \in C_{\alpha(\omega)}$ can vanish at some points. Thus for all points $y \in D$

(2.4.4) $$\alpha(s[\omega](y)) \geqslant \alpha(\omega),$$

and the equality holds apart from a closed subset in D.

Thus the orders $\alpha(s[\omega](y))$ of geometric sections at points can jump on closed subsets. Nevertheless the subspaces $Gr_V^\beta \mathscr{H}^{(0)}(y) \subset C^\beta(y)$ generated by $s(\omega, \alpha(\omega))$ behave well. Indeed, by Varchenko's theorem (II.8.9.11) the spectral numbers n_α remain constant under μ-const deformations. Hence by (II.8.1.4″) the dimensions of spaces

(2.4.5) $$\dim Gr_V^\beta \mathscr{H}^{(0)}(y) = n_\alpha + n_{\alpha+1} + \dots + n_\beta = f_\lambda^\beta,$$
$$\beta = \alpha + n - p, \ -1 < \alpha \leqslant 0,$$

also remain constant. Thus we have:

(2.4.6) *Proposition* The vector spaces $Gr_V^\beta \mathscr{H}^{(0)}(y) \subset C^\beta(y)$ form a subfibration in C^β, and the sheaf of holomorphic sections of this is a locally free subsheaf $Gr_V^\beta \mathscr{H}^{(0)} \subset C^\beta$.

(2.4.7) *Corollary.* For any β there exist relative $(n+1)$-forms $\omega_{1,\beta}, \ldots, \omega_{m,\beta}$, $m = m(\beta) = \dim Gr_V^\beta \mathscr{H}^{(0)}(y) = f_\lambda^p$, on some neighborhood U_{y_0} of a point $y_0 \in D$ such that $\alpha(\omega_{j,\beta}) = \alpha(s[\omega_j, \beta](y)) = \beta$ for all $y \in U_{y_0}$, and the main parts $s(\omega_{j,\beta}, \beta)$ form a basis of $Gr_V^\beta \mathscr{H}^{(0)}$.

(2.4.8) *Corollary* For any $(n+1)$-form $\omega_0 \in \Omega^{n+1}_{\mathbb{C}^{n+1},0}$ in the neighborhood of a point $y_0 \in D$ there exists a relative $(n+1)$-form ω such that $s[\omega](y_0) = s[\omega_0]$ and $\alpha(s[\omega](y_0)) = \alpha(\omega)$.

Indeed, if $\alpha(s[\omega_0]) = \beta$, then it is sufficient to take ω generating the class $s[\omega_0]$ in $Gr_V^\beta \mathscr{H}^{(0)}$ at the point y_0.

(2.4.9) *Remark* The vector spaces $Gr_V^\alpha \mathscr{H}^{(0)}(y) \simeq \{s(\omega, \alpha)(y) | \alpha(s[\omega](y_0)) = \alpha\}$ are generated by the main (leading) components of geometric sections. One can consider larger vector spaces $\{s(\omega, \alpha)(y)\}$ generated by components of order α (not necessarily leading order) of all geometric sections. The dimensions of these spaces unlike $\dim Gr_V^\alpha \mathscr{H}^{(0)}(y)$ can jump. Their behavior is related to the Bernstein polynomials.

3 The period map

Before considering the 'natural' period map defined by the MHS on a vanishing cohomology we will first consider the period map defined by M. Saito [Sa7] via embeddings of Brieskorn lattices $\mathscr{H}^{(0)}(y) \subset \mathscr{M}$ because such an embedding defines the corresponding MHS.

3.1 Identification of meromorphic connections in a μ-const family of singularities

Let $F = F_Y \colon (\mathbb{C}^{n+1}, 0) \times Y \to (\mathbb{C}, 0) \times Y$ be a m-parametric μ-const deformation of a singularity $f \colon (\mathbb{C}^{n+1}, 0) \to (\mathbb{C}, 0)$ with base Y, $\dim Y = m$, not necessarily local, $f = f_{y_0}$, $y_0 \in Y$. Let

(3.1.1) $$F = F_Y \colon \mathscr{X}_Y \to S \times Y$$

be the corresponding 'Milnor fibration' arranged locally as in (1.1.3).

If the base Y is simply connected, then the elementary sections (1.1.3)

$$s[A, \beta] = t^\beta t^N A(t, y),$$

where $A(t, y) \in H_\lambda$ are multivalued horizontal sections of the fibration $\underline{H}$, make it possible to identify all meromorphic connections $\mathcal{M}(y) = \oplus_\alpha \hat{C}_\alpha(y)$, $y \in Y$, and all their homogeneous components $C_\alpha(y)$. The sections $s[A, \beta]$ are horizontal along y, $(\partial/\partial y_j) \cdot s[A, \beta] = 0$, and using these we can form a basis of sections at every point y. Then the canonical isomorphism $\mathcal{M}(y_1) \tilde{\to} \mathcal{M}(y_2)$ takes an element, which has some coordinates relative to the 'elementary' basis, to the element with the same coordinates relative to this basis. More explicitly, if $A_1, \ldots, A_{\mu_\lambda}$ is a basis of H_λ and $s(y)$ is a section of the sheaf C_β, $\mathrm{e}^{-2\pi i\beta} = \lambda$, then $s(y)$ can be written in the form

$$s(y) = \sum_{j=1}^{\mu_\lambda} g_j(y) s[A_j, \beta] = \sum_{j=1}^{\mu} t^\beta t^N g_j(y) A_j(t, y) \tag{3.1.2}$$

and we can consider $s(y)$ as a family of elements of the space $C^\beta(y_0)$ with coordinates $g_j(y)$ relative to the basis $s[A_j, \beta](y_0)$. Thus we have the canonical isomorphisms $\mathcal{M}(y_1) \tilde{\to} \mathcal{M}(y_2)$ for any two points y_1 and $y_2 \in Y$, which are isomorphisms of $\mathbb{C}\{t\}[t^{-1}]$-modules commuting with ∂_t (or isomorphisms of $\mathbb{C}[\partial_t]$-modules). Under the isomorphisms the $V^\cdot$-filtration goes to the $V^\cdot$-filtration and, in particular, $\mathcal{L}(y_1) \tilde{\to} \mathcal{L}(y_2)$. Further we shall assume that a point y_0 is chosen, and we'll often write $\mathcal{M}(y_0)$, $\mathcal{L}(y_0)$, etc., simply as $\mathcal{M}$, $\mathcal{L}$, etc.

If the base Y is not simply connected, then for any $\gamma \in \pi_1(Y, y_0)$ we have the automorphism h_γ: $\mathcal{M}(y_0) \to \mathcal{M}(y_0)$ defined by extending sections along γ. In this case to identify the connections $\mathcal{M}(y)$ we have either to consider $\mathcal{M}(y_0)$ up to automorphisms h_γ, or to turn to the universal covering $\tilde{Y} \to Y$ (to consider the canonical fibre in the sense of §2, chapter II).

3.2 The period map defined by the embedding of Brieskorn lattices

Assume that we have a μ-const deformation (3.1.1). Then we have the Brieskorn lattice $\mathcal{H}^{(0)}(y)$ at every point $y \in Y$. Taking into account the canonical identification $\mathcal{M}(y) \tilde{\to} \mathcal{M}(y_0) = \mathcal{M}$ we obtain a family of Brieskorn lattices $\mathcal{H}^{(0)}(y) \subset \mathcal{M}$ in the meromorphic connection $\mathcal{M}$ associated with the deformation (3.1.1). By (II.8.3.7) we have

$$\mathcal{L} = V^{>-1} \supset \mathcal{H}^{(0)}(y) \supset V^{n-1} \tag{3.2.1}$$

and the embedding $\mathcal{H}^{(0)}(y) \subset \mathcal{M}$ is completely defined by the embedding of finite-dimensional spaces

$$\mathcal{H}^{(0)}(y)/V^{n-1} \subset V^{>-1}/V^{n-1} =: V. \tag{3.2.2}$$

Let us find the dimensions of these spaces. Since $\sum_{-1<\alpha\leqslant 0}\dim C_\alpha = \mu$ and $\sum_{-1<\alpha\leqslant n-1}\dim C_\alpha = n\mu$, we have

$$\dim V = \sum_{-1<\alpha<n-1} \dim C_\alpha = n\mu - \mu_1,$$

where $\mu_1 = \dim C_n = \dim H^n(X_\infty, \mathbb{C})_1$.

From the symmetry of spectral numbers $n_\alpha = n_{n-1-\alpha}$ (II.8.3.3) it is easy to obtain (one can see this visually in the picture on p. 107) the following lemma:

(3.2.4) *Lemma*

$$\dim \mathscr{H}^{(0)}(y)/V^{n-1} = \tfrac{1}{2}(n\mu - \mu_1) = \tfrac{1}{2}\dim V.$$

■

Denote by $\Pi' = Grass(k, V)$ the Grassman manifold of k-dimensional subspaces in a vector space V, $\dim V = 2k = n\mu - \mu_1$. Then Π' is the 'space of periods' containing all Brieskorn lattices $\mathscr{H}^{(0)}(y) \bmod V^{n-1}$ of singularities f_y μ-equivalent to the singularity f. We obtain the *period map*

(3.2.5) $$\Phi\colon Y \to \Pi',\ y \mapsto \mathscr{H}^{(0)}(y)/V^{n-1}$$

associated with the deformation F_Y of the singularity f *defined by the embedding of Brieskorn lattices*. By (2.4.6) Φ is a holomorphic map. In general, Φ is multivalued if Y is not simply connected.

(3.2.6) The space Π' is too large. We can construct a more 'economical' space $\Pi \subset \Pi'$ containing all Brieskorn lattices. Let $Sp(f) = \sum n_\alpha\cdot(\alpha)$ be the spectrum of the singularity f. Let

$$\alpha_{\min} = \min\{\alpha | n_\alpha \neq 0\},\ \alpha_{\max} = \max\{\alpha | n_\alpha \neq 0\}$$

be the smallest and the largest spectral numbers, respectively. It follows from (II.8.3.4) that $\alpha_{\min}$ and $\alpha_{\max} \in (-1, n)$, and from (II.8.3.3) that $\alpha_{\min} + \alpha_{\max} = n - 1$. In the same way as in (II.8.3.7) we can prove a refinement.

(3.2.7) *Lemma*

$$V^{>-1} \supset V^{\alpha_{\min}} \supset \mathscr{H}^{(0)}(y) \supset V^{>\alpha_{\max}-1} \supset V^{n-1}.$$

Proof By (II.8.1.4″) we have $Gr_V^\beta \mathscr{H}^{(0)}(y) = 0$ for $\beta < \alpha_{\min}$. Hence $\mathscr{H}^{(0)}(y) \subset V^{\alpha_{\min}}$. Furthermore, $Gr_V^\beta \mathscr{H}^{(0)}(y) = C_\beta$ for $\beta > \alpha_{\max} - 1$

because otherwise there exists a spectral number $\geqslant \beta + 1 > \alpha_{\max}$. This yields $\mathscr{H}^{(0)}(y) \supset V^{>\alpha_{\max}-1}$. ■

(3.2.8) It follows from (3.2.7) that $\Phi(Y) \subset \Pi \subset \Pi'$, where Π is the Grassman manifold of vector subspaces in $V^{\alpha_{\min}}/V^{>\alpha_{\max}-1}$ of dimensions $\dim \mathscr{H}^{(0)}(y)/V^{>\alpha_{\max}-1}$.

(3.2.9) To calculate the period map Φ: $Y \to \Pi$ explicitly, we have:

(i) To calculate the spectrum $Sp(f)$ and to consider the vector space $V = V^{\alpha_{\min}}/V^{>\alpha_{\max}-1} = \oplus_{\alpha_1 \leqslant \alpha \leqslant \beta_1} C_\alpha$, where $\alpha_1 = \alpha_{\min}$, and $\beta_1 = \alpha_{\max} - 1$.

(ii) To take enough relative $(n+1)$-forms $\omega_1(y), \ldots, \omega_1(y)$ so that corresponding geometric sections $s[\omega_1], \ldots, s[\omega_1]$ define the position of the subspace $\mathscr{H}^{(0)}(y)/V^{>\beta_1} = \langle s[\omega_1], \ldots, s[\omega_1]\rangle$ in V. The problem now is to learn how to find the vector functions $s[\omega_j](y)$. Here we give two methods which we'll use in what follows: (1) if F_Y is a semiquasihomogeneous deformation (2.3.6), we can use proposition (2.3.10) to decompose $s[\omega]$ into the Taylor series; (2) one can find differential equations such that the vector functions $s[\omega_j](y)$ are their solutions, and then solve them to find these functions. If one succeeds in finding the vector functions $s[\omega_j](y)$, then in order to write explicit formulae for the period map, the Plücker coordinates ξ of a point in Π, corresponding to the subspace $\mathscr{H}^{(0)}(y)/V^{>\beta_1}$, have to be expressed as functions of y.

Now we examine some examples in detail.

3.3 Example: the period map for E_{12} singularities

Let us consider a two-dimensional singularity f: $(\mathbb{C}^3, 0) \to (\mathbb{C}, 0)$ defined by the function $f(x, y, z) = x^3 + y^7 + z^2$. This is a quasihomogeneous unimodular singularity of type E_{12} in the classification of V. I. Arnol'd. The normal form of singularities E_{12} is ([AGV])

(3.3.1) $$E_{12}: x^3 + y^7 + axy^5 + z^2.$$

This function defines a semiquasihomogeneous μ-const deformation of the singularity f with base $Y = \mathbb{C} \ni a$. We want to calculate the period map Φ associated with this deformation.

The singularity f is quasihomogeneous of degree 1 with variables of weights $w_0 = \frac{1}{3}$, $w_1 = \frac{1}{7}$, $w_2 = \frac{1}{2}$. We have $\mathrm{wt}(xy^5) = \frac{22}{21}$. By (II.8.4.8) the forms

(3.3.2) $$\omega_{k,1} = x^k y^1 \omega_0,\ \omega_0 = \mathrm{d}x \wedge \mathrm{d}y \wedge \mathrm{d}z,\ 0 \leqslant k \leqslant 1,\ 0 \leqslant l \leqslant 5,$$

form a basis of the $\mathcal{O}_{S,0}$-module $\mathcal{H}^{(0)}$, and the numbers

$$\{\alpha(\omega_{k,1}) = \tfrac{1}{3}(k+1) + \tfrac{1}{7}(l+1) - \tfrac{1}{2},\ 0 \leqslant k \leqslant 1,\ 0 \leqslant l \leqslant 5\}$$

form the spectrum of the singularity f. If we order the spectral numbers $\alpha_i = k_i/42$, $1 \leqslant i \leqslant 12$, the numerators k_i give the set

$$\{-1, 5, 11, 13, 17, 19, 23, 25, 29, 31, 37, 43\}.$$

Thus, $\alpha_{\min} = \alpha_1 = -\frac{1}{42}$ ($\alpha_1 = \alpha(s[\omega_0])$), and $\alpha_{\max} = \frac{43}{42}$ and $\beta_1 = \alpha_{\max} - 1 = \frac{1}{42}$. The spectrum $Sp(f)$ is symmetric wrt the point $(n-1)/2 = \frac{1}{2}$.

From (II.7.2.6) we schematically visualise $\mathcal{M}$ and the inclusions $\mathcal{H}^{(0)}(a) \subset \mathcal{M}$ with the following picture

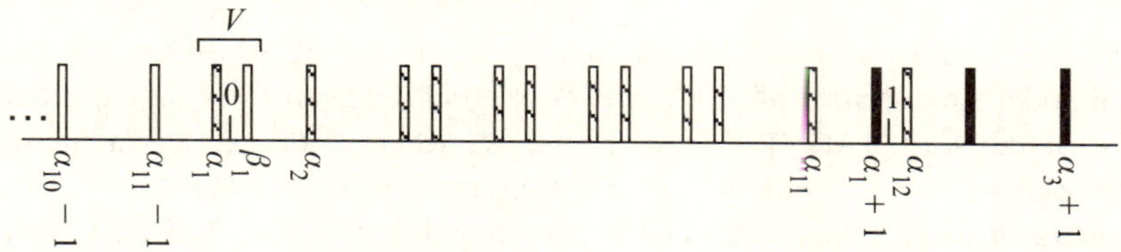

Here $\mathcal{M} = \hat{\oplus} C_\alpha$, where α runs 12 progressions $\beta_1 + n$, $\alpha_2 + n$, $\ldots$, $\alpha_{11} + n$, $\alpha_1 + n + 1$, $n \in \mathbb{Z}$, and the root subspaces C_α are depicted as columns of height $\dim C_\alpha = 1$. Those parts of columns corresponding to $Gr_V^\beta \mathcal{H}^{(0)} \subset C_\alpha$ are hatched and those corresponding to $\partial_t^{-1} Gr_V^{\beta-1} \mathcal{H}^{(0)} = Gr_V^\beta \mathcal{H}^{(-1)} \subset Gr_V^\beta \mathcal{H}^{(0)}$ are shaded.

In this case from (3.2.8) the space V is

(3.3.3) $$V = C_{\alpha_1} \oplus C_{\beta_1},\ \dim V = 2,$$

and the space $\mathcal{H}^{(0)}(a)/V^{>\alpha_{\max}-1}$ is one-dimensional. Therefore, the Grassman manifold $\Pi = \mathbb{P}^1$ is a projective line.

The embedding of a Brieskorn lattice $\mathcal{H}^{(0)}(a) \subset \mathcal{M}$ is defined by the geometric sections $s[\omega_0](a)$,

$$\mathcal{H}^{(0)}(a) = \mathbb{C}\cdot(s(\omega_0, \alpha_1)(a) + s(\omega_0, \beta_1)(a)) \oplus V^{>\beta_1}$$

and the point $\Phi(a) \in \mathbb{P}^1$ corresponds to the line

(3.3.4) $$\mathbb{C}\cdot(s(\omega_0, \alpha_1)(a) + s(\omega_0, \beta_1)(a)) \subset C_{\alpha_1} \oplus C_{\beta_1} = V.$$

To calculate $s(\omega_0, \alpha_1)(a)$ and $s(\omega_0, \beta_1)(a)$, we use their decompositions into the Taylor series (2.3.11). In this case $m = 1$ and monomial $M = M_1 = xy^5$. By (2.3.13) we have

(3.3.5) $$s(\omega_0, \alpha_1)(a) = s[\omega_0](0),$$

because $\alpha(\omega_0) = \nu(\omega_0)$. To decompose $s(\omega_0, \beta_1)(a)$, we have to find i from (2.3.9) for $\alpha = \beta_1$, $b_1 = \nu(M) - 1 = \nu(xy^5) - 1 = \frac{1}{21}$, i.e. from the equation $\frac{1}{42} = bi - \frac{1}{42}$. This gives $i = 1$, and from (2.3.11) we have

(3.3.6) $$s(\omega_0, \beta_1)(a) = -\partial_t s[xy^5\omega_0](0)\cdot a.$$

We take the vectors

$$b_0 = s[\omega_0](0) \in C_\alpha \text{ and } b_1 = -\partial_t s[xy^5\omega_0](0) \in C_{\beta_1}$$

as a basis of the space $V = C_{\alpha_1} \oplus C_{\beta_1}$. The line $\mathbb{C}\cdot(\xi_0 b_0 + \xi_1 b_1) \subset V$ corresponds to the point $(\xi_0 \colon \xi_1) \in \mathbb{P}^1$. It then follows from (2.3.4), (3.3.5), and (3.3.6) that the period map is given by the formula

(3.3.7) $$\Phi\colon Y = \mathbb{C} \to \mathbb{P}^1 = \Pi,\ a \mapsto (1\colon a).$$

If $\Pi_0\colon \xi_0 \neq 0$ is the open subset $\Pi_0 = \mathbb{C} \subset \mathbb{P}^1$ with coordinate $\xi = \xi_1/\xi_0$, then $\Phi\colon Y \to \Pi_0 \subset \Pi$ is simply the identical map, $\xi = a$.

(3.3.8) *The general case* The singularity E_{12} is one of the 14 exceptional unimodal singularities of V. I. Arnol'd. Normal forms of these singularities are of the form [AGV] $f_0(x, y, z) + a\cdot M$, where $f_0(x, y, z)$ is a quasi-homogeneous monomial, and M is a monomial of weight $\mathrm{wt}(M) > 1$. The spectra of these singularities can be found in Table (II.8.8.1). For all these singularities we have: all $\dim C_\alpha = 1$, $\alpha_{\min} = \alpha_1 < 0$, $\alpha_{\max} = \alpha_\mu > 1$ and $\alpha_i > \beta_1 = \alpha_\mu - 1$ for $i = 2, \ldots, \mu$. The remaining calculations and the result are the same as for the singularity E_{12}. These calculations are carried out in [He1].

3.4 The period map for hyperbolic singularities $T_{p,q,r}$

The unimodal (two-dimensional) singularities of the series $T_{p,q,r}$ have the following normal form [AGV]

(3.4.1) $$f_a = x^p + y^q + z^r + axyz,\ a \neq 0,$$

where

(3.4.2) $$\rho =: \frac{1}{p} + \frac{1}{q} + \frac{1}{r} < 1.$$

This is a μ-const deformation with the one-dimensional base $Y = \mathbb{C}^* = \mathbb{C}\backslash\{0\} \ni a$. Let us find the corresponding period map (see [He1]).

The singularities f_a are not quasihomogeneous, but they are non-degenerate with respect to the Newton boundary. For them $\mu = p + q + r - 1$, and the monomials

(3.4.3) $$1,\ xyz,\ x^k (0 < k < p),\ y^k (0 < k < q),\ z^k (0 < k < r),$$

form a basis of the Jacobian algebra Q_{f_a}, and the forms

(3.4.4) $$M_i\omega_0,\ \omega_0 = \mathrm{d}x \wedge \mathrm{d}y \wedge \mathrm{d}z,\ 1 \leqslant i \leqslant \mu,$$

where M_i are the monomials (3.4.3), represent basis vectors of the space $\Omega_f = \mathscr{H}^{(0)}(a)/\mathscr{H}^{(-1)}(a)$. From (II.8.5.8) we obtain that

$$(3.4.5)\qquad \alpha([\omega_0](a)) = 0,\ \alpha(s[xyz\omega_0](a)) = 1,$$
$$\alpha([x^k\omega_0](a)) = \frac{k}{p},\ \alpha([y^k\omega_0](a)) = \frac{k}{q},\ \alpha([z^k\omega_0](a)) = \frac{k}{r},$$

and the spectral numbers are

$$(3.4.6)\qquad \mathrm{Sp}(f_a) = \left\{0;\ 1;\ \frac{k}{p}(0 < k < p),\ \frac{k}{q}(0 < k < q),\ \frac{k}{r}(0 < k < r)\right\}.$$

It follows from this that $\dim C_0 = \dim H^n(X_\infty, \mathbb{C})_{\lambda=1} = 2$. A diagram based on (II.7.2.6) is of the form (the example shown is for the singularity $T_{3,3,4}$):

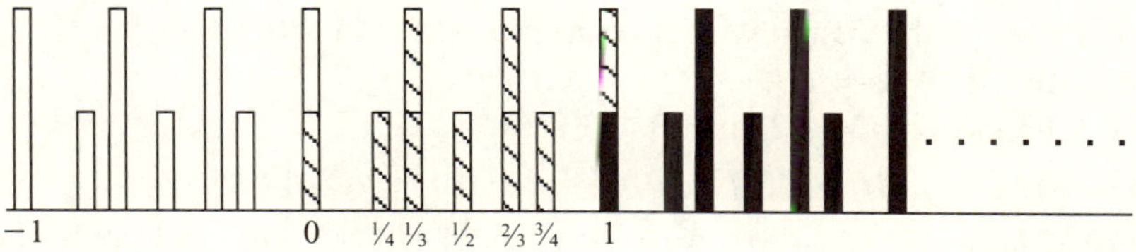

For the singularity f_a we have $\alpha_1 = \alpha_{\min} = 0$, $\alpha_{\max} = 1$, $\beta_1 = \alpha_{\max} - 1 = 0 = \alpha_1$. Therefore, $V = V^{\alpha_1}/V^{>\beta_1} = C_0$,

$$(3.4.7)\qquad V^0 \supset \mathscr{H}^{(0)}(a) \supset V^{>0},\ \mathscr{H}^{(0)}(a) = \mathbb{C}\cdot s(\omega_0, 0)(a) \oplus V^{>0},$$

and the embedding $\mathscr{H}^{(0)}(a) \subset \mathscr{M}$ is defined by the embedding of the one-dimensional space $\mathbb{C}\cdot s(\omega_0, 0)(a)$ into the two-dimensional space $V = C_0$.

To find $s(\omega_0, 0)(a)$, we will obtain a differential equation which has this vector function as its solution. Since $\partial f_a/\partial a = xyz$, we have for $\omega_0 = \mathrm{d}x \wedge \mathrm{d}y \wedge \mathrm{d}z$ from (2.2.8) that

$$(3.4.8)\qquad \partial_a s[\omega_0](a) = -\partial_t s[xyz\omega_0](a),$$
$$\partial_a^2 s[\omega_0](a) = -\partial_t^2 s[(xyz)^2\omega_0](a).$$

We now have to find a relation between the derivatives of the function $s(\omega_0, 0)$. Actually we already have a differential equation: since $s(\omega_0, 0) \in C_0$, C_0 is a root subspace of the operator $t\partial_t$ and $\dim C_0 = 2$, we have

$$(3.4.9)\qquad (t\partial_t)^2 s(\omega_0, 0)(a) = 0.$$

It now remains to express the operator

$$(3.4.10)\qquad (t\partial_t)^2 = \partial_t^2 t^2 - 3\partial_t t + 1$$

as ∂_a and ∂_a^2 by means of (3.4.8) and to use that the function $f_a = t$ in $\mathscr{M}$. To do this we employ an obvious identity which follows from (3.4.1) and (3.4.2),

$$(3.4.11)\qquad f_a = \frac{1}{p}xf'_{a,x} + \frac{1}{q}yf'_{a,y} + \frac{1}{r}zf'_{a,z} + a(1-\rho)xyz.$$

To express $\partial_t ts[\omega_0](a)$ we multiply (4.4.11) by ω_0 and apply the operator ∂_t to the corresponding geometric sections

(3.4.12)

$$\partial_t ts[\omega_0](a) = \partial_t s[f_a\omega_0](a) = \partial_t s\left[\left(\frac{1}{p}xf_{a,x} + \ldots + a(1-\rho)xyz\right)\omega_0\right](a).$$

We then have

(3.4.13) $$\partial_t s[xf'_{a,x}\omega_0](a) = s[\omega_0](a)$$

and analogous equalities for the derivatives wrt y and z. This follows from the identity $xf'_x\omega_0 = \mathrm{d}f \wedge \eta$, where $\eta = x\,\mathrm{d}y \wedge \mathrm{d}z$, and from the rule of calculation of the Gauss–Manin connection (I.5.2.2), $\partial_t s[xf'_x\omega_0] = s[\mathrm{d}\eta]$, $\mathrm{d}\eta = \mathrm{d}x \wedge \mathrm{d}y \wedge \mathrm{d}z = \omega_0$.

It follows from (3.4.12) and (3.4.13) that

(3.4.14) $$\partial_t ts[\omega_0](a) = \rho s[\omega_0](a) + a(1-\rho)\partial_t s[xyz\omega_0](a).$$

To express $\partial_t^2 t^2 s[\omega_0](a)$, we multiply (3.4.11) by $f_a\omega_0$ and apply the operator ∂_t,

(3.4.15)

$$\begin{aligned}\partial_t t^2 s[\omega_0](a) &= \partial_t s[f_a^2\omega_0](a)\\ &= \partial_t s\left[\left(\frac{1}{p}xf_af'_{a,x} + \ldots + \frac{1}{r}zf_af'_{a,z} + a(1-\rho)f_axyz\right)\omega_0\right](a).\end{aligned}$$

The terms on the right-hand side (except the last one) are transformed by means of the formula

(3.4.16) $$\begin{aligned}\partial_t s[xf'_{a,x}f_a\omega_0](a) &= s[f_a\omega_0 + xf'_{a,x}\omega_0](a)\\ &= (t+\partial_t^{-1})s[\omega_0](a),\end{aligned}$$

which follows from the formulae $xf_af'_{a,x}\omega_0 = \mathrm{d}f_a \wedge (f_ax\,\mathrm{d}y \wedge \mathrm{d}z)$, and $\mathrm{d}(f_ax\,\mathrm{d}y \wedge \mathrm{d}z) = f_a\omega_0 + xf'_{a,x}\omega_0$. We substitute f_a from (3.4.11) in the last term on the right-hand side of (3.4.15) and apply the formula

(3.4.17) $$\partial_t s[x(xyz)f'_{a,x}\omega_0](a) = 2s[xyz\omega_0](a),$$

which follows from the formulae $x(xyz)f'_{a,x}\omega_0 = \mathrm{d}f \wedge (x^2yz\,\mathrm{d}y \wedge \mathrm{d}z)$, and $\mathrm{d}(x^2yz\,\mathrm{d}y \wedge \mathrm{d}z) = 2xyz\omega_0$. Applying (3.4.16) and (3.4.17), we obtain from (3.4.15) that

$$\begin{aligned}\partial_t t^2 s[\omega_0](a) = \rho(t+\partial_t^{-1})s[\omega_0](a) &+ a(1-\rho)(2\rho s[xyz\omega_0](a)\\ &+ a(1-\rho)\cdot\partial_t s[(xyz)^2\omega_0](a)).\end{aligned}$$

Applying ∂_t to both sides of the equality, we obtain

(3.4.18)

$$\partial_t^2 t^2 s[\omega_0](a) = \rho\partial_t ts[\omega_0](a) + \rho s[\omega_0](a) + 2a(1-\rho)\rho\partial_t s[xyz\omega_0](a)$$
$$+ a^2(1-\rho)^2\partial_t^2 s[(xyz)^2\omega_0](a).$$

Substituting (3.4.14) and (3.4.18) into (3.4.10), we obtain

(3.4.19) $\partial_t^2 t^2 s[\omega_0](a)$
$$= (1-\rho^2)\{a^2\partial_t^2 s[(xyz)^2\omega_0](a) - 3a\partial_t s[xyz\omega_0](a) + s[\omega_0](a)\}.$$

Finally taking (3.4.8) into account, we rewrite the differential equation (3.4.9) for the root component $s(\omega_0, 0)(a)$ in the following form

(3.4.20) $$a^2\partial_a^2 s(\omega_0, 0)(a) + 3a\partial_a s(\omega_0, 0)(a) + s(\omega_0, 0)(a) = 0$$

or

(3.4.21) $$(a\partial_a + 1)^2 s(\omega_0, 0)(a) = 0.$$

The differential equation (3.4.21) has the general solution $c_1 a^{-1} + c_2 a^{-1}\ln a$. Taking into account the initial conditions satisfied by $s(\omega_0, 0)(a)$ for $a = 1$, we obtain that

(3.4.22) $$s(\omega_0, 0)(a) = b_1 a^{-1} + b_2 a^{-1}\ln a,$$

where $b_1 = s(\omega_0, 0)(1)$, $b_2 = b_1 + \partial_c s(\omega_0, 0)(1)$, are vectors in the space $C_0 = V$.

To summarize: the period space for the singularities $T_{p,q,r}$ is the projective line $\mathbb{P}^1$ of lines in C_0. Take vectors b_1 and b_2 as the basis in C_0, and let $(\xi_0: \xi_1)$ be coordinates of the line $\xi_0 b_1 + \xi_1 b_2$. Then the period map

$$\Phi: Y = \mathbb{C}^* \to \mathbb{P}^1 = \Pi$$

takes

$$a \mapsto (a^{-1}: a^{-1}\ln a) = (1: \ln a).$$

The map $\Phi: \mathbb{C}^* \to \mathbb{C} \subset \mathbb{P}^1$ is multivalued since $\mathbb{C}^*$ is not simply connected. If $\tilde{Y} = \mathbb{C} \to \mathbb{C}^* = Y$, $\tilde{a} \mapsto a = e^{\tilde{a}}$, is the universal covering of Y, then the map $\tilde{\Phi}: \tilde{Y} \to \mathbb{C} \subset \mathbb{P}^1$ is the identity map, $\xi = \tilde{a}$.

3.5 The period map for simply-elliptic singularities

The simply-elliptic singularities can be characterized from the point of view of their desingularization in the following way. They are normal two-dimensional singularities such that the exceptional curve of their minimal resolution is a smooth elliptic curve E. The hypersurface singularities among them are the singularities $\tilde{E}_8$, $\tilde{E}_7$, $\tilde{E}_6$ These are those for which $(E^2) = -1$, -2 and -3 respectively. In Arnol'd's classification of critical points of functions they are the unimodal singularities $J_{10} = T_{2,3,6}$, $X_9 = T_{2,4,4}$, $P_8 = T_{3,3,3}$ respectively.

The R-equivalent (or $\mathcal{K}$-equivalent) classes of these singularities correspond exactly to the isomorphic classes of elliptic curves E defined by the j-invariant. Each elliptic curve can be written in the Legendre normal form

$$z^2 = y(y-1)(y-\lambda),\ \lambda \in \mathbb{C}\backslash\{0, 1\}, \tag{3.5.1}$$

and its j-invariant is $j = j(\lambda) = \frac{4}{27}(\lambda^2 - \lambda + 1)^3/\lambda^2(\lambda - 1)^2$. Each of the singularities $\tilde{E}_8$, $\tilde{E}_7$, $\tilde{E}_6$ has the corresponding Legendre normal form f_λ, where $\lambda \in Y = \mathbb{C}\backslash\{0, 1\}$. However, to calculate the period map it is convenient to use another normal form. We'll review in detail the case of the singularity $\tilde{E}_7$ for which the calculations are the most simple. The results for the singularities $\tilde{E}_6$ and $\tilde{E}_8$ are completely analogous [He1].

(3.5.2) The simply-elliptic two-dimensional singularities of type $\tilde{E}_7$ have the Legendre normal form

$$yx(y-x)(y-\lambda x) - z^2,\ \lambda \in \mathbb{C}\backslash\{0, 1\},$$

and the corresponding stable equivalent one-dimensional singularities

$$g_\lambda = yx(y-x)(y-\lambda x).$$

For the construction of the period map we'll use another normal form:

$$f_a(x, y) = x^4 + ax^2y^2 + y^4,\ a \in \mathbb{C}\backslash\{-2, 2\}. \tag{3.5.3}$$

The singularities f_a and g_λ are R-equivalent for $a = 4\lambda - 2$. Form (3.5.3) is more convenient for calculations because the parameter a appears only in one monomial.

Let us calculate the period map for the μ-const deformation (3.5.3) with the one-dimensional base $Y = \mathbb{C}\backslash\{-2, 2\}$.

The singularities $f_a(x, y)$ are quasihomogeneous with weights $w_0 = w_1 = \frac{1}{4}$, the Milnor number μ is equal to 9, a $\mathbb{C}$-basis of $\mathcal{H}^{(0)}(a)/\mathcal{H}^{(-1)}(a)$ is given by the forms

$$x^ky^l\omega_0,\ \text{where } \omega_0 = \mathrm{d}x \wedge \mathrm{d}y,\ 0 \leqslant k,\ l \leqslant 2, \tag{3.5.4}$$

and the spectral numbers coincide with the orders $\alpha(s[x^ky^l\omega_0](a))$ of the geometric sections of these forms, and by (II.8.4.8) these orders coincide with the Newton orders $\nu(x^ky^l\omega_0)$. Hence,

$$\begin{aligned} Sp(f_a) &= \{\tfrac{1}{4}(k+1) + \tfrac{1}{4}(l+1) - 1 | 0 \leqslant k,\ l \leqslant 2\} \\ &= \{-\tfrac{1}{2}, -\tfrac{1}{4}, -\tfrac{1}{4}, 0, 0, 0, \tfrac{1}{4}, \tfrac{1}{4}, \tfrac{1}{2}\} \end{aligned} \tag{3.5.5}$$

A diagram based on (II.7.2.6) is of the following form:

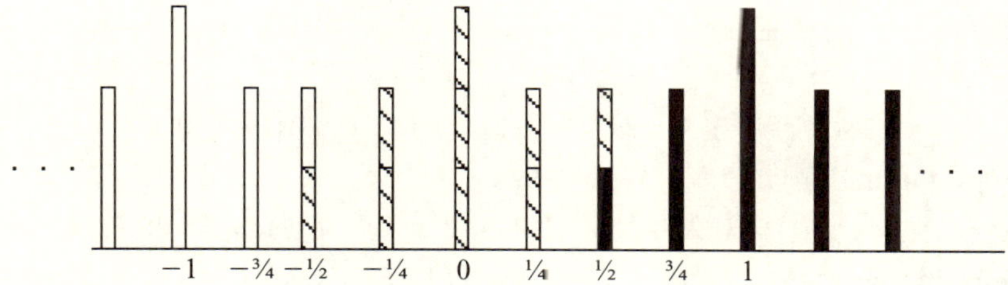

$\alpha_1 = \alpha_{\min} = -\frac{1}{2}$, $\beta_1 = \alpha_{\max} - 1 = -\frac{1}{2} = \alpha_1$. Hence, $V = V^{\alpha_1}/V^{>\beta_1} = C_{-1/2}$, $\dim C_{-1/2} = 2$, and

(3.5.6) $V^{-1/2} \supset \mathscr{H}^{(0)}(a) \supset V^{>-1/2}$, $\mathscr{H}^{(0)}(a) = \mathbb{C}\, s[\omega_0](a) \oplus V^{>-1/2}$.

Note that the geometric sections $s[\omega_0](a)$ are homogeneous $s[\omega_0](a) = s(\omega_0, -\frac{1}{2})(a)$. It follows from Taylor's formula (3.3.11), $s[\omega_0](y) = \sum_{i\geqslant 0}(-\partial_t)^i s[x^2y^2)^i\omega_0](0)(y^i/i!)$, because all $(\partial_t)^i s[(x^2y^2)^i\omega_0](0) \in C_{-1/2}$.

Thus the embedding $\mathscr{H}^{(0)}(a) \subset \mathscr{M}$ is defined by the disposition of the line $\mathbb{C}\cdot s[\omega_0](a) \subset C_{-1/2} = V$. The period space $\Pi = \mathbb{P}^1$ and the period map is

(3.5.7) $\Phi\colon Y = \mathbb{C}\backslash\{-2, 2\} \to \mathbb{P}^1,\ a \mapsto [\mathbb{C}\cdot s[\omega_0](a)]$

To determine $s[\omega_0](a)$ we have to find a differential equation which has this vector function as its solution. Since $\partial f_a/\partial a = x^2y^2$, we have from (2.2.8) that

(3.5.8) $\partial_a s[\omega_0](a) = -\partial_t s[x^2y^2\omega_0](a),$

and

$$\partial_a^2 s[\omega_0](a) = \partial_t^2 s[(x^2y^2)^2\omega_0](a).$$

We want to find relations between these derivatives. To do this, we express $(x^2y^2)^2\omega_0$ in $x^2y^2\omega_0$ and ω_0 in $\mathscr{H}^{(0)}(a)$. First, we can write x^2y^2 from the equation $f_a = t$ (3.5.3), and $(x^2y^2)^2$ from equation (3.5.3) squared

(3.5.9) $f_a^2 = (x^4 + y^4)^2 + 2ax^2y^2(x^4 + y^4) + a^2(x^2y^2)^2.$

Second, we use relations which hold for the Brieskorn lattice $\mathscr{H}^{(0)}(a) = \Omega_X^2/\mathrm{d}f \wedge \mathrm{d}\mathscr{O}_X$ (I.5.3.3) in the case of one-dimensional singularities ($n = 1$): for any function $g \in \mathbb{C}\{x, y\}$ we have $\mathrm{d}f_a \wedge \mathrm{d}g \equiv 0 \bmod \mathrm{d}f_a \wedge \mathrm{d}\mathscr{O}_X$, i.e.

(3.5.10) $(f_{a,x}\cdot g_y - f_{a,y}\cdot g_x)\omega_0 = 0$ in $\mathscr{H}^{(0)}(a)$.

Applying this for $g = x^5y,\ xy^5$, we obtain relations in $\mathscr{H}^{(0)}(a)$

(3.5.11)
$$(x^8 - 5x^4y^4)\omega_0 = 0,$$
$$(5x^4y^4 - y^8)\omega_0 = 0.$$

Subtracting the second relation from the first, we get

$$(x^8 + y^8 - 10x^4y^4 - 2ax^2y^2(x^4 + y^4))\omega_0 = 0,$$

or

$$(x^4 + y^4)^2 - 12x^4y^4 - 2ax^2y^2(x^4 + y^4))\omega_0 = 0.$$

From this and from (3.5.9) we obtain

$$f_a^2\omega_0 = ((12 + a^2)x^4y^4 + 4ax^2y^2(x^4 + y^4))\omega_0 = 0.$$

Substituting $x^4 + y^4 = f_a - ax^2y^2$ from (3.5.3) gives

$$f_a^2\omega_0 = ((12 - 3a^2)x^4y^4 + 4af_ax^2y^2)\omega_0 = 0,$$

or

(3.5.12) $$x^4y^4\omega_0 = \frac{f_a^2}{12 - 3a^2} - \frac{4a}{12 - 3a^2} f_a x^2y^2\omega_0 \bmod \mathrm{d}f_a \wedge \mathrm{d}\mathcal{O}_X.$$

Now from (3.5.8) and (3.5.12) we obtain

(3.5.13)

$$\partial_a^2 s[\omega_0](a) = \partial_t^2 t^2 \frac{1}{12 - 3a^2} s[\omega_0](a) - \partial_t^2 t \frac{4a}{12 - 3a^2} s[x^2y^2\omega_0](a).$$

Since the singularities $\tilde{E}_7$ are quasihomogeneous, the monodromy T is semisimple. Hence the root spaces C_α are eigensubspaces of the operator $t\partial_t$ with eigenvalues α. From the commutator relation $[\partial_t, t] = 1$ $\partial_t t = t\partial_t + 1$, it follows that the operator $\partial_t^2 t^2 = \partial_t(\partial_t t)t = \partial_t(t\partial_t + 1)t = (\partial_t t)^2 + \partial_t t$ is equal to $(-\frac{1}{2})^2 + \frac{1}{2} = \frac{3}{4}$ on the space $C_{-1/2}$, and the operator $\partial_t^2 t = \partial_t(t\partial_t + 1)$ is equal to $\frac{3}{2}\partial_t$ on the space $C_{1/2}$. Finally, since the geometric sections $s[\omega_0](a) = s(\omega_0, -\frac{1}{2})(a) \in C_{-1/2}$ and $s[x^2y^2\omega_0](a) \in C_{1/2}$ are homogeneous, relation (3.5.13) is rewritten in the form

$$\partial_a^2 s[\omega_0](a) = \frac{1}{12 - 3a^2}\frac{3}{4}s[\omega_0](a) - \frac{4a}{12 - 3a^2}\frac{3}{2}\partial_t s[x^2y^2\omega_0](a),$$

or, from (3.5.8), in the form

(3.5.14) $$(4 - a^2)\partial_a^2 s[\omega_0](a) = \tfrac{1}{4}s[\omega_0](a) + 2a\partial_a s[\omega_0](a).$$

Returning to the Legendre parameter $\lambda = \frac{1}{2} + \frac{1}{4}a$ ($a = 4\lambda - 2$, $\partial_a = \frac{1}{4}\partial_\lambda$, $4 - a^2 = -16\lambda(\lambda - 1)$), we obtain that the function

$$s[\omega_0](a) = s[\omega_0](4\lambda - 2) =: u(\lambda)$$

is a solution of the differential equation

(3.5.15) $$\lambda(\lambda - 1)\partial_\lambda^2 u + (2\lambda - 1)\partial_\lambda u + \tfrac{1}{4}u = 0.$$

This is the hypergeometric differential equation with $\alpha = \beta = \frac{1}{2}$, $\gamma = 1$.

(3.5.16) Now let A_1, A_2 be a $\mathbb{Z}$-basis in $H^1(X_\infty, \mathbb{C})_{-1}$, and let $A_1(t, a)$, $A_2(t, a)$ be the corresponding multivalued horizontal sections of the fibration H_{-1}. Then the elementary sections $e_1 = s[A_1, -\frac{1}{2}](a)$, $e_2 = s[A_2, -\frac{1}{2}](a)$ from (1.3.1) form a basis in $C_{-1/2}$. Now if $s[\omega_0](a) = I_1(a)e_1 + I_2(a)e_2$ is a representation of $s[\omega_0](a)$ in this basis, where $I_1(a)$, $I_2(a)$ are (multivalued) holomorphic functions on $Y = \mathbb{C}\backslash\{-2, 2\}$, then the period map (3.5.7) is given by the formula

$$\Phi(a) = (I_1(a) \colon I_2(a)).$$

Since the basis e_1, e_2 is horizontal, $\partial_a e_1 = \partial_a e_2 = 0$, and $s[\omega_0](a)$ is a solution of (3.5.15), we have that $I_1(a) = I_1(4\lambda - 2)$, $I_2(a) = I_2(4\lambda - 2)$ are also solutions of the hypergeometric equation (3.5.15). Thus, the period map Φ is given by hypergeometric functions. The classical theory of such functions enables us to obtain more detailed information about Φ.

(3.5.17) The same results occur for singularities of types $\tilde{E}_8$ and $\tilde{E}_6$. Analogous but more tedious calculations were done by Hertling [He1] who showed that as in the case of $\tilde{E}_7$ the function $s[\omega_0](\lambda)$ satisfies the hypergeometric equation (3.5.15). We refer the reader to [He1] for the details.

It is useful to compare the period map Φ, obtained above, with the period map of the modular family of elliptic curves, described in detail in §9.3 of [Br-K].

(3.5.18) Hertling [He1] has also calculated the period map for the bimodal singularity $W_{1,0}$ (and other bimodal singularities in [He2]), which has the normal form

$$f_{a,b}(x, y) = x^4 + y^6 + ax^2y^3 + bx^2y^4,\ a \neq 2, -2.$$

This is a two-parametric μ-const deformation with the base $Y = (\mathbb{C}\backslash\{-2, 2\}) \times \mathbb{C}$. Hertling showed that in this case the 'true' period space Π is a two-dimensional manifold which is a Stein fibration over $\mathbb{P}^1$ with a Chern class equal to 2. The calculation of the period map leads to the hypergeometric equation with $\alpha = \frac{1}{2}, \beta = \frac{1}{3}, \gamma = \frac{11}{12}$.

3.6 The period map defined by MHS on the vanishing cohomology

A natural development of the ideas related to the period map of a family of non-singular projective manifolds leads to the period map which in the case of the μ-const deformation (1.1.3) $F \colon \mathscr{X} \to S \times Y$ of isolated singularities $f_y = F(., y)$ sends a point $y \in Y$ to the mixed Hodge structure on $H^n(\mathscr{X}(y_\infty), \mathbb{Z})$,

$$y \mapsto \text{MHS on } H^n(\mathscr{X}(y)_\infty, \mathbb{Z}).$$

However, we began with the period map defined by embedding of Brieskorn lattices because the MHS is defined by the embedding $\mathscr{H}^{(0)}(y) \subset \mathscr{M}$ and the $V^{\cdot}$-filtration on $\mathscr{M}$ by virtue of (II.7.2.1). The weight filtration of the MHS is defined by the monodromy (i.e. by topology) and it is unchanged under μ-const deformations. We haven't considered yet the question of the polarization of the MHS on the vanishing cohomology [S3, V10, Ka2]. Disregarding the question of the variation of the MHS (see the review [B-Z]), we mean the period map here to be a map which sends a point y to the Hodge filtration $F^{\cdot}(y)$ of the MHS on $H^n(\mathscr{X}(y)_\infty, \mathbb{Z})$.

(3.6.1) Defining the period space $\Pi' = \{\mathscr{H}^{(0)}(y) | V^{>-1} \supset \mathscr{H}^{(0)}(y) \supset V^{n-1}\} = Grass(k, V)$ in (3.2.5) and (3.2.8) we viewed $\mathscr{H}^{(0)}(y)$ only as a vector space in $\mathscr{M}$. However, $\mathscr{H}^{(0)}(y)$ is a $\mathbb{C}\{t\}$-module invariant wrt the operator ∂_t^{-1} acting on $\mathscr{L} = V^{>-1}$ ($\mathscr{H}^{(0)}(y)$ is a $\mathbb{C}\{t\}\{\{\partial_t^{-1}\}\}$-module), and there is a $V^{\cdot}$-filtration on $\mathscr{M}$. Let Π be the subset of vector subspaces W in $\mathscr{M}$, $V^{>-1} \supset W \supset V^{n-1}$, i.e. $W \in \Pi^{\cdot}$, such that W is a $\mathbb{C}\{t\}\{\{\partial_t^{-1}\}\}$-module, and such that the dimensions

(3.6.2) $$m_\alpha = \dim Gr_V^\alpha W, \quad \alpha \in \mathbb{Q},$$

are given.

Recall (II.7.2.1.) that the Hodge filtration $F^{\cdot}(y)$ on $H^n(\mathscr{X}(y)_\infty, \mathbb{C}) \overset{\psi}{\leftarrow} \mathscr{L}(y)/t\mathscr{L}(y) = \oplus_{-1<\alpha\leqslant 0} C_\alpha(y)$ is defined by $Gr_V^{\cdot}\mathscr{H}^{(0)}(y)$. Namely, by (II.7.2.3) we have

$$F^p C_\alpha = \partial_t^{n-p} Gr_V^{\alpha+n-p} \mathscr{H}^{(0)}$$

and

(3.6.3) $$f_\lambda^p = \dim F^p C_\alpha = m_{\alpha+n-p}.$$

(3.6.4) Let $\overline{\Pi}$ be the manifold of flags $F^{\cdot}$ for which the dimensions $f^p = \dim F^p$ are defined by equalities (3.6.3) as m_β. Here $\overline{\Pi}$ is the *period space for Hodge filtrations*. The rule (II.7.2.1) defines the flag $F^{\cdot}(W)$ in terms of $W \in \Pi$, and therefore defines a holomorphic map $\Pi \to \overline{\Pi}$.

(3.6.5) *Definition* Let f_y, $y \in Y$, be a μ-const deformation of an isolated singularity. The map

$$Y \xrightarrow{\overline{\Phi}} \overline{\Pi}, \quad y \mapsto F^{\cdot}(y),$$

where $F^{\cdot}(y)$ is the Hodge filtration on $H^n(\mathscr{X}(y)_\infty, \mathbb{Z})$, is called the period map defined by MHS or, for short, the *MHS-period map*.

We have a commutative diagram

$$\begin{array}{ccc} Y & \xrightarrow{\Phi} & \Pi \\ & \overline{\Phi}\searrow & \downarrow \\ & & \overline{\Pi} \end{array}.$$

The period map $\overline{\Phi}$, is a holomorphic map by (2.4.6) in the same way of Φ is.

(3.6.7) The Hodge filtration $F^{\cdot}(y)$ is a rougher invariant than $\mathscr{H}^{(0)}(y)$. In going from the embedding of the Brieskorn lattice $\mathscr{H}^{(0)}(y) \subset \mathscr{M}$ to the Hodge filtration $F^{\cdot}(y)$, i.e in going from $\mathscr{H}^{(0)}(y)$ to $Gr^{\cdot}_V \mathscr{H}^{(0)}(y) \subset Gr^{\cdot}_V \mathscr{M}$, we lose information. As an example let us look at the period map $\overline{\Phi}$ for the μ-const deformation (3.3.1) of the singularity E_{12}. For E_{12} all spaces C_α are one-dimensional, $\dim C_\alpha(y) = 1$. Hence $F^p C_\alpha(y)$ is either 0 or the whole space $C_\alpha(y)$. Thus $F^p(y)$ does not vary and $\overline{\Phi}(Y)$ is one point, and at the same time Φ: $Y = \mathbb{C} \to \Pi$ is an embedding by (3.3.7).

This example is a particular case of the behavior of period maps for μ-const deformations of quasihomogeneous singularities which we'll consider in the next section.

(3.6.8) *Remark* The information lost (when we go from $\mathscr{H}^{(0)}(y)$ to $Gr^{\cdot}_V \mathscr{H}^{(0)}(y)$) is expressed by M. Saito in terms of some linear maps $c_{\beta,\alpha}$ in [Sa7].

4 The infinitesimal Torelli theorem

4.1 The $V^{\cdot}$-filtration on Jacobian algebra. The necessary condition for μ-const deformation

Let

$$f_y(x) = F_Y(x, y) \colon (\mathbb{C}^{n+1}, 0) \times (\mathbb{C}^m, 0) \to (\mathbb{C}, 0) \tag{4.1.1}$$

be an m-parametric deformation of an isolated singularity $f_0(x) = f$: $(\mathbb{C}^{n+1}, 0) \to (\mathbb{C}, 0)$. There is a canonical map of the tangent space $T_0 Y$ to the base of deformation $Y = \mathbb{C}^m$ at the point $0 \in Y$ to the Jacobian algebra $Q_f = \mathscr{O}_{\mathbb{C}^{n+1},0}/J_f$ (the Kodaira–Spencer map),

$$T_0 Y \to Q_f, \quad \frac{\partial}{\partial y_j} \mapsto \left.\frac{\partial F_Y}{\partial y_j}\right|_{y=0} \bmod J_f. \tag{4.1.2}$$

The deformation (4.1.1) is miniversal if this map is an isomorphism. The miniversal deformation is usually obtained in the following way [AGV].

We choose functions (monomials) $M_1(x) = 1,\ M_2(x), \ldots, M_\mu(x) \in \mathscr{O}_{\mathbb{C}^{n+1},0}$ representing a basis of Q_f, and consider the deformation

$$f_\lambda(x) = F(x, \lambda) \colon (\mathbb{C}^{n+1}, 0) \times (\mathbb{C}^\mu, 0) \to (\mathbb{C}, 0),$$

defined by the function

$$F(x, \lambda) = f(x) + \lambda_1 + \lambda_2 M_2(x) + \ldots + \lambda_\mu M_\mu(x). \tag{4.1.3}$$

The base $\Lambda = \mathbb{C}^\mu$ of the miniversal deformation is identified with the Jacobian algebra as in (4.1.2)

$$T_0\Lambda \rightrightarrows Q_f, \quad \frac{\partial}{\partial y_j} \mapsto M_j(x). \tag{4.1.4}$$

Any deformation (4.1.1) is induced (modulo isomorphism) from the miniversal one by a base change $\varphi\colon Y \to \Lambda,\ \lambda = \lambda(y),\ F_Y(x, y) = F(x, \lambda(y))$. The map of the tangent spaces $\mathrm{d}\varphi(0)\colon T_0 Y \to T_0\Lambda$,

$$\frac{\partial}{\partial y_j} \mapsto \mathrm{d}\varphi(0)\left(\frac{\partial}{\partial y_j}\right) = \frac{\partial \lambda_1}{\partial y_j}(0)\frac{\partial}{\partial \lambda_1} + \ldots + \frac{\partial \lambda_\mu}{\partial y_j}(0)\frac{\partial}{\partial \lambda_\mu}, \tag{4.1.5}$$

is interpreted under identification (4.1.4) as the map (4.1.2)

$$\frac{\partial}{\partial y_j} \mapsto \frac{\partial F_Y}{\partial y_j}\bigg|_{y=0} = \sum_k \frac{\partial F}{\partial \lambda_k}\frac{\partial \lambda_k}{\partial y_j}(0) = \sum_k \frac{\partial \lambda_k}{\partial y_j}(0) M_k(x).$$

Now we introduce the $V^\cdot$-filtration on the Jacobian algebra $Q_f \simeq T_0\Lambda$. There is a $V^\cdot$-filtration on the Brieskorn lattice $\mathscr{H}^{(0)} \subset \mathscr{M}$ induced from the meromorphic connection $\mathscr{M}$ (II.6.3.2). The filtration on $\mathscr{H}^{(0)}$ induces the quotient filtration (II.8.4.5) on

$$\Omega_f = \mathscr{H}^{(0)}/\mathscr{H}^{(-1)} \simeq \Omega^{n+1}_{\mathbb{C}^{n+1},0}/\mathrm{d}f \wedge \Omega^n_{\mathbb{C}^{n+1},0}.$$

The filtration on Ω_f defines a filtration on End Ω_f, where Ω_f is a Q_f-algebra. The multiplication by an element of Q_f defines the endomorphism of Ω_f. We define the *$V^\cdot$-filtration* on Q_f, the filtration induced from End Ω_f,

$$V^\alpha Q_f = \{g \in Q_f | g V^\beta \Omega_f \subset V^{\beta+\alpha}\Omega_f, \forall \beta\}. \tag{4.1.6}$$

Now let (4.1.1) be a μ-const deformation as in (1.1.1).

(4.1.7) *Theorem* [V-Ch] If $f_y(x)$ (4.1.1) is a m-parametric μ-const deformation of a singularity f with base Y, then

$$\mathrm{d}\varphi(T_0 Y) \subset V^1 T_0\Lambda = V^1 Q_f,$$

i.e.

$$\frac{\partial f_y}{\partial y_j}_{|y=0} \in V^1 Q_f, \quad j = 1, \ldots, m. \tag{4.1.8}$$

Proof. Let an element $[\omega_0] \in \Omega_f$ be represented by an $(n+1)$-form ω_0. Since $\mathscr{H}^{(-1)} = \partial^{-1}(\mathscr{H}^{(0)})$, we have by (II.8.4.5) to prove that for $(\partial f_y/\partial y_j)_{|y=0}$ there exists an $(n+1)$-form η such that

$$\alpha\left(s\left(\frac{\partial f_y}{\partial y_j}_{|y=0}\omega_0\right) + \partial^{-1}s[\eta]\right) \geqslant \alpha(s[\omega_0]) + 1.$$

Let $\omega = \omega(y)$ be a relative $(n+1)$-form such that condition (2.4.7),

$$s[\omega](0) = s[\omega_0] \text{ and } \alpha(s[\omega](y_0)) = \alpha(w),$$

holds in a neighborhood of the point $0 \in Y$. Then

$$\alpha(\partial_{y_j}s[\omega](y) \geqslant \alpha(s[\omega](0)).$$

Apply the operator ∂_t^{-1} to formula (2.2.6),

$$\partial_{y_j}s[\omega](y) = s\left[\frac{\partial\omega}{\partial y_j}\right](y) - \partial_t s\left[\frac{\partial f_y}{\partial y_j}\omega\right](y).$$

We obtain

$$s\left[\frac{\partial f_y}{\partial y_j}\omega\right](y) - \partial_t^{-1}s\left[\frac{\partial\omega}{\partial y_j}\right](y) = -\partial_t^{-1}(\partial_{y_j}s[\omega](y)).$$

Hence

$$\alpha\left(s\left[\frac{\partial f_y}{\partial y_j}_{|y=0}\omega_0\right] - \partial_t^{-1}s\left[\frac{\partial\omega}{\partial y_j}\right](0)\right) \geqslant \alpha(s[\omega](0)) + 1 = \alpha(s[\omega_0]) + 1.$$

■

(4.1.9) *Corollary* If $f(x)$ is a quasihomogeneous isolated singularity, then singularities $f_y(x)$ of its μ-const deformation are quasihomogeneous or semiquasihomogeneous.

This follows from the fact that the V-filtrations on $\mathscr{H}^{(0)}$, Ω_f and Q_f coincide with the Newton filtration defined by weights $w_0, \ldots, w_n$ (II.4.4.7), and $V^1Q_f = \{\overline{g} \in Q_f | \nu(g) \geqslant 1\}$, and $\mathscr{V}(f) = 1$.

4.2 Calculation of the tangent map of the period map. The horizontality of the MHS-period map

Let $f_y(x)$ be an m-parametric μ-const deformation as in (3.1.1), and Φ: $Y \to \Pi'$, $\Pi' = Grass(k, V)$, be the period map defined by the embedding of Brieskorn lattices (3.2.5).

(4.2.1) As is well known, the tangent space to the Grassmann manifold Π' at a point $[W] \in \Pi'$ where $W \subset V$ is a subspace of dimension dim $W = k$, is identified with the space

$$T_{[W]}\Pi' = \mathrm{Hom}(W, V/W)$$

in the following way. If a vector $\xi \in T_{[W]}\Pi'$ is represented by a curve $W(t)$, $W(0) = W$, in Π', then the homomorphism $\xi\colon W \to V/W$ corresponding to ξ takes a vector $v \in W$ to the vector $\xi(v) = (\mathrm{d}/\mathrm{d}t)v(t)|_{t=0} \bmod W$, where $v(t)$ is a curve in V such that $v(0) = v$, $v(t) \in W(t)$.

Now let us calculate the tangent map

$$\mathrm{d}\Phi(0)\colon T_0 Y \to T_{[W]}\Pi',$$

where $W = \mathscr{H}^{(0)}(0)/V^{n-1} \subset V = V^{>-1}/V$.

(4.2.2) *Lemma* If $\xi = \Sigma c_j \partial/\partial y_j \in T_0 Y$, then

$$\mathrm{d}\Phi(0)(\xi) = -\partial_t \frac{\partial f_y}{\partial \xi}_{|y=0},$$

i.e. if an element of W is represented by the geometric section $s[\omega_0]$, $\omega_0 \in \Omega^{n+1}_{\mathbb{C}^{n+1},0}$, then

$$\mathrm{d}\Phi(0)\colon s[\omega_0] \mapsto -\partial_t s\left[\frac{\partial f_y}{\partial \xi}_{|y=0}\omega_0\right]. \tag{4.2.3}$$

Proof By (4.2.1) to calculate $\mathrm{d}\Phi(0)(\xi)$ we have to take $v(y) \in \mathscr{H}^{(0)}(y)$, and then to differentiate $(\partial v/\partial \xi)_{|y=0}$. Let us take the section $v(y) = s[\omega_0](y) = \mathrm{Res}\,[\omega_0/(f_y(x) - t)]$. Then by (2.2.9) we have

$$\left.\frac{\partial s[\omega_0](y)}{\partial \xi}\right|_{y=0} = \left(\sum_{j=1}^{m} c_j \partial_{y_j} s[\omega_0]\right)(0)$$

$$= -\partial_t s\left[\sum_{j=1}^{m} c_j \frac{\partial f_y}{\partial y_j}_{|y=0}\omega_0\right](0) = -\partial_t s\left[\frac{\partial f_y}{\partial \xi}_{|y=0}\omega_0\right].$$

■

(4.2.4) *The tangent map to the period map defined via MHS* Now let $\Pi \subset \Pi'$ be the submanifold defined in (3.6.1), $\overline{\Pi}$ be the flag manifold, and $\Phi\colon Y \to \overline{\Pi}$ be the MHS-period map from (3.6.5). The tangent map $\mathrm{d}\overline{\Phi}$ is calculated in the same way as $\mathrm{d}\Phi$ in (4.2.2). Since $\overline{\Pi} \subset \Pi^{n}_{p=1}$ *Grass*(f^p, H), where $H = H^n(X_\infty, \mathbb{C}) = F^0$, we have $T_*\overline{\Pi} \subset \oplus^{n}_{p=1}$ $\mathrm{Hom}\,(F^p, H/F^p)$, where $* \in \overline{\Pi}$ corresponds to the flag $F^{\cdot}(0)$. The Hodge filtration $F^{\cdot}$ on H was defined in (II.7.2.1) by the filtration on $\mathscr{L}/t\mathscr{L} = \oplus_{-1<\alpha\leqslant 0} C_\alpha$ and by the isomorphism $\mathscr{L}/t\mathscr{L} \rightleftarrows H$, where

$$F^pH_\lambda \simeq F^pC_\alpha = \partial_t^{n-p} Gr_V^{\alpha+n-p}\mathscr{H}^{(0)} \xleftarrow{\partial_t^{n-p}} Gr_V^{\alpha+n-p}\mathscr{H}^{(0)} \subset C_{\alpha+n-p},$$
$$-1 < \alpha \leqslant 0.$$

Therefore, to calculate

$$\text{(4.2.5)} \qquad d\overline{\Phi}(0): \quad T_0Y \to T_*\overline{\Pi},$$

we can take $s[\omega_0] \in V^{\alpha+n-p}\mathscr{H}^{(0)}$, the section $s[\omega_0](y) = \text{Res}\,[\omega_0/(f_y(x)-t)]$, and apply formula (4.2.3) to it. By (4.1.8) we have $s[(\partial f_y/\partial\xi)_{|y=0}\omega_0] \in V^{\alpha+n-p+1}\mathscr{H}^{(0)} \bmod V^{\alpha+n-p+1}\mathscr{H}^{(-1)}$, and hence $d\overline{\Phi}(0)$ takes F^p to $F^{p-1}/F^p \subset H/F^p$.

We define the *horizontal subfibration* $T^h\overline{\Pi}$ of the tangent fibration $T\overline{\Pi}$ to the flag manifold $\overline{\Pi}$ by the condition: at the point $* \in \overline{\Pi}$ corresponding to a flag $F^\cdot$ we have

$$T_*^h\overline{\Pi} \subset \bigoplus_{p+1}^{n} \text{Hom}(F^p, F^{p-1}/F^p) = \bigoplus_{p+1}^{n} \text{Hom}(F^p/F^{p+1}, F^{p-1}/F^p)$$
$$= \bigoplus_{p+1}^{n} \text{Hom}(Gr_F^p, Gr_F^{p-1}).$$

We have then obtained the following:

(4.2.6) *Proposition* The MHS-period map $\overline{\Phi}$: $Y \to \overline{\Pi}$ of a μ-const deformation f_y is horizontal, i.e. $d\overline{\Phi}$ takes values in $T^h\overline{\Pi}$.

This means that for the variation of MHS defined by a μ-const deformation, Griffiths' transversality condition $\nabla(\mathscr{F}^p) \subset \Omega_Y^1 \otimes_{\mathscr{O}_Y} \mathscr{F}^{p-1}$ holds.

4.3 The infinitesimal Torelli theorem

Let $D_\mu \subset \Lambda$ be the μ-const stratum in the base of miniversal deformation of a singularity f. Then D_μ can be viewed as an analog of the moduli spaces of projective varieties. Let Φ (or $\overline{\Phi}$) be the period map $D_\mu \to \Pi$. Assertions about the injectivity of Φ, i.e. assertions stating that a singularity is uniquely determined by its periods, are called Torelli (type) theorems. We saw in (3.6.7) that in general $\overline{\Phi}$ is not an embedding. We'll show that Φ is an embedding locally at non-singular points of D_μ.

(4.3.1) *Theorem* ([S7]) If f_y is an m-parametric μ-const deformation (3.1.1) of an isolated singularity $f = f_0$ such that the images of $(\partial f_y/\partial y_1)_{|y=0}, \ldots, (\partial f_y/\partial y_\mu)_{|y=0}$ in Q_f are linear independent, i.e. by (4.1.5) $d\varphi(0)$ is a monomorphism for the map φ: $Y \to \Pi'$ inducing f_y from the miniversal deformation, then the tangent map $d\Phi(0)$ of the period map Φ: $Y \to \Pi'$ defined by the embedding of Brieskorn lattices, is also a monomorphism, and hence, Φ is locally an embedding.

Proof If $\xi = \sum c_j(\partial/\partial y_j) \in T_0 Y$, and $\xi \neq 0$, then $d\varphi(0)(\xi) = (\partial f_y/\partial \xi)_{|y=0} \neq 0$ in Q_f, i.e. $(\partial f_y/\partial \xi)_{|y=0} \notin J_f$. Consequently, for $\omega_0 \in \mathscr{H}^{(0)}$ the product $(\partial f_y/\partial \xi)_{|y=0}\omega_0 \notin \mathscr{H}^{(-1)} \subset \mathscr{H}^{(0)}$ or $s[(\partial f_y/\partial \xi)_{|y=0}\omega_0] \notin \mathscr{H}^{(-1)}$ if $\mathscr{H}^{(0)}$ is viewed as a submodule of $\mathscr{M}$. But $\mathscr{H}^{(-1)} = \partial_t^{-1}\mathscr{H}^{(0)}$. Hence $\partial_t s[(\partial f_y/\partial \xi)_{|y=0}\omega_0] \notin \mathscr{H}^{(0)}$, i.e. this element is not equal to zero in $V/W = V^{>-1}/\mathscr{H}^{(0)}(0)$. Thus, $d\Phi(0)(\xi) \neq 0$ by (4.2.3).

In particular, if D_μ is the μ-const stratum in the base of miniversal deformation,

(4.3.2) $$D_\mu \subset \Lambda, \quad D_\mu^0 = D_\mu \backslash \operatorname{Sing} D_\mu,$$

then Φ: $D_\mu^0 \to \Pi$ locally is an embedding.

(4.3.3) In [Sa8] M. Saito defined the period map Φ: $Y \to \Pi$ for the case of a μ-const deformation F_Y: $(\mathbb{C}^{n+1}, 0) \times Y \to (\mathbb{C}, 0) \times Y$ with a not necessarily non-singular base Y. To avoid problems with the topological triviality of μ-const deformations (1.1.3)–(1.1.5) M. Saito uses Deligne's sheaves of vanishing cycles (II.8.11.8). Let $X \subset \mathbb{C}^{n+1}$ be an open subset containing 0, Y be a reduced analytic space, and $F = F_Y$: $X \times Y \to \mathbb{C}$ be a holomorphic function. Assume that $f_y = F_{|X\times\{y\}}$: $X \times \{y\} \to \mathbb{C}$ has an isolated singularity, and $\operatorname{Sing} f_y = \{0\}$, $f_y(0) = 0$. Then F is a μ-const deformation of an isolated singularity with base Y. Instead of the fibration $\underline{H}$ on $(\mathbb{C}\backslash\{0\}) \times Y$ in (1.2.1) M. Saito considers Deligne's sheaf of vanishing cycles $\varphi_f \mathbb{C}_{X\times S}$ which is concentrated on $Y = \{0\} \times Y$. This is a constant sheaf, and its fibres at points $y \in Y$ are a vanishing cohomology of singularities f_y. This enables us to identify the vanishing cohomology at different points $y \in Y$. Again using Deligne's sheaves M. Saito shows how to glue the Gauss–Manin differential systems of singularities f_y to a locally constant sheaf on Y. Operating in the category of D-modules he proves theorem (4.3.1), and its generalization to the case of an arbitrary base Y as a corollary of this theorem.

(4.3.4) *Theorem* [Sa8] If $D_\mu \subset \Lambda$ is the μ-const stratum in the base of miniversal deformation of an isolated singularity f, then the period map Φ: $D_\mu \to \Pi$ has zero-dimensional fibres, and by restricting to an open subset in D_μ we can assume that the fibres of Φ are finite.

4.4 The period map in the case of quasihomogeneous singularities

(4.4.1) *The* μ-const *stratum*. Let $f(x)$ be a quasihomogeneous singularity of degree 1 with weights $w_0, \ldots, w_n$. Let $M_1(x) = 1, M_2(x), \ldots, M_\mu(x)$ be a monomial basis in Q_f and $F(x, \lambda)$ be the miniversal deformation (4.1.3). The quasihomogeneity of $f(x)$ means that the morphism f: $(\mathbb{C}^{n+1}, 0) \to (\mathbb{C}, 0)$, $t = f(x)$, is equivariant with respect to the action of $\mathbb{C}^* \ni \tau$ on $\mathbb{C}^{n+1}$, $\tau \circ (x_0, \ldots, x_n) = (\tau^{w_0} x_0, \ldots, \tau^{w_n} x_n)$, and on $\mathbb{C}$, $\tau(t) = \tau t$. We can make the deformation (4.1.3), $F(x, \lambda)$: $\mathbb{C}^{n+1} \times \mathbb{C}^\mu \to \mathbb{C}$, equivariant also if we define weights of λ_j such that $\mathrm{wt}\,\lambda_j M_j(x) = 1$, i.e. we put

(4.4.2)
$$\nu_j = \mathrm{wt}\,\lambda_j = 1 - \mathrm{wt}\,M_j(x),$$

and define the action of $\mathbb{C}^*$ on the base $\mathbb{C}^\mu = \Lambda$ by the rule $\tau \circ (\lambda_1, \ldots, \lambda_n) = (\tau^{\nu_1}\lambda_1, \ldots, \tau^{\nu_\mu}\lambda_\mu)$.

We can decompose $\mathbb{C}^\mu = \Lambda = \Lambda^+ \times \Lambda^0 \times \Lambda^-$ into a product according to the weights of the $\mathbb{C}^*$-action. The deformations from Λ^0 are quasihomogeneous and are defined by the diagonal monomials, and the deformation from Λ^- are defined by the upper diagonal monomials. By corollary (4.1.8) ([V6]) for the quasihomogeneous singularity the μ-const stratum

$$D_\mu = \{0\} \times U^0 \times \Lambda^-,$$

where U^0 is an open subset in Λ^0, is therefore non-singular.

Let us consider the period map Φ and the MHS-period map $\overline{\Phi}$ for the stratum D_μ. Let π: $U^0 \times \Lambda^- \to U^0$ be the projection, and j: $U^0 = U^0 \times \{0\} \to U^0 \times \Lambda^-$ be the closed embedding. We have the diagram

(4.4.3)
$$\begin{array}{ccc} U^0 \times \Lambda^- = D_\mu & \xrightarrow{\Phi} & \Pi \\ \pi\downarrow\uparrow j & \searrow \overline{\Phi} & \downarrow \\ U^0 & \to & \overline{\Pi} \end{array}$$

By theorem (4.3.1) Φ is locally an embedding. In [S7] it is shown that the Hodge filtration $F^\cdot$ on $H^n(X_\infty, \mathbb{C})$ is unchanged under upper diagonal deformations, i.e. when we add upper diagonal monomials to a quasihomogeneous singularity. Hence the period map $\overline{\Phi}$ is constant on the fibres of π, and consequently, Φ factors through π. In particular, if the dimension of a fibre is greater than 0, then the Torelli theorem is false for $\overline{\Phi}$. On the other hand, if a singularity is quasihomogeneous, then on passing $\mathscr{H}^{(0)} \mapsto Gr^\cdot_V \mathscr{H}^{(0)}$ we do not lose any information, and the maps Φ and $\overline{\Phi}$ are equivalent on $U^0 \underset{j}{\hookrightarrow} D_\mu$. We thus obtain the following result of M. Saito:

(4.4.4) *Theorem* [Sa8] If D_μ is the μ-const stratum in the base of the miniversal deformation of a quasihomogeneous singularity, then there is a projection π: $D_\mu \to U^0$ to a closed submanifold $U^0 \subset D_\mu$ such that the MHS-period map $\overline{\Phi}$: $D_\mu \to \Pi$ is constant on fibres of π, and the map $\overline{\Phi}$ is locally an embedding on U^0.

5 The Picard–Fuchs singularity and Hertling's invariants

In this section we'll be concerned with the 'global' Torelli problem. In §3 for a μ-const deformation f_y, $y \in Y$, of isolated singularities we defined the period map Φ: $Y \to \Pi$ to the Grassman manifold Π, taking a point y to the Brieskorn lattices $\mathscr{H}^{(0)}$ mod V^{n-1} in the vector space $V^{>-1}/V^{n-1}$ (or in a more 'economical' space). If the base Y is not simply connected, then the map Φ, in general, is multivalued. So before going into the question of the injectivity of Φ, we ought to consider the analog of Griffiths' modular space D/Γ, i.e. to take the quotient of Π by an appropriate group. There is another reason for this. In singularity theory singularities are usually considered within equivalence relation. For example, the R-equivalence is considered for function germs. This leads to the equivalence relation '$\underset{R}{\sim}$' on Y. To have a map from $Y/\underset{R}{\sim}$ to Π, again we have to take a quotient of Π by some group.

5.1 The Picard–Fuchs singularity PFS(f) according to Varchenko

(5.1.1) *The Picard–Fuchs equation and the period matrix* Let f: $(\mathbb{C}^{n+1}, 0) \to (\mathbb{C}, 0)$ be a germ of a holomorphic function with an isolated singularity. With a misuse of language we will not differentiate between a holomorphic $(n+1)$-form ω and the geometric section $s[\omega]$ determined by it. Let $\omega_1, \ldots, \omega_\mu$ be a basis in the lattice $\mathscr{H}^{(0)}$ which we'll write as a line $\overline{\omega} = (\omega_1, \ldots, \omega_\mu)$. Let $\Gamma(t)$ be the matrix of connection coefficients of the Gauss–Manin connection wrt the basis $\overline{\omega}$. Then $\Gamma(t)$ is the matrix of the operator ∂_t wrt the basis $\overline{\omega}$,

$$\partial_t\overline{\omega} = \overline{\omega}\Gamma(t),\ \partial_t\omega_j = \sum_i \Gamma_{ij}(t)\omega_i.$$

The system of linear differential equations

$$y' = A(t)y, \text{ where } A(t) = \Gamma^t(t),$$

is called by Varchenko [AGV] the *Picard–Fuchs equation* of the singularity f wrt the basis $\overline{\omega}$ (I.8.1.5). This system has the following meaning. It expresses in coordinates the condition of horizontality of a section $\gamma = \overline{\omega}^* y$ of the sheaf $\mathscr{H}^* = \underline{H}_* \otimes \mathscr{O}_{S'}$ wrt the dual connection ∇^* wrt the basis $\overline{\omega}^* = (\omega_1^* \dots \omega_\mu^*)$ dual to the basis $\overline{\omega}$. Here $y = (y_1 \dots y_\mu)^t$ is the column of coordinates of γ wrt the basis $\overline{\omega}^*$.

If $\gamma = \gamma(t)$ is a family of homology classes, i.e. a horizontal section of the sheaf $\mathscr{H}^*$, then the integrals

$$I_1(t) = \int_{\gamma(t)} \frac{\omega_1}{\mathrm{d}f}, \dots, I_\mu(t) = \int_{\gamma(t)} \frac{\omega_\mu}{\mathrm{d}f}$$

give a solution of the Picard–Fuchs equation (I.8.1.4.). If $\gamma_1, \dots, \gamma_\mu$ is a basis of horizontal sections, then the period matrix

$$\Omega(t) = \Omega_{ij}(t), \text{ where } \Omega_{ij} = \int_{\gamma_j} \frac{\omega_i}{\mathrm{d}f} = \langle \omega_i, \gamma_j \rangle,$$

is the fundamental matrix of solutions of the Picard–Fuchs equation.

Let H^* be the vector space of solutions of the Picard–Fuchs equation. H^* can be regarded as the space of multivalued horizontal sections of the homological fibration $\underline{H}_*$. The homology with $\mathbb{Z}$-coefficients defines a lattice $H^*_{\mathbb{Z}} \subset H^*$. The lattice $H^*_{\mathbb{Z}}$ is invariant wrt the monodromy M.

(5.1.2) The Picard–Fuchs equation, i.e. the matrix $A(t)$, depends on the choice of the basis $\overline{\omega}$. The change to a new basis transforms this matrix as follows. If $\bar{\tilde{\omega}} = (\tilde{\omega}_1 \dots \tilde{\omega}_\mu)$ is a new basis in $\mathscr{H}^{(0)}$ and $C(t) \in GL(\mu, \mathscr{O}_{S,0})$ is the transition matrix to this basis, $\bar{\tilde{\omega}} = \overline{\omega} C(t)$, then the matrix $Q(t) = (C^t(t))^{-1}$ is the transition matrix to the basis $\bar{\tilde{\omega}}^*$, $\bar{\tilde{\omega}}^* = \overline{\omega}^* Q(t)$. If $\tilde{y}$ is the column of new coordinates, $\gamma = \overline{\omega}^* y = \bar{\tilde{\omega}}^* \tilde{y}$, then the new coordinates are related to the old ones by

$$y = Q\tilde{y}.$$

The substitution $y = Q\tilde{y}$ transforms the equation $y' = Ay$ into the equation

$$\tilde{y}' = \tilde{A}\tilde{y}, \text{ where } \tilde{A} = Q^{-1}AQ - Q^{-1}Q'.$$

(5.1.3) *Definition* A system $y' = A(t)y$ with a lattice $W_{\mathbb{Z}} \subset W$ in the space of solutions W of this system, invariant wrt the monodromy M, is called the *framed system*. The lattice $W_{\mathbb{Z}}$ is called the *framing of the system*.

(5.1.4) *Definition* Two framed systems $(A(t), W_{\mathbb{Z}})$ and $(\tilde{A}(t), \tilde{W}_{\mathbb{Z}})$ are called *equivalent* if there exists a substitution of coordinates $y = Q\tilde{y}$, $Q \in GL(\mu, \mathscr{O}_{S,0})$ which transforms the system $y' = A(t)y$ to $\tilde{y}' = \tilde{A}(t)\tilde{y}$, and the lattice $W_{\mathbb{Z}} \subset W$ to $\tilde{W}_{\mathbb{Z}} \subset \tilde{W}$, $W_{\mathbb{Z}} = Q\tilde{W}_{\mathbb{Z}}$. *The singularity of a framed system* is the equivalence class of this system.

(5.1.5) *Definition The Picard–Fuchs singularity PFS(f)* of a singularity f is the singularity of the framed system $(A(t), W_{\mathbb{Z}})$, where $A(t) = \Gamma^t(t)$, and $\Gamma(t)$ is the matrix of connection coefficients wrt a basis $\omega_1, \ldots, \omega_\mu \in \mathscr{H}^{(0)}$, and $W_{\mathbb{Z}} = H^*_{\mathbb{Z}} \subset H^*$.

(5.1.6) The Picard–Fuchs singularity and the period matrix. Choose a basis $\overline{\gamma} = (\gamma_1, \ldots, \gamma_\mu)$ in $W_{\mathbb{Z}}$ and write coordinates of the basis vectors $\gamma_j \in W_{\mathbb{Z}}$ to columns of a matrix Ω. We can call the matrix Ω the period matrix of the framed system $(A(t), W_{\mathbb{Z}})$.

In fact to give a framed system is to give its period matrix. The matrix $A(t)$ and the lattice $W_{\mathbb{Z}}$ are uniquely determined by $\Omega(t)$. Indeed, since $\Omega(t)$ is the fundamental matrix of solutions, we have $\Omega' = A\Omega$ and hence $A(t) = \Omega'(t)\Omega^{-1}(t)$. The lattice $W_{\mathbb{Z}}$ is the lattice generated by the columns of the matrix $\Omega(t)$.

The matrix Ω depends on the choice of the bases $\overline{\omega}$ and $\overline{\gamma}$. The passage to new bases $\bar{\tilde{\omega}} = \overline{\omega}C$, $C \in GL(\mu, \mathscr{O}_{S,0})$, and $\bar{\tilde{\gamma}} = \overline{\gamma}U$, $U \in GL(\mu, \mathbb{Z})$, transforms the matrix Ω to the matrix

$$\tilde{\Omega} = C^t\Omega U = Q^{-1}\Omega U$$

(one can obtain this very quickly by writing Ω as a product of matrices $\Omega = \overline{\omega} \circ \gamma$, where $\omega_i \circ \gamma_j = \langle \omega_i, \gamma_j \rangle$).

Thus, we can regard the singularity of a framed system $(A(t), W_{\mathbb{Z}})$ as the equivalence class of its period matrix wrt the action of group $GL(\mu, \mathscr{O}_{S,0}) \times GL(\mu, \mathbb{Z})$.

In [AGV] Varchenko proposed the following conjecture.

Varchenko's conjecture The Picard–Fuchs singularities of non-R-equivalent germs of functions with isolated singularities are different at least locally, i.e. when the germs of functions lie sufficiently close to each other.

We wish to reformulate this conjecture in terms of the period map. For this we need to introduce an equivalence relation on the set of Brieskorn lattices (to introduce an action of a group on the manifold Π).

5.2 *The Hertling invariant* $Her_1(f)$

There are two reasons for introducing an equivalence relation on the set of Brieskorn lattices. The first is connected with the period map of families of singularities with non-simply-connected bases, and the second with the construction of invariants of R-equivalence classes of singularities.

(5.2.1) The horizontal monodromy. Let $F = F_Y: (\mathbb{C}^{n+1}, 0) \times Y \to (\mathbb{C}, 0) \times Y$ be a μ-const deformation of an isolated singularity $f = f_{y_0}$, $y_0 \in Y$. In the case of a simply connected base Y we defined in (3.2.5) and (3.6.1) the period map

$$\Phi: Y \to \Pi,\ y \mapsto \mathscr{H}^{(0)}(y) \subset \mathscr{M} = \mathscr{M}(y_0),$$

and also the MHS-period map

$$\Phi: Y \to \overline{\Pi},\ y \mapsto F^{\cdot}(y),$$

where $F^{\cdot}(y)$ is a flag in the vector space $H = H^n(X_\infty(y_0), \mathbb{C})$.

If Y is not simply connected, then for any $\gamma \in \pi_1(Y, y_0)$ there appears the monodromy transformation $\rho(\gamma)$ of H, which we'll call (to distinguish it from T) the *horizontal monodromy* corresponding to γ. We obtain a representation ρ of the group $\pi_1(Y, y_0)$ on the vanishing cohomology H of the singularity f_{y_0}. If Y is the base of the miniversal family, i.e. if $Y = D^0_\mu \subset \Lambda$ is the manifold of non-singular points of the μ-const stratum in the base of this family, then we can call the group $G_h = \rho(\pi_1(D^0_\mu))$ the *group of horizontal monodromy* of the singularity f.

The horizontal monodromy commutes with the monodromy T (this follows from the group $\pi_1(S^1 \times S^1) = \mathbb{Z}^2$ being abelian) and, obviously, preserves the lattice $H_\mathbb{Z} \subset H$. Hence $G_h \subset G_1$, where

(5.2.2) $$G_1 = \{g \in GL(H_\mathbb{Z}) | gT = Tg\}$$

is the subgroup of integral automorphisms of H commuting with the monodromy T.

The group G_1 naturally acts on the set of flags in H and, consequently, acts on the manifold $\overline{\Pi}$. In order to be able to define the period map in the case of a non-simply-connected base Y, we can consider the quotient $\overline{\Pi}/G_1$, as is usually done in the theory of period maps (II.2.3.3).

We can proceed similarly in the case of the period map defined by embedding of Brieskorn lattices. For $\gamma \in \pi_1(Y, y_0)$ this is the monodromy transformation $\rho(\gamma)$ of the space $\mathscr{M} = \mathscr{M}(y_0)$. We can define the action of the group G_1 on $\mathscr{M}$ and, consequently, on Π. We'll return to this presently.

(5.2.3) The invariants of R-equivalence classes. The point $\Phi(y) \in \Pi$ or $\overline{\Phi}(y) \in \overline{\Pi}$ in the period space is not an invariant of the R-equivalence class of a singularity f_y. Let us consider how these invariants transform under a change of coordinates.

Let f_1 and $f_2: (\mathbb{C}^{n+1}, 0) \to (\mathbb{C}, 0)$ be two R-equivalent singularities. This means that there exists a holomorphic change of coordinates $g: (\mathbb{C}^{n+1}, 0) \to (\mathbb{C}^{n+1}, 0)$ such that $f_1 = f_2 \circ g$, i.e. $f_1(x) = f_2(g(x))$. The morphism g induces an isomorphism of the Milnor fibrations $X_1 \to S$ and

$X_2 \to S$ which commutes with the monodromy T and preserves other topological invariants (i.e. preserves the intersection form, the bilinear Seifert form, transforms one distinguished basis to another distinguished basis). Now if f_1 and f_2 belong to a μ-const deformation, $f_1 = f_{y_1}$, $f_2 = f_{y_2}$, then under canonical isomorphisms $H(y_1) \simeq H \simeq H(y_2)$ (respectively, $\mathscr{M}(y_1) \simeq \mathscr{M} \simeq \mathscr{M}(y_2)$) the filtrations $F^{\cdot}(y_1)$ and $F^{\cdot}(y_2)$ (respectively, the Brieskorn lattices $\mathscr{H}^{(0)}(y_1)$ and $\mathscr{H}^{(0)}(y_2)$), in general, do not coincide, and again we see that there exists an automorphism $g \in G_1$ such that $F^{\cdot}(y_1) = g(F^{\cdot}(y_2))$. Thus, under the map $\overline{\Phi}_1 \colon Y \to \overline{\Pi}/G_1$ the points $\overline{\Phi}_1(y_1)$ and $\overline{\Phi}_1(y_2)$ coincide, and $\overline{\Phi}_1(y)$ gives the real invariant of R-equivalence class of the singularity $f_y(x)$.

(5.2.4) Now we consider the action of the group G_1 on $\mathscr{M}$. Recall that we have constructed (II.6.4.8) an embedding of the vanishing cohomology vector space in the meromorphic connection

$$\psi \colon H^n(X_\infty, \mathbb{C}) = H \rightrightarrows \bigoplus_{-1<\alpha\leqslant 0} C_\alpha \subset \mathscr{M}.$$

The automorphisms $g \in GL(H)$ are transferred by means of ψ to $\oplus_{-1<\alpha\leqslant 0} C_\alpha$, $g \mapsto \psi \circ g \circ \psi^{-1}$, and then are extended to $\mathbb{C}\{t\}[t^{-1}]$-automorphisms of $\mathscr{M}$. We obtain an inclusion $GL(H) \to \operatorname{Aut}\mathscr{M}$, $g \mapsto \psi(g)$. We showed (II.6.6.4) that $g \in GL(H)$ commutes with the monodromy T if and only if $\psi(g)$ commutes with ∂_t, i.e. $\psi(g)$ is a $\mathbb{C}[\partial_t]$-module automorphisms of $\mathscr{M}$. Let $\Gamma_1 = \psi(G_1)$ be the subgroup of $\operatorname{Aut}\mathscr{M}$ corresponding to $G_1 \subset GL(H)$. The group Γ_1 preserves C_α and hence acts on the vector space $V = V^{>-1}/V^{n-1}$ and, consequently, acts on the Grassman manifold Π.

(5.2.5) Hertling defined the Picard–Fuchs singularity in [He1] as follows:

Definition The *Hertling invariant* $Her_1(f)$ of a singularity f is the set

$$Her_1(f) = \{\gamma(\mathscr{H}^{(0)}(y)) | \gamma \in \Gamma_1\},$$

i.e. the 'orbit' of its Brieskorn lattice in $\mathscr{M}$ wrt the group Γ_1.

(5.2.6) *Lemma* The two invariants of a singularity f, the Picard–Fuchs singularity $PFS(f)$ and the Hertling invariant $Her_1(f)$, coincide.

This is almost a tautology, since both these invariants are given by the period matrix $\Omega(t)$ to within the equivalence relation. If $\gamma(\mathscr{H}^{(0)}(y)) \in Her_1(f)$, then choosing a basis $s_1, \ldots, s_\mu$ in $\gamma(\mathscr{H}^{(0)}(y))$ and a basis $\delta_1, \ldots, \delta_\mu$ in the space H^* of multivalued horizontal sections of the holomorphic fibration $\underline{H}_*$, we obtain the matrix $(\langle s_i(t), \delta_j(t)\rangle)$, which determines an element of $PFS(f)$.

(5.2.7) Now if $f_y(x)$ is a μ-const family of isolated hypersurface singularities, parametrized by a manifold Y, we obtain the map

$$\Phi_1 \colon Y \to \Pi/\Gamma_1,$$

and we can regard the Hertling invariant $Her_1(f_y)$ of the singularity $f_y(x)$ as the point $\Phi_1(y) \in \Pi/\Gamma_1$ in quotient space.

(5.2.8) Denote by '$\sim_R$' the equivalence relation on Y given by the R-equivalence relation of functions.

$$y_1 \underset{R}{\sim} y_2 \Leftrightarrow \text{the germs } f_{y_1}(x) \text{ and } f_{y_2}(x) \text{ are } R\text{-equivalent.}$$

By (5.2.3) we obtain the map

$$\Phi_1 \colon Y/\underset{R}{\sim} \to \Pi/\Gamma_1$$

(we are not interested here in the structure of quotients as analytic spaces). Now we can formulate the Varchenko's conjecture as follows.

Conjecture The map Φ_1 is (locally) injective.

5.3 The Hertling invariants $Her_2(f)$ and $Her_3(f)$

(5.3.1) As was shown by Hertling's calculations [He1] for unimodular singularities the local form of Varchenko's conjecture holds, but the global form doesn't always.

Example Let us consider the μ-const deformation of the singularity E_{12} from (3.3.1). Let $m_0 = \min(m|T^m = 1) = 42$ be the order of the monodromy, and $m_{\mathrm{R}} = m_0/2 = 21$. Let $\Phi \colon Y = \mathbb{C} \to \mathbb{P}^1 = \Pi$, $a \mapsto (1 \colon a)$, be the monodromy map (3.3.7). It is obvious that if $(a_1/a_2)^{m_R} = 1$, then f_{a_1} and f_{a_2} are R-equivalent. The converse is also true. Thus, f_{a_1} and f_{a_1} are R-equivalent $\Leftrightarrow (a_1/a_2)^{m_R} = 1$, and at the same time, as Hertling showed,

$$PFS(f_{a_1}) = PFS(f_{a_2}) \Leftrightarrow \left(\frac{a_1}{a_2}\right)^{2m_R} = 1.$$

Consequently, for the singularity E_{12} the period map $\Phi_1 \colon Y/\underset{R}{\sim} \to \Pi/\Gamma_1$ is two-sheeted, for every Picard–Fuchs singularity there are exactly two R-equivalence classes of singularities with given PFS.

(5.3.2) Thus, the invariant $PFS(f) = Her_1(f)$ is sufficiently rough, in the sense that it does not separate the R-equivalence classes. This led Hertling to consider finer invariants, which were, as before, invariants of R-equivalence classes, but which separated the R-equivalence classes. We have to consider subgroups of G_1 which respect more topological properties than G_1 does. For $g_* \in GL(H_n(X_\infty, \mathbb{Z}))$ we denote by $g^* \in GL(H^n(X_\infty, \mathbb{Z}))$ the automorphism of the dual Milnor lattice $H^n(X_\infty, \mathbb{Z})$.

Put

$$G_2 = \{g^* \in G_1 | g_* \text{ preserves the intersection form on } H_n(X_\infty, \mathbb{Z})\},$$
$$G_3 = \{g^* \in G | g_* \text{ preserves the bilinear Seifert form}\}.$$

Since (see [AGV]) the Seifert form determines the intersection form and the monodromy, we have $G_1 \supset G_2 \supset G_3$. Denote by

$$\Gamma_3 \subset \Gamma_2 \subset \Gamma_1 \subset \operatorname{Aut} \mathscr{M}$$

the corresponding groups of automorphisms of the meromorphic connection $\mathscr{M}$.

(5.3.3) *Definition* Hertling defined the invariants $Her_i(f)$, $i = 1, 2, 3,$ of a singularity f as the sets

$$Her_i(f) = \{\gamma(\mathscr{H}^{(0)}) | \gamma \in \Gamma_i\},$$

i.e. the 'orbit' of its Brieskorn lattice in $\mathscr{M}$ wrt the group Γ_i.

Thus, the invariants $Her_i(f)$, $i = 1, 2, 3,$ are the invariants of R-equivalence classes of increasing fineness,

$$\mathscr{R}(f) = \mathscr{R}(g) \Rightarrow Her_3(f) = Her_3(g) \Rightarrow Her_2(f) = Her_2(g) \Rightarrow Her_1(f) = Her_1(g).$$

From the point of view of the period map of a family of singularities $f_y(x)$, we can consider the following commutative diagram of quotients and maps

$$\begin{array}{ccccccc} Y & \longrightarrow & Y/\underset{R}{\sim} & & & & \\ \Phi\downarrow & & \Phi_3\downarrow \; \Phi_2 \searrow & & \Phi_1 \searrow & & \\ \Pi & \longrightarrow & \Pi/\Gamma_3 & \longrightarrow & \Pi/\Gamma_2 & \longrightarrow & \Pi/\Gamma_1 \end{array}$$

We can regard the Hertling invariant $Her_i(f_y)$ of the singularity $f_y(x)$ as the point $\Phi_i(y) \in \Pi/\Gamma_i$, $i = 1, 2, 3,$ in the quotient space of the period space.

5.4 Hertling's results

(5.4.1) In [He1] and [He2] Hertling considered the invariants Her_1 and Her_2 for all uni- and bimodal singularities. He proved the following theorem.

Theorem

(i) The invariant Her_2 separates the R-equivalence classes (i.e. Φ_2 is injective) for all uni- and bimodal singularities, possibly with the exceptions of the series $Z_{1,14k}$, $S_{1,10k}$, $S^{\#}_{1,10k}$, $k \in \mathbb{N}$, in the classification of Arnol'd (see [AGV]).

(ii) The invariant Her_1 determines the R-equivalence classes for the simple-elliptic singularities $\tilde{E}_6$, $\tilde{E}_7$, $\tilde{E}_8$, for the hyperbolic singularities $T_{p,q,r}$ and for the exceptional unimodal singularities E_{13}, Z_{11}, Z_{12}, Z_{13}, Q_{11}, W_{13}, S_{11}, S_{12}.

(iii) For the remaining exceptional singularities E_{12}, E_{14}, Q_{10}, Q_{12}, W_{12}, U_{12}, there are two R-equivalence classes of singularities with equal invariants Her_1.

For the series $Z_{1,14k}$, $S_{1,10k}$, $S^{\#}_{1,10k}$ the methods used are not sufficient to calculate Her_2 explicitly. However, Hertling believes that Her_2 also determines the R-equivalence classes for them.

Hertling's results are formulated in more detail for the unimodal singularities as follows:

(5.4.2) For the 14 exceptional unimodal singularities we can take the μ-const deformation $f_a(x)$ in the normal form

$$f_a = f_0(x, y, z) + aM,$$

where f_0 is a quasihomogeneous polynomial of weight 1, $\mathrm{wt}(M) > 1$, $a \in \mathbb{C}$.

Let m_0 be the order of monodromy T, and

$$m_R = \begin{cases} m_0, & \text{if } m_0 \text{ is odd,} \\ \dfrac{m_0}{2}, & \text{if } m_0 \text{ is even.} \end{cases}$$

Two singularities f_{a_1} and f_{a_2} are R-equivalent, i.e. $\mathscr{R}(f_{a_1}) = \mathscr{R}(f_{a_2})$ if and only if $(a_1/a_2)^{m_R} = 1$.

Theorem

(i) $Her_2(f_{a_1}) = Her_2(f_{a_2}) \Leftrightarrow (a_1/a_2)^{m_R} = 1 \Leftrightarrow \mathscr{R}(f_{a_1}) = \mathscr{R}(f_{a_2})$.

(ii) For eight of exceptional unimodal singularities E_{13}, Z_{11}, Z_{12}, Z_{13},

Q_{11}, W_{13}, S_{11}, S_{12} we have

$$Her_1(f_{a_1}) = Her_1(f_{a_2}) \Leftrightarrow \left(\frac{a_1}{a_2}\right)^{m_R} = 1.$$

(iii) For the singularities E_{12}, E_{14}, Q_{10}, Q_{12}, W_{12}, and U_{12} we have

$$Her_1(f_{a_1}) = Her_1(f_{a_2}) \Leftrightarrow \left(\frac{a_1}{a_2}\right)^{2m_R} = 1.$$

(5.4.3) For the hyperbolic singularities $T_{p,q,r}$ consider the μ-const deformation given in the normal form (3.4.1).

Theorem

$$Her_1(f_{a_1}) = Her_1(f_{a_2}) \Leftrightarrow \left(\frac{a_1}{a_2}\right)^{\mathrm{lcm}(p,q,r)} = 1 \Leftrightarrow \mathscr{R}(f_{a_1}) = \mathscr{R}(f_{a_2}).$$

(5.4.4) The simple-elliptic singularities $\tilde{E}_6$, $\tilde{E}_7$, $\tilde{E}_8$ can be given in, for example, the Legendre normal form, see (3.5.2). The R-equivalence class of such a singularity is determined by the j-invariant of the corresponding elliptic curve.

Theorem

$$Her_1(f_{\lambda_1}) = Her_1(f_{\lambda_2}) \Leftrightarrow j(\lambda_1) = j(\lambda_2) \Leftrightarrow \mathscr{R}(f_{\lambda_1}) = \mathscr{R}(f_{\lambda_2}).$$

(5.4.5) The results obtained by Hertling enabled him to modify Varchenko's conjecture as follows:

Conjecture The invariant Her_2 (or Her_3) of an isolated hypersurface singularity determines its R-equivalence class.

The methods applied by Hertling, which use the classification of singularities and explicit calculations (which even in the case of bimodal singularities are very cumbersome and difficult), cannot be applied to the proof of this conjecture. It is not yet clear how to prove this conjecture in a general form.

References

[A] A'Campo: *Invent. Maths*, 1973, **20**, 147–69.1

[Ar] Arnol'd, V. I.: On some problems in singularity theory. Geom. and Anal., papers dedicated to the memory of V. K. Patody, 1981, 1–10.

[AGV] Arnold, A. I., Gusein-Zade, S. M., Varchenko, A. N.: *Singularities of Differentiable Maps*, Vols. 1 and 2, Monographs in Mathematics 82 and 83, Birkhäuser, Boston, 1985 and 1988.

[Br] Brieskorn, E.: Die Monodromie der isolierten Singularitäten von Hyperflächen, *Manuscripta Math.*, 1970, 2, 103–61.

[Br-k] Brieskorn, E., Knörrer, H.: *Ebene algebraische Kurven*, Birkhäuser, 1981.

[B-Z] Brylinski, J.-L., Zucker, S., An overview of recent advances in Hodge theory, *Encyclopaedia of Math. Sciences*, VI NITI, 1991, vol. 69, pp 48–165.

[B-G] Buchweitz-Greuel, *Invent. Maths*, 1980, **58**, 241–81.

[C-L] Coddington and Levinson, *Theory of Ordinary Differential Equations*, McGraw Hill, New York, 1955.

[Da] Danilov, V. I.: Newton polyhedra and vanishing cohomology, *Funct. Anal. Appl.*, 1979, **13**, No. 2, 32–47 (in Russian)

[D1] Deligne, P.: *Equations Differentielles a Points Singular Regulier*, Lect. Notes in Math., Springer, Berlin, 1970, vol. 163.

[D2] Deligne, P.: Theorie de Hodge, II, *Publ. Math. IHES*, 1971, **40**, 5–58.

[D3] Deligne, P.: Theorie de Hodge, III, *Publ. Math. IHES*, 1974, **44**, 5–77.

[D-Di] Deligne, P., Dimca, A.: Filtrations de Hodge et par l'ordre du pôle pour les hypersurfaces singularities, *Ann. scient. Ec. Norm. Sup.*, 4 serie. 1990, **23**, 645–56.

[Di1] Dimca, A.: On the Milnor filtrations of weighted homogeneous polynomials, *Compos. Math.*, 1990, **76**, No 1–2, 19–47.

[Di2] Dimca, A.: *Differential Forms and Hypersurface Singularities*, Lect. Notes Math., Springer, Berlin, 1991, No 1462, pp. 122–53.

[Di3] Dimca, A.: *Singularities and Topology of Hypersurfaces*, Springer Verlag, 1992, 263pp.

[DS] Doorn, van, M. G. M., Steenbrink, J. H. M.: A supplement to the monodromy theorem, *Abh. math. Semin. Hamburg*, 1989, **59**, 225–33.

[E] Elkik, R.: Singularites rationnelles et deformations, *Invent. Math.*, 1978, **47**, 139–48.

[G-M] Gelfand, S. I., Manin, Yu. I.: *Homological Algebra*, in series Itogi nauki i techniki, seria Seuremennie problemy matematiki, VINITI, 1989, vol. 38, 238 pp.

[Go] Goryunov, V. V. Adjacencies of spectra of certain singularities, *Vestnik MGU*, ser math., 1981, **4**, 19–22.

[G1] Greuel, G. -M.: Der Gauss–Manin–Zusammenhang isolierter Singularitäten von vollständigen Durchschnitten, *Math. Ann.* 1975, **214**, 235–66.

[Gr] Griffiths, P.: On the periods of certain rational integrals I, II, *Ann. Math.*, 1969, **90**, 460–541.

[H] Hamm, H.: *Zur analytischen und algebraischen Beschreibung der Picard–Lefschetz–Monodromie*, Habilitationsschrift, Göttingen, 1974, 106pp.

[He1] Hertling, C.: Analytische invarianten bei den unimodularen und bimodularen Hyperflächensingularitäten. Dissertation Bonner Mach. Schriften 250, Bonn 1992.

[He2] Hertling, C.: Ein Torelli-Satz für die unimodalen und bimodalen Hyperflächensingularitäten, *Math. Ann.*, 1995, **302**, 359–94.

[Ka1] Karpishpan, Ya.: Pole order filtration on the cohomology of algebraic links, *Compos. Math.* 1999, **78**, No 2, 213–26.

[Ka2] Karpishpan, Ya.: Torelli theorems for singularities, *Invent. math.*, 1990, **100**, 97–141.

[K-Ku] Kulikov, Vic. S., Kurchanov, P. F.: Complex algebraic varieties: periods of integrals, Hodge structures. *Encyclopaedia of Math. Sciences*, 1989, **36**, 5–231.

[Ko] Kouchnirenko, A; *Invent. Math.*, 1976, **32**, 1–31.

[L] Lê Dung Trang: Sur les noeuds algebriques, *Compos. Math.*, 1972, **25**, 281–321.

[L-R] Lê Dung Trang, Ramanujam, C. P.: The invariance of Milnor's number implies the invariance of topological type, *Amer. J. Math.*, 1976, **98**, 67–78.

[M1] Malgrange, B.: Integrales asymptotiques et monodromie, *Ann. Sci Ecole Norm. Super.*, 1974, **7**, 405–430.

[M2] Malgrange, B.: Letter to the editors, *Invent. Math*, 1973, **20**, 171–2.

[Mi] Milnor, J.: *Singular Points of Complex Hypersurfaces*, Annals of Math. Studies, Princeton Univ. Press, Princeton, 1968, vol 61.

[N] Navarro, A.: Sur la théorie de Hodge-Deligne, *Invent. Math.*, 1987, **90**, 11–76.

[Ph1] Pham, F.: *Singularités des Systèmes Différentiels de Gauss–Manin*, Progress in Math., Birkhäuser, Boston, 1979, vol. 2.

[Ph2] Pham, F.: Structures de Hodge mixtes assosiées à un germe de fonction à point critique isolé, *Asterisque*, 1983, No 101–102, 268–285.

[SK] Saito, K.: Period mapping associated to a primitive form, *Publ. Res. Inst. Math. Sci.*, 1983, **19**, No 3, 1231–64.

[Sa1] Saito, M.: Gauss–Manin system and mixed Hodge structure, *Proc. Japan Acad.*, 1982, **58**, Ser. A, 29–32.

[Sa2] Saito, M.: Supplement to 'Gauss–Manin system and mixed Hodge structure', *Astérisque*, 1983, No 101-102, 320–31.

[Sa3] Saito, M.: On the exponents and the geometric genus of an isolated hypersurface singularity, *Proc. Symp. Pure Math. AMS*, 1983, **40**, Pt. 2, 465–72.

[Sa4] Saito, M.: Hodge filtrations on Gauss–Manin systems. I, *J. Fac. Sci. Univ. Tokyo*, 1984, Sec. IA, **30**, No 3, 489–98.

[Sa5] Saito, M.: Hodge structure via filtered D-modules, *Astërisque*, 1985, No 130, 342–51.

[Sa6] Saito, M.: Exponents and Newton polyhedra of isolated hypersurface singularities, *Math. Ann.*, 1988, **281**, 411–17.

[Sa7] Saito, M.: On the structure of Brieskorn lattice, Ann. Inst. Fourier, 1989, **39**, No 1, 27–72.

[Sa8] Saito, M.: Period mapping via Brieskorn modules, *Bull. Soc. Mat. Fr.*, 1991, **119**, No 2, 141–71.

[Sa9] Saito, M.: On Steenbrink's conjecture, *Math. Ann.*, 1991, **289**, No 4, 703–16.

[Sa10] Saito, M.: Mixed Hodge modules and applications, *Proc. Int. Congr.* Math., Kyoto, Japan, 1990. Math. Soc. Japan, 1991, 725–34.

[Sc1] Scherk, J.: On the Gauss–Manin connection of an isolated hypersurface singularity, *Math. Ann.*, 1978, **238**, 23–32.

[Sc2] Scherk, J.: On the monodromy theorem for isolated hypersurface singularities, *Invent. Math.*, 1980, **58**, 289–301.

[Sc3] Scherk, J.: A note on two local Hodge filtrations, *Proc. Symp. Pure Math. A.M.S.,* **10**, 1983, Part 2, 473–7.

[Sc4] Scherk, J.: Isolated singular points and the Gauss–Manin connection, PhD thesis, University of Oxford, 1977.

[Sc-S] Scherk, J. and Steenbrink, J. H M.: On the mixed Hodge structure on the cohomology of the Milnor fibre, *Math. Ann.*, 1985., **271**, No 4, 641–65.

[Sm] Schmid, W.: Variation of Hodge structure: The singularities of the period mapping, *Invent. Math.*, 1973, **22**, 211–319.

[Se] Sebastiani, M.: Preuve d'une conjecture de Brieskorn, *Manuscr. Math.*, 1970, 2, No 2, 301–6.

[Se-T] Sebastiani, M. Thom, R.: Un resultat sur le monodromie. *Invent. Math.*, 1971, **13**, No 1–2, 90–6.

[Si] Siersma, D: The monodromy of a series of hypersurface singularities, *Comment. Math. Helv.*, 1990, **65**, No 2, 181–97.

[S1] Steenbrink, J. H. M.: Limits of Hodge structures, *Invent. Math.*, 1975, **31**, No 3, 229–57.

[S2] Steenbrink, J. H. M.: Intersection form for quasihomogeneous singularities, *Compos. Math.*, 1977, **34**, No 2, 211–23.

[S3] Steenbrink, J. H. M.: Mixed Hodge structure on the vanishing cohomology, in *Real and Complex Singularities, Oslo 1976*, Alphen aan den Rijn, Oslo 1977, pp. 525–63.

[S4] Steenbrink, J. H. M.: Mixed Hodge structures associated with isolated singularities, *Proc. Symp. Pure Math. AMS*, 1983, **40**, Part 2, 513–36.

[S5] Steenbrink, J. H. M.: Semicontinuity of the singularity spectrum, *Invent. Math.*, 1985, **79**, No 3, 557–65.

[S6] Steenbrink, J. H. M.: The spectrum of hypersurface singularities, *Astërisque*, 1989, No 179–80, 163–84.

[S7] Steenbrink, J. H. M.: *Applications of Hodge Theory to Singularities*, Proc. Int. Congress of Math., Kyoto, Japan, 1990. Math. Soc. Japan, 1991, pp. 569–76.

[Sr] Straten van D.: *On the Betti numbers of the Milnor fibre of a certain class of hypersurface singularities.* Lect. Notes Math., Springer, Berlin, 1987, No 1273, pp. 203–20.

[V1] Varchenko, A. N.: Gauss–Manin connection of isolated singular point and Bernstein polynomial, *Bull. Sci. math.*, 2nd series, 1980, **104**, 205–23.

[V2] Varchenko, A. N.: Asymptotics of holomorphic forms define mixed Hodge structure, *Dokl. Acad. Nauk SSSR*, 1980, **225**, No 5, 1035–38 (in Russian) [English transl.: *Sov. Math. Dokl.*, 1980, **22**, 772–5].

[V3] Varchenko, A. N.: Asymptotic Hodge structure in vanishing cohomology, *Izv. Akad. Nauk SSSR, Ser. Mat.*, 1981, **45**, 540–91 (in Russian) [English transl.: *Math. USSR Izv.*, 1982, **18**, 469–512].

[V4] Varchenko, A. N.: On the monodromy operator in vanishing cohomology and the operator of multiplication by f in the local ring, *Dokl. Acad. Nauk SSSR*, 1981, **260**, No 2, 272–6 (in Russian) [English transl.: *Sov. Math. Dokl.*, 1981, **24**, 248–52].

[V5] Varchenko, A. N.: The complex exponent of a singularity does not change along strata μ =const, *Funct. Anal. Appl.*, 1982, **16**, No 1, 1–12.

[V6] Varchenko, A. N.: A lower bound for the codimension of strata μ =const in terms of the mixed Hodge structure, *Vestnik MGU, Mat. Mech.*, 1982, No 6, 28–31 (in Russian) [English transl.: *Moscow Univ. Math. Bull.*, 1982, **37**, 30–3].

[V7] Varchenko, A. N.: The spectrum and falling to pieces of a critical point of a function, *Dokl. Acad. Nauk SSSR*, 1983, **270**, No 2, 267–70 (in Russian).

[V8] Varchenko, A. N.: On semicontinuity of the spectrum and an upper bound for the number of singular points of projective hypersurfaces, *Dokl. Acad. Nauk SSSR*, 1983, **270**, No 6, 1294–7 (in Russian).

[V9] Varchenko, A. N.: Asymptotics of integrals and Hodge structures, in *Modern Problems of Mathematics*, VINITI, 1983, vol. 22, 130–66 (in Russian).

[V10] Varchenko, A. N.: On the local residue and the intersection form on the vanishing cohomology, *Izy. Akad. Nauk SSSR, Ser. Mat.*, 1985, **49**, No 1, 32–54 (in Russian) [English transl.: *Math. USSR Izv.*, 1986, **26**, 31–52].

[V-Ch] Varchenko, A. N., Chmutov, S. V.: Tangent cone to the stratum μ =const, *Vestnik MGU, Mat. Mech.*, 1985, No 1, 6–9 (in Russian) [English transl.: *Moscow Univ. Math. Bull*, 1985, **40**, 7–12].

[V-Kh] Varchenko, A. N., Khovanskii, A. G.: Asymptotics of integrals over vanishing cycles and the Newton polyhedron, *Dokl. Acad. Nauk SSSR*, 1985, **283**, No 3, 521–5 (in Russian) [English transl.: *Sov. Math. Dokl.*, 1985, **32**, 122–7].

[Y] Yomdin, I. N.: Complex varieties with singularities of dimension one, *Siberia Math. J.*, 1974, **15**, 1061–82.

Index